To Barry

From one my [illegible]

POSTURE AND MOVEMENT

Posture and Movement

Editors

Richard E. Talbott, Ph.D.
Department of Physiology
University of Oregon
Health Science Center
Portland, Oregon

Donald R. Humphrey, Ph.D
Department of Physiology
Emory University School
of Medicine
Atlanta, Georgia

Raven Press ▪ New York

Raven Press, 1140 Avenue of the Americas, New York, New York 10036

Made in the United States of America

Library of Congress Cataloging in Publication Data

Symposium on Posture and Movement, Anaheim,
Calif., 1977.

Includes bibliographical references and index.
1. Posture--Congresses. 2. Sensory-motor
integration--Congresses. 3. Man--Attitude and
movement--Congresses. I. Talbott, Richard E.
II. Humphrey, Donald R. III. Title.
[DNLM: 1. Movement–Congresses. 2. Posture–
Congresses. WE103.3 S989p 1977]
QP301.S955 1977 612'.76 77-85515
ISBN O-89004-259-4

This book is dedicated to
Professor John M. Brookhart,
with the respect and affection
of his colleagues

Preface

Perspective, or point of view, is critical in science, for it serves the role of a loosely formulated hypothesis. It is the little-considered constraint that each experimentalist carries into the laboratory. The present volume contains a series of chapters in which the authors have, to varying degrees, sought to clearly define what are the perspectives, constraints, or limitations characterizing particular experimental approaches to the understanding of neural control of posture and movement. Further, there is within these pages an attempt to look forward and define the variables that must be measured in order to adequately test the assumptions implicit in a particular perspective regarding nervous system function.

The neurosciences constitute an immense body of experimental facts so that no single volume could possibly address more than a few. Posture is but one end of a spectrum of movements. Thus, this volume is really about one very broad topic, namely, movements made by mammals.

Particular aspects of that subject have been addressed. The neuronal circuitry and response properties measurable from neural elements constitute one major theme in this volume since it is essential to relate the internal manifestations of movement to the external parameters of that movement. The qualities of particular sensory inputs, what they are measuring, and how they are incorporated into a movement is another major theme. A third theme presented concerns the characteristics of the controlled device: How do the dynamics of the musculoskeletal system affect or determine the sort of control problem that a nervous system must solve? Another theme dealt with here is the careful specification of performance so that explicit models of that performance and the likely characteristics of components yielding that performance can be determined. The final theme presented is the recognition that more open-mindedness is needed in this field of research. Many procedures have been used and will be used, but it is essential that experimental tests be formulated in ways that will facilitate translation of the facts from one level of study into useful constraints upon explanation at another level.

This volume is like a sampler; it presents a specimen of neuroscientists' collective skill at defining and deciphering a difficult problem. The volume also reveals the value of synergy, that is, the joint working or cooperation needed in science to solve particular problems, just as that concept was originally used to describe the full range of cooperation among the muscles of the whole body to produce some movement.

The volume will be of interest to those neurologists and neuroscientists involved in motor systems. It will also be of interest to a broader segment of neuroscientists who are concerned with the historical and future perspective of this branch of science.

Richard E. Talbott
Donald R. Humphrey

Acknowledgment

The symposium on Posture and Movement, held in Anaheim, California, was organized in part to honor Professor John M. Brookhart. Professor Brookhart's many contributions to the neurosciences and his broad perspective on that field made it appropriate to assemble a group of scientists who likewise could discuss the broader issues critical to successful study of neural control of posture and movement.

Contents

Contributors

Neal H. Barmack: *Department of Ophthalmology, Neurological Sciences Institute, Good Samaritan Hospital and Medical Center, Portland, Oregon 97210*

Marc D. Binder: *Department of Physiology, College of Medicine, Arizona Health Sciences Center, University of Arizona, Tucson, Arizona 85724*

Emilio Bizzi: *Department of Psychology, Massachusetts Institute of Technology, Cambridge, Massachusetts 92139*

Barry R. Botterman: *Department of Physiology, College of Medicine, Arizona Health Sciences Center, University of Arizona, Tucson, Arizona 85724*

John M. Brookhart: *Department of Physiology, School of Medicine, University of Oregon Health Science Center, Portland, Oregon 97201*

Vernon B. Brooks: *Department of Physiology, University of Western Ontario, London, Ontario, NA65C1 Canada*

James C. Houk: *Department of Physiology, Northwestern University Medical School, Chicago, Illinois 60611*

Donald R. Humphrey: *Department of Physiology, Emory University School of Medicine, Atlanta, Georgia 30322*

William S. Levine: *Electrical Engineering Department, University of Maryland, College Park, Maryland 20742*

Lewis M. Nashner: *Neurological Sciences Institute, Good Samaritan Hospital and Medical Center, Portland, Oregon 97210*

Lloyd D. Partridge: *Department of Physiology and Biophysics, University of Tennessee Center for thc Health Sciences, Memphis, Tennessee 38163*

Andres Polit: *Department of Psychology, Massachusetts Institute of Technology, Cambridge, Massachusetts 02139*

Douglas G. Stuart: *Department of Physiology, College of Medicine, Arizona Health Sciences Center, University of Arizona, Tucson, Arizona 85724*

Richard E. Talbott: *Department of Physiology, University of Oregon Health Science Center, School of Medicine, Portland, Oregon 97201*

Carlo A. Terzuolo: *Laboratory of Neurophysiology, University of Minnesota, Minneapolis, Minnesota 55455*

Paulo Viviani: *Laboratoire du Physiologie de Travail du CNRS, Department de Physiologie Neurosensorielle, 75005 Paris, France*

Marjorie Woollacott: *Neurological Sciences Institute, Good Samaritan Hospital and Medical Center, Portland, Oregon 97210*

Laurence R. Young: *Man Vehicle Laboratory, Department of Aeronautics and Astronautics, Massachusetts Institute of Technology, Cambridge, Massachusetts 02139*

Felix E. Zajac, III: *Electrical Engineering Department, University of Maryland, College Park, Maryland 20742*

Posture and Movement, edited by
R. E. Talbott and D. R. Humphrey.
Raven Press, New York © 1979.

Ferrier, the Synergy Concept, and the Study of Posture and Movement

Richard E. Talbott*

Department of Physiology, School of Medicine, University of Oregon Health Sciences Center, Portland, Oregon 97201

INTRODUCTION

David Ferrier was 33 years old when he published the first edition of his book entitled *The Functions of the Brain* (66). Ferrier had just recently completed his historic series of experiments at the West Riding Asylum in which he had used faradic stimulation to explore the brain of monkeys, cats, dogs, and many other species. *The Functions of the Brain* was, in large measure, a compilation of those studies. But the book was also more than that. It was primarily Ferrier's attempt to assimilate known facts about the brain into some kind of coherent picture. His introduction needs no modification to be relevant even 100 years later. Ferrier wrote in 1876: "There is, perhaps, no subject in physiology of greater importance and general interest than the functions of the brain, and there are few which present to experimental investigation conditions of greater intricacy and complexity. No one who has attentively studied the results of the labors of the numerous investigators in this field of research, can help being struck by the want of harmony, and even positive contradictions, among the conclusions which apparently the same experiments and the same facts have led to in different hands."

In 1906, Sherrington dedicated his book, *The Integrated Action of the Nervous System,* to Ferrier, but the philosophical orientation adopted by Sherrington was quite different. Sherrington (67, p. 7) stated:

> The reflex-arc is the unit mechanism of the nervous system when that system is regarded in its integrative function. The unit reaction in nervous integration is the reflex, because every reflex is a complete act of integration. The nervous synthesis of an individual from what without it were a mere aggregation of commensual organs resolves itself into coordination by reflex action. But, though the unit reaction in the integration is a reflex, not every reflex is a unit reaction, since some reflexes are compounded of simpler reflexes. Coordination, therefore, is in part the compounding of reflexes.

Later, Creed et al. (15, p. v) presented this intriguing assessment of the scope of a study of reflexes as "an attempt to trace general principles of coordination

* At present, in private practice of law in Portland, Oregon.

observable in them." Creed et al. (15) continued, "In its variety of 'reflex-figure' and 'pattern,' the problem dealt with here is that of physiological production and execution, and not that of biological significance." When Sherrington's book was reissued in 1947, he had modified his view of reflex studies somewhat. Sherrington (67, p. xiii) stated, "The volume here reprinted concerns itself predominantly with the type of motor behaviour which is called 'reflex'; it might give the impression that in reflex behaviour it saw the most important and far-reaching of all types of 'nerve' behavior. That is in fact not so. But reflex action presents certain advantages for physiological description. It can be studied free from complication with the psyche." Commenting on Descartes' view of animals as machines that interact, Sherrington (67, p. xiv) continued,

> His wheelwork animals geared into a turning universe. Cat, dog, horse, etc. in his view had no thoughts, no ideas; they were trigger-puppets which events in the circumambient universe touched-off into doing what they do. It was a view less strange than might seem from this condensed epitome. But it lets us feel Descartes can never have kept an animal pet. Experiment today does, however, put within reach of the observer a puppet-animal which conforms largely with Descartes' assumptions. In the more organized animals of the vertebrate type the shape of the central nerve-organ allows a simple operation to reduce the animal to the Descartes condition. An overlying outgrowth of the central nerve-organ of the head can be removed under anesthesia, and on the narcosis passing off the animal is found to be a Cartesian puppet; it can execute certain acts but is devoid of mind.

SUMMATION OF REFLEXES

A commonly shared credo among neurophysiologists who study posture seems to derive from a view of nervous system organization put forth by Magnus (43,44). The assumption made by Magnus was that the several reflexes should add algebraically to produce the final postural attitude. Further, it was assumed that inferences about particular reflex mechanisms could not be evaluated by removing those reflexes one at a time. The constituent reflexes had to be studied in isolation. To do that experimentally it was necessary to start with an animal in the zero condition, by which Magnus (44) meant that all centers or mechanisms concerned with complex functions had to be paralyzed, that is, removed. Under such an experimental scheme the labyrinthine reflexes, the optical righting reflexes, tactile righting reflexes, and so forth, would be studied only after all other identifiable reflex mechanisms had been eliminated. The behavior of the whole animal was then taken to be the result of an algebraic summation of the constituent reflexes such that, it was argued (44), "good pictures and statues give the impression of being natural if the body is represented according to the laws of attitudinal reflexes."

Young (83) has criticized Ferrier for discarding the study of psychologically meaningful issues in favor of physiologically measurable events. But the criticism is not totally warranted. Ferrier was quite concerned with large, integrated actions. In the 1886 edition of *The Functions of the Brain,* Ferrier repeatedly stresses the coordinate activities of the nervous system. For example, Ferrier (25, p. 77) stated his belief that it is "probable that each segment of the cervical and lumbar enlargements of the spinal cord, whence nerves proceed to the limbs, is a centre of coordinated synergic muscular movements, of a character adapted to the habits and requirements of the animal in its ordinary modes of activity." Again, Ferrier (25, p. 200) wrote, "Every form of active muscular exertion necessitates the simultaneous cooperation of an immense assemblage of synergic movements throughout the body to secure steadiness and maintain the general equilibrium." Those views of Ferrier written in 1886 were derived from the data available at that time coupled with the philosophy of the time, as Young (83) has argued. It seems apparent, though, that the views expressed by Ferrier had much more psychological relevance than the long sequence of experiments which follow from the reflex-oriented approaches advocated later by Sherrington and Magnus.

MOVEMENT AS A MORPHOLOGICAL ENTITY

Irrespective of one's view of what mind is, certainly the warning by Efron (19) is pertinent to those who attempt to study higher nervous system functions. Efron (19) noted explicitly the problem of excluding the concept of consciousness when studies of higher organisms are undertaken. Similarly, the philosophical orientation of Beritoff seems to acknowledge that necessity for including consciousness. Beritoff (6, at p. 4) stated, "It is important to recognize the type of behavior of the higher vertebrates which is developed entirely from individual experience and depends upon the activity of the central nervous system functions reorganized in the process of biological adaptation." Beritoff went on to describe his concept of image-driven behavior, which is not built up of a collection of elemental conditioned responses. He further developed the idea of a spatial orientation for such image-driven behaviors.

Bernstein (7) also recognized the necessity of viewing larger units of behavior as the elements of nervous system function. He and his followers developed the concept of movements as morphological objects; movements were not to be viewed as "chains of details but as structures, which are differentiated into details; they are structurally whole" (7, p. 69).

Greene [(33), see also (70) at p. 352] in the introduction to the volume *Models of the Structural-Functional Organization of Certain Biological Systems* commented on the research derived from the views of Bernstein. "The first stage of a motor task, according to this view, is the coming together of joints and muscles into a small number of closely knit groups with a resultant reduction in the number of degrees of freedom of the motor apparatus, since the higher centers have only to select, and possibly tune, these groups" (33, p. xviii). A major problem, in Greene's view, arises from our limited ability to recognize the significant informational units of movement. Gelfand et al. (30) further developed some of the concepts of Bernstein. They used the concept of synergy in much the same way as Ferrier (25) did; a synergy is a class of movements having similar kinematics. Hence, "for realization of a movement it is enough to control a small number of independent parameters, although the number of muscles participating in the movement can be large" (30, p. 331). Gelfand et al. (30) present a good analogy of their conception of synergy and movement:

> The basic synergies and the simplest neurophysiological mechanism enumerated here form, so to say, "a dictionary of movements." Using this analogy, we can say that the efforts of the muscles are the letters of the language of movements, and the synergies combine these letters into words, the number of which are much less than simply the number of combinations of letters. Moreover, the wealth of this dictionary provides a variety of admissible movements. The majority of motor problems arising before an organism are contained in the scope of this dictionary and only in exceptional cases does there arise a need for its enrichment (30, p. 332).

A slightly different statement of the same philosophical point of view can be derived from the ethological approach to animal behavior. For example, Ploog (61) has stressed the importance of studying species-specific behavior so that the nervous system is performing those functions for which it was shaped during evolution. Obviously, the appropriate stimuli to use in such studies would be "natural" stimuli, which are taken to include those stimuli which are linked to responses of the whole organism. Thus, in that frame of reference, a punctate skin indentation would not be particularly significant to most animals (56). Similarly, Uttal (80) has argued that activity recorded from the nervous system can only be taken as an indication of some coding process if it can be demonstrated that the recorded signal is causally related to some behavioral manifestation. Essentially, that situation is what Mountcastle (55) had to contend with when it was found that single cells in cortical area SI did not seem to alter their response patterns while a monkey's behavior in response to the stimuli did change.

COMBINED PHYSIOLOGICAL-PSYCHOLOGICAL STUDIES

If one adopts a philosophy of experimentation on higher nervous systems such as that outlined in the preceding paragraphs, one could easily be led to conclude either that only some form of psychological experiment would yield valid results, thus calling forth the necessity for a review of the immense psychological literature, or that the limitations of certain forms of prior neurophysiological research should be acknowledged and that other approaches, not yet fully developed, but holding a promise of combining the best of physiological rigor with the best of psychologically meaningful questions should be pursued. The latter route has more immediate appeal. An exhaustive compilation of such approaches will not be attempted, for the range of possibilities is quite wide. Some examples will suffice to indicate them.

The volume edited by Gelfand et al. (29) contains a good summary of some of the more recent Russian experiments on posture. The common thread seems to be a view of function incorporating a minimizing principle, e.g., a reduction in the degrees of freedom and a distribution of control to local levels, meaning the spinal cord. Another thread relates to the concept of synergy, which can be taken to mean task-oriented response with variable component mixtures. Thus, a movement, in this view, is not just a synergy of the muscles acting directly on a limb but the synergy must incorporate the other muscle activity which sets the body such that the movement can take place without disturbing the overall postural stability. In a related fashion, contact placing by a cat is preceded by a generalized postural adjustment (46,47,36). That important observation and what it says about nervous system organization would be totally lost if the overall reactions of the animal had not been monitored and just the movement of the stimulated paw had been noted. Studies with a restricted focus [e.g., (3)] must necessarily yield incomplete and possibly misleading pictures of neuronal organization. Likewise, care must be taken not to assume that the mechanism for some observed phenomena such as contact placing is organized more cephalad than perhaps it really is (28). The concept of spinally organized synergies as propounded by Ferrier (25) still may apply for some forms of movement.

Experiments utilizing behaving animals or humans must clearly be organized in a way which will permit a broad view of what the experimental subject is doing. Greenwood and Hopkins (34) found that a generalized startle reaction could most likely account for the earlier observations of Melvill Jones and Watt (48,49). Melvill Jones and Watt had proposed a reflex mechanism originating in the otolith which would serve to cushion a fall from a height. The difference in conclusions reached resulted directly from the wider view of the behavioral problem taken by Greenwood and Hopkins (34). The observation by Rossignol and Melville Jones (62) should likewise be reassessed to determine whether their proposed audio-spinal influence is exclusively or even primarily aimed at the lumbar spinal cord to influence control of leg movement.

Considerable knowledge about nervous system organization related to behavior can be gained from accurate biomechanical analysis of movement and posture. Viviani and Terzuolo (81) and Soechting (68) adopted such a point of view, though it is possible that the particular movement they chose to study, namely, a rapid arm flexion after overcoming a restraining load followed by a rapid arrest of that motion, imposes some fairly artificial constraints on the synergies available to move an arm in some more biologically useful pursuit. Stern, Jr. (69) has carefully incorporated known muscle properties in his theoretical model of limb dynamics. Stern provided a set of predictions and variations of those predictions which could readily be checked experimentally, although, apparently, data in quite the right form are not yet available. The basic premise of Stern was that physiological attributes of muscle must significantly determine the nature of movement. An analysis of the actual forces generated by a walking animal (45) or human (20) coupled with a fairly complete analysis of muscle activity throughout the body (75,76,77,21,22,23) would go far toward assessing the validity and the generality of the synergy concept. Such measurements would also provide the data needed to assess the correspondence between actual animal motion and models of motion derived from the basic physical principles used to describe the motion of bodies [see (79)]. Similarly, a quantitative assessment of quiet stance (12) and of the physiological parameters associated with perturbation of such stance (52,11,72,13) can lead to explicit dynamic models for postural control as well as to the possibility of bridging the gap between animal models of visually affected posture (73) and the human illusion of motion due to visual stimuli (8).

The pattern of activity exhibited during walking should and has provided a fruitful model for assessing nervous system mechanisms during a biologically useful task (35). A critical variable in experiments on walking is whether the phenomenon studied by means of treadmill-induced movement is really the same as that studied when free motion over the ground is permitted. There seems to be agreement that the two tasks are different and that the measured parameters of movement are, therefore, different (50,32,82,57). The details of the differences are now emerging, but a full exposition of them must await more agreement among the investigators involved as to just which measurements are to be used for across-condition comparisons. Certainly it is clear that the total pattern of afferent input is different in the two experimental conditions; for example, the relative motion of the visual surround as a function of overground stepping is totally different from the usual static visual field encountered during treadmill stepping (82). The posture of the animal, whether crouched and apprehensive or erect and relaxed, has a definite bearing on the observed measures of performance (50). Whatever deviations treadmill performance has from "normal" overground progression, the use of treadmill-associated stepping at least allows certain questions to be asked concerning the neural mechanisms involved. As an example, the studies of Lissin et al. (42), Forssberg et al. (27), and Duysens and Pearson (18) seem to establish the context-dependent effects of a stimulus delivered to a moving limb in that the expected "reflex" reactions due to such a stimulus could be reversed according to the position of the limb at the time the stimulus was delivered.

NONTRADITIONALLY STUDIED SOURCES OF AFFERENT INFORMATION

In the future thought should be directed toward the issue of ensembles of afferent information available to the nervous system as the basis for generating reactions by the organism. To that end, it is suggestive that there are some regions of the vertebrate nervous system which have largely been ignored as sources of information potentially useful for determining a behavioral response.

One body region richly supplied with a variety of receptors was recognized by Ferrier (24) as potentially playing a major role in sensations of vertigo, namely the viscera. The role of the viscera as a source of a complex set of signals integrated into an overall posture has not been adequately studied, though present evidence apparently rules out a role for the visceral receptors in the causation of motion sickness (51). Even Ferrier (25) dropped the idea from his second edition. Yet, it is well established that some of the receptors located in the viscera send signals to the cortex and that the visceral afferents may converge onto pathways usually thought to serve as transmission pathways for skin-derived signals. For example, it has long been established that the Pacinian corpuscles of the cat mesentery contribute a portion of the largest afferent fibers found in the splanchnic nerves (65,38,31). The mesenteric Pacinian corpuscles are not the only deformation-sensitive receptors in the viscera (59,41). Another example of a visceral mechanoreceptor is the movement receptor located near the bifurcation of the mesenteric arteries on the serosal surface of the small intestine (9,53). Similar, slowly adapting mechanoreceptors have been located in the pancreas and spleen (53). At both the spinal cord level (63,64,26,17) and at central levels, particularly cortical areas SI, SII, and orbital cortex (58,40), visceral afferents have been shown to converge onto neural elements which also are affected by signals from somatosensory receptors. In addition to potential convergence, the existence of rapidly conducting pathways from the viscera to the somatosensory cortex (1,2,16,5) would seem to demand an expansion and a reassessment of the columnar organization, as propounded by Mountcastle (54), to include afferents derived from regions other than just skin and deep tissues restricted primarily to the limbs. Thus, modality-pure columns representing the extensive set of afferents located within the thoracic and abdominal cavities should also be sought as a necessary prediction from the columnar hypothesis (78).

Several potential functions might be served by the wide variety of visceral receptors (59,41,60), but little work has concentrated on their potential influence on the general orientational state of an animal. It is suggestive in this latter context to consider the influence exerted by this large set of receptors on behavior, as shown by the prompt arousal reaction (37) and the related dysynchronization of the EEG (14) which can be induced by exposing solely the abdominal regions of an animal to a brief dose of X-rays. A mechanism for such prompt behavioral and EEG phenomena is not available, but it has been suggested that one possible signal to the nervous system which is engendered by the X-rays could be radiation-induced changes in the overall level of receptor input from the viscera (71,74). Such a change in level of overall sensory input referable to the viscera may partly explain the positional reflexes which Magnus (43) ascribed to skin contact but which Ito and Sanada (39) have attributed to shifts in the position of the visceral contents with changes in position. The experiments of Ito and Sanada (39) seem to suggest that a powerful reciprocal reflex effect is produced via the splanchnic nerves, most likely from mesenteric receptors which change their activity as a function of a change in position. Similarly, the data of Anand and Pillai (4) suggest that an increase in the total level of afferent activity from the stomach can lead to suppression of hypothalamic feeding center activity and an augmentation of hypothalamic satiety center neurons (41).

CONCLUSIONS

The overall orientation of this chapter has been that animals do something on the basis of the information they gather via a widely distributed array of receptors, that what the animals do is of some biological utility to the animal under normal circumstances, and that the task for neurophysiologists is to determine how the animal processes the information about its environment to generate appropriate behavior. Clearly, the depth of presentation was curtailed in favor of outlining a broader perspective of the interacting problems connected with such an endeavor. The paper by Brodal (10), in which he describes the effects of his own stroke upon his behavior, provides a fitting conclusion to this attempt at integrating a broader perspective for neurophysiological investigations of posture and movement.

> The specificity of the morphological organization of the nervous system is amazing. The more we know about the structure of the brain, the more we realize how this specificity can be traced to the minutest levels. It is the rule, rather than the exception, that even a small nucleus can be subdivided into parts or territories which differ with regard to cytoarchitecture, glial architecture, vasoarchitecture, fiber connexions, synaptic arrangements and by its chemistry. This, of course, is the morphological counterpart of the localization of function in the central nervous system. . . . From this point of view, the entire nervous system may be considered to be composed of a multitude of minor units, each with its particular morphological organization and its specific task. . . . If one focusses attention on the fiber connexions of the central nervous system one is struck by another feature of its organization: The multiplicity of connexions. . . . A scrutiny of the fiber connexions of the central nervous system as a whole leaves one with the conviction that there are morphological possibilities for an impulse from a certain part of the brain to be transmitted along circumvential routes of varying complexity to virtually every other part of the central nervous system! These multifarious interconnections between structures presumably provide possibilities for cooperation and integration of function between them and make it increasingly difficult and unjustified to describe a certain part of the brain as simply "motor," "visual," etc. . . . On the one hand, a multitude of small units in the central nervous system can be identified which all have their particular structure and function, on the other hand they are all more or less amply interconnected and dependent on each other. These views lead to the conclusion that destruction of even a localized part of the brain will cause consequences for several functions in addition to those which are more specifically dependent on the region damaged. In other words, for optimal, perfect function, we need the whole brain (10, pp. 687–688).

In a similar vein 100 years ago Ferrier perceptively wrote: "An animal with brain intact exhibits a varied spontaneity of action not, at least immediately, conditioned by present impressions on its organs of sense. When the hemispheres are removed, all the actions of the animal become the immediate and necessary response to the form and intensity of the stimulus communicated to its afferent nerves" (24, p. 40). Ferrier later discussed this same proposition in these terms: "Though the phenomena of reflex action are manifested by the spinal cord and its segments when completely separated from the brain, yet under normal conditions they are not dissociated, and it is therefore not likely that the laws and conditions of reflex action are the same under both circumstances. When the spinal centers form only a part of the whole cerebrospinal system, stimuli which excite reflex action also affect the whole nervous system, cerebral as well as spinal" (25, p. 69).

The essence of the view espoused here is captured by these statements, since what has been presented here is a constellation comprised of a multitude of elements. For too long the elements have captured the attention of experimentalists in neurophysiology. It is time to start viewing the constellation in full splendor. The chapters in this volume contain ample evidence of the ferment to be found in present-day neurophysiology. The shift from a focus on elements of the nervous system to the more complex aspects of its integrative functions can clearly be gleaned from the materials presented.

SUMMARY

The molecular view of the nervous system organization has dominated conceptualization of its functions. The idea that higher levels of organization may not be inferable from more elemental components of that system is a view to be explored more fully in the future. The units of behavior may not correspond with the units of neuronal activity studied, and such a disparity in focus may frustrate acquisition of an understanding about the behavior. In terms of control, it may be that the higher levels of nervous system organization operate by minimizing the number of degrees of freedom. Not every motor unit of every muscle need be controlled by the central nervous system even though a circuit for doing so may be experimentally demonstrated. Biologically useful behaviors do not consist of isolated actions of single neurons or single muscles or single limbs. The entire complex of the body participates in posture and motion, thus necessitating an expanded focus for studies about nervous system function. Part of that expanded focus should include explicit recognition that manifold sources of afferent activity play upon the several levels of the nervous system in order to provoke some functionally integrated response.

REFERENCES

1. Amassian, V. E. (1951): Cortical representation of visceral afferents. *J. Neurophysiol.*, 14:433–444.
2. Amassian, V. E. (1951): Fiber groups and spinal pathways of cortically represented visceral afferents. *J. Neurophysiol.*, 14:445–460.
3. Amassian, V. E., Weiner, H., and Rosenblum, M. (1972): Neural systems subserving the tactile placing reaction: A model for the study of higher level control of movement. *Brain Res.*, 40:171–178.

4. Anand, B. K., and Pillai, R. V. (1967): Activity of single neurons in the hypothalamic feeding centers: Effect of gastric distension. *J. Physiol.,* 192:63–78.
5. Aubert, M., and Pietri, N. (1965): Aires de projections du nerf splanchnique sur le cortex cerebral du chat anesthesie au chloralose. *J. Physiol.,* 57:218–219.
6. Beritoff, J. S. (1965): *Neural Mechanisms of Higher Vertebrate Behavior,* translated and edited by W. T. Liberson. Little, Brown and Co., Boston.
7. Bernstein, N. (1967): *The Co-ordination and Regulation of Movements.* Pergamon Press, Oxford.
8. Berthoz, A., Pavard, B., and Young, L. R. (1975): Perception of linear horizontal self-motion induced by peripheral vision (linearvection). Basic characteristics and visual-vestibular interactions. *Exp. Brain Res.,* 23(5):471–489.
9. Bessou, P., and Perl, E. R. (1966): A movement receptor of the small intestine. *J. Physiol.,* 182:404–426.
10. Brodal, A. (1973): Self-observations and neuro-anatomical considerations after a stroke. *Brain,* 96:675–694.
11. Brookhart, J. M., Mori, S., and Reynolds, P. J. (1970): Postural reactions to two directions of displacement in dogs. *Am. J. Physiol.,* 218:719–725.
12. Brookhart, J. M., Parmeggiani, P. L., Petersen, W. A., and Stone, S. A. (1965): Postural stability in the dog. *Am. J. Physiol.,* 208:1047–1057.
13. Brookhart, J. M., and Talbott, R. E. (1974): The postural response of normal dogs to sinusoidal displacement. *J. Physiol.,* 243:287–307.
14. Cooper, G. P., and Kimeldorf, D. J. (1964): Electroencephalographic desynchronization in irradiated rats with transected spinal cords. *Science,* 143:1040–1041.
15. Creed, R. S., Denny-Brown, D., Eccles, J. C., Liddell, E. G. T., and Sherrington, C. S. (1932): *Reflex Activity of the Spinal Cord.* Clarendon Press, Oxford.
16. Downman, C. B. B. (1951): Cerebral destination of splanchnic afferent impulses. *J. Physiol.,* 113:434–441.
17. Duda, P. (1964): Facilitatory and inhibitory effects of splanchnic afferentation on somatic reflexes. *Physiol. Bohemoslov.,* 13:137–141.
18. Duysens, J., and Pearson, K. G. (1976): The role of cutaneous afferents from the distal hindlimb in the regulation of the step cycle of thalamic cats. *Exp. Brain Res.,* 24:245–255.
19. Efron, R. (1966): The conditioned reflex: A meaningless concept. *Perspect. Biol. Med.,* 9:488–514.
20. Elftman, H. (1939): Forces and energy change in the leg during walking. *Am. J. Physiol.,* 125:339–366.
21. Engberg, I. (1964): Reflexes to foot muscles in the cat. *Acta Physiol. Scand.* 62:(Suppl. 235).
22. Engberg, I., and Lundberg, A. (1962): An electromyographic analysis of stepping in the cat. *Experientia,* 18:174–176.
23. Engberg, I., and Lundberg, A. (1969): An electromyographic analysis of muscular activity in the hindlimb of the cat during unrestrained locomotion. *Acta Physiol. Scand.,* 75:614–630.
24. Ferrier, David (1876): *The Functions of the Brain.* G. P. Putnams Sons, New York.
25. Ferrier, David (1886): *The Functions of the Brain,* 2nd ed. Smith, Elder and Co., London.
26. Fields, H. L., Partridge, L. D., Jr., and Winter, D. L. (1970): Somatic and visceral receptive field properties of fibers in ventral quadrant white matter of the cat spinal cord. *J. Neurophysiol.,* 33:827–837.
27. Forssberg, H., Grillner, S., and Rossignol, S. (1975): Phase dependent reflex reversal during walking in chronic spinal cats. *Brain Res.,* 85:103–107.
28. Forssberg, H., Grillner, S., and Sjostrom, A. (1974): Tactile placing reactions in chronic spinal kittens. *Acta Physiol. Scand.,* 92(1):114–120.
29. Gelfand, I. M., Gurfinkel, V. S., Fomin, S. V., and Tsetlin, M. L., editors (1971): *Models of the Structural-Functional Organization of Certain Biological Systems.* The MIT Press, Cambridge, Mass.
30. Gelfand, I. M., Gurfinkel, V. S., Tsetlin, M. L., and Shik, M. L. (1971): Some problems in the analysis of movements. In: *Models of the Structural-Functional Organization of Certain Biological Systems,* edited by I. M. Gelfand, V. S. Gurfinkel, S. V. Fomin, and M. L. Tsetlin, pp. 329–345. The MIT Press, Cambridge, Mass.
31. Gernandt, B., and Zotterman, Y. (1946): Intestinal pain: An electrophysiological investigation on mesenteric nerves. *Acta Physiol. Scand.,* 12:56–72.

32. Goslow, G. E., Jr., Reinking, R. M., and Stuart, D. G. (1973): The cat step cycle: Hind limb joint angles and muscle lengths during unrestrained locomotion. *J. Morphol.*, 141:1–42.
33. Greene, P. H. (1971): Introduction. In: *Models of the Structural-Functional Organization of Certain Biological Systems*, edited by I. M. Gelfand, V. S. Gurfinkel, S. V. Fomin, and M. L. Tsetlin. The MIT Press, Cambridge, Mass.
34. Greenwood, R., and Hopkins, A. (1976): Muscle responses during sudden falls in man. *J. Physiol.*, 253:507–518.
35. Grillner, S. (1975): Locomotion in vertebrates: Central mechanisms and reflex interaction. *Physiol. Rev.*, 55(2):247–304.
36. Haour, R., Massarino, R., Massion, J., and Swett, J. E. (1976): A device used for study of postural reactions in the quadruped. *EEG Clin. Neurophysiol.*, 40(4):427–431.
37. Hunt, E. L., and Kimeldorf, D. J. (1963): Arousal reactions with a brief partial- and whole-body X-ray exposure. *Nature*, 200:536–538.
38. Irving, J. T., McSwiney, B. A., and Suffolk, S. F. (1937): Afferent fibres from the stomach and small intestine. *J. Physiol.*, 89:407–420.
39. Ito, T., and Sanada, Y. (1965): Location of receptors for righting reflexes acting upon the body in primates. *Jpn. J. Physiol.*, 15:235–242.
40. Korn, H. (1969): Splanchnic projection to the orbital cortex of the cat. *Brain Res.*, 16:23–38.
41. Leek, B. F. (1972): Abdominal visceral receptors. In: *Enteroreceptors, Handbook of Sensory Physiology, Vol. III/1*, edited by E. Neil. Springer-Verlag, Berlin.
42. Lissin, V. V., Frankenstein, S. I., and Rechtmann, M. B. (1973): The influence of locomotion on flexor reflex of the hindlimb in cat and man. *Exp. Neurol.*, 38:180–183.
43. Magnus, R. (1926): On the co-operation and interference of reflexes from other sense organs with those of the labyrinth. *Laryngoscope*, 36:701–712.
44. Magnus, R. (1926): Some results of studies in the physiology of posture. *Lancet*, 211:531–536, 585–588.
45. Manter, J. T. (1938): The dynamics of quadrupedal walking. *J. Exp. Biol.*, 15:522–540.
46. Massion, J., and Smith, A. M. (1973): Activity of ventrolateral thalamic neurons related to posture and movement during contact placing responses in the cat. *Brain Res.*, 61:400–406.
47. Massion, J., and Smith, A. M. (1974): Ventrolateral thalamic neurons related to posture during a modified placing reaction. *Brain Res.*, 71:353–359.
48. Melvill Jones, G., and Watt, D. G. D. (1971): Observations on the control of stepping and hopping movements in man. *J. Physiol.*, 219:709–728.
49. Melvill Jones, G., and Watt, D. G. D. (1971): Muscular control of landing from unexpected falls. *J. Physiol.*, 219:729–738.
50. Miller, S., Van der Burg, J., and Van der Meche, F. G. A. (1975): Co-ordination of movements of the hindlimbs and forelimbs in different forms of locomotion in normal and decerebrate cats. *Brain Res.*, 91:217–237.
51. Money, K. E. (1970): Motion sickness. *Physiol. Rev.*, 50:1–39.
52. Mori, S., and Brookhart, J. M. (1968): Characteristics of the postural reactions of the dog to a controlled disturbance. *Am. J. Physiol.*, 215:339–348.
53. Morrison, J. F. B. (1973): Splanchnic slowly adapting mechanoreceptors with punctate receptive fields in the mesentery and gastrointestinal tract of the cat. *J. Physiol.*, 233:349–361.
54. Mountcastle, V. B. (1957): Modality and topographic properties of single neurons of cat's somatic sensory cortex. *J. Neurophysiol.*, 20:408–434.
55. Mountcastle, V. B. (1976): The world around us: Neural command functions for selective attention. (The F. O. Schmitt lecture in Neuroscience, 1975). *Neurosciences Res. Prog. Bull. (Suppl.)*, 14:1–47.
56. Nathan, P. W. (1976): The gate-control theory of pain. A critical review. *Brain*, 99:123–158.
57. Nelson, R. C., Dillman, C. J., Logasse, P., and Bickett, P. (1972): Biomechanics of overground versus treadmill running. *Medicine and Science in Sports*, 4(4):233–240.
58. Newman, P. P. (1962): Single unit activity in the viscero-sensory areas of the cerebral cortex. *J. Physiol.*, 160:284–297.
59. Newman, P. P. (1974): Visceral afferent functions of the nervous system. *Monographs of the Physiological Society, No. 25.* Edward Arnold, London.
60. Paintal, A. S. (1973): Vagal sensory receptors and their reflex effects. *Physiol. Rev.*, 53(1):159–227.

61. Ploog, D. (1971): The relevance of natural stimulus patterns for sensory information processes. *Brain Res.,* 31:353–359.
62. Rossignol, S., and Melvill Jones, G. (1976): Audio-spinal influence in man studied by the H-reflex and its possible role on rhythmic movements synchronized to sound. *EEG Clin. Neurophysiol.,* 41:83–92.
63. Selzer, M., and Spencer, W. A. (1969): Convergence of visceral and cutaneous afferent pathways in the lumbar spinal cord. *Brain Res.,* 14:331–348.
64. Selzer, M., and Spencer, W. A. (1969): Interactions between visceral and cutaneous afferents in the spinal cord: Reciprocal primary afferent fiber depolarization. *Brain Res.,* 14:349–366.
65. Sheehan, D. (1933): The afferent nerve supply of the mesentery and its significance in the causation of abdominal pain. *J. Anat. (Lond.),* 67:233–249.
66. Sherrington, C. S. (1928): Sir David Ferrier, 1843–1928. *Proc. R. Soc. Lond.* [Biol.] 103B:VIII–XVIII.
67. Sherrington, Sir Charles S. (1906): *The Integrative Action of the Nervous System,* 1947 ed., Yale University Press, New Haven.
68. Soechting, J. F. (1973): Modeling of a simple motor task in man: Motor output dependence on sensory input. *Kybernetic,* 14(1):25–34.
69. Stern, J. T., Jr. (1974): Computer modelling of gross muscle dynamics. *J. Biomechanics,* 7(5):411–428.
70. Szentagothai, J., and Arbib, M. A. (1974): Conceptual models of neural organization. *Neurosci. Res. Program Bull.,* 12(3).
71. Talbott, R. E. (1970): The effect of beta radiation on mesenteric Pacinian corpuscle response. *Radiation Res.,* 42(3):539–551.
72. Talbott, R. E. (1973): Postural control: A quantitative study of nervous system functions in the dog. In: *Control of Posture and Locomotion,* edited by R. B. Stein, K. G. Pearson, R. S. Smith, and J. B. Redford, pp. 273–289. Plenum Press, New York.
73. Talbott, R. E. (1974): Modification of the postural response of normal dogs by blindfolding. *J. Physiol.,* 243:309–320.
74. Talbott, R. E., Jones, D. C., and Kimeldorf, D. J. (1969): Alteration of a small-intestine mechanoreceptor response by beta radiation. *Exp. Neurol.,* 23:366–377.
75. Tokuriki, M. (1973): Electromyographic and joint-mechanical studies in quadrupedal locomotion. I. Walk. *Jpn. J. Vet. Sci.,* 35:433–446.
76. Tokuriki, M. (1973): Electromyographic and joint-mechanical studies in quadrupedal locomotion. II. Trot. *Jpn. J. Vet. Sci.,* 35:525–533.
77. Tokuriki, M. (1974): Electromyographic and joint-mechanical studies in quadrupedal locomotion. III. Gallop. *Jpn. J. Vet. Sci.,* 36:121–132.
78. Towe, A. L. (1975): Notes on the hypothesis of columnar organization in somatosensory cerebral cortex. *Brain, Behav. Evol.,* 11(1):16–47.
79. Townsend, M., and Seireg, A. (1972): Optimal trajectories and controls for systems of coupled rigid bodies. *J. Engrg. for Industry,* 94:472–482.
80. Uttal, W. R. (1967): Evoked brain potentials: Signs or codes. *Perspect. Biol. Med.,* 10:627–639.
81. Viviani, P., and Terzuolo, C. A. (1973): Modeling of a simple motor task in man: Intentional arrest of an ongoing movement. *Kybernetic,* 14:35–62.
82. Wetzel, M. C., Atwater, A. E., Wait, J. V., and Stuart, D. G. (1975): Neural implications of different profiles between treadmill and overground locomotion timings in cats. *J. Neurophysiol.,* 38:492–501.
83. Young, Robert M. (1970): *Mind, Brain and Adaptation in the Nineteenth Century: Cerebral Localization and Its Biological Context from Gall to Ferrier.* Clarendon Press, Oxford.

Posture and Movement, edited by
R. E. Talbott and D. R. Humphrey.
Raven Press, New York © 1979.

Motor Programs Revisited

Vernon B. Brooks

Department of Physiology, University of Western Ontario, London, Ontario, Canada, N6A 5C1

INTRODUCTION

What do we mean by "motor programs"? One definition could be that: "Motor programs are communications in the central nervous system that are based on past experience and that can generate postural adjustments and movements." This definition is less specific than that put forward by Keele (85) ten years ago: ". . . a set of muscle commands that are structured before a movement sequence begins, and that allows the entire sequence to be carried out uninfluenced by peripheral feedback." The present formulation invites, even mandates, consideration of different levels of complexity, yet it allows that some "programmed" movements can be influenced by sensory input under some circumstances. I will try to clarify this and to illustrate how some brain parts may participate, with strong emphasis on strategies of movements. The sequence to be followed is listed below. A brief summary is appended to each topic in the text.

1. Hierarchies of Motor Programs
2. Use of Programs in Different Kinds of Movements
   a. Simple movements of moderate speed
   b. Ballistic movements
3. Properties of Motor Programs
   a. How separate are the neural substrates for hierarchical levels?
   b. What movement properties are programmed?

It is my particular pleasure on this occasion to base my presentation on some experiments begun 10 years ago, (and their sequelae), that were first published in the *Journal of Neurophysiology* (14,147), after generous, kindly, and careful correspondence by the editor, Professor J. M. Brookhart.

HIERARCHIES OF MOTOR PROGRAMS

Motor programs are not all of one kind. Anyone watching a person or animal go about their daily tasks, let alone perform standardized motor tests, comes away with the impression that *general decisions* to act come first, followed by selection of *strategies* of how to act, that are resolved into "*set*" also based on motor learning, leading to *preparations for movement execution,* and finally *combinations of tactical detail.* Subjects seem to estimate the muscular efforts needed for a movement by perception of the situation or from memory. These estimates refer to many parts of the body in concert: trunk and head, as well as limbs. Thus "high-level" programs are likely to "set" many task-related processes, so that movements can be made in the *contexts* of the intention as well as the expected degree of readiness of being able to carry it out, with or without adjustments (see Use of Programs, below). These hierarchies of programs, going from the general to the particular, are not envisaged to generate movements by simple streams of instructions in a "top-down" sequence of brain structures, but rather by multiple feedback and feedforward interactions, whose neural substrates are touched on below, and whose integration with feedback-controlled events is discussed under Movement Properties, below.

Bernstein (5) anticipated a hierarchical motor organization from the peripheral to the conscious level, more than 40 years ago. He supported this view with the observation that the degrees of freedom increased from a minimum for infants as coordination was learned, and that the shape of learned skilled movements was independent of alterations in their size or of the motor structure involved in implementing the task, and with mathematical analysis of the degrees of freedom for each level of motor control. He argued that a corresponding hierarchical structure in the localization of motor skill must exist in the brain. As he put it, ". . . the central effectors achieve coordination of movements only by plastically reacting to the totality of signals from the afferent field, adapting the impulses transmitted to the situation that actually obtains. . . ." The relation between intent and execution has been discussed in the psychological literature for about 100 years. Some examples of this older work will be juxtaposed to recent findings to illustrate the evolution of modern Neuroscience. R. S. Woodworth examined how human subjects executed movements whose intended end-points were well rehearsed. The following quote from his monograph published in 1899 (157) is poignant today: "Some doubt may arise whether my work is really psychological or physiological. The field of voluntary movement undoubtedly lies, like the field of sensation, in the borderland between the two sciences. On which side of the border the present studies lie, I have not thought it worth while to attempt to decide."

The main thrust of the present contribution is to stress how programs for learned movements of laboratory animals may derive from *past experience,* and that results obtained with animals resemble and fit observations made on human subjects. This sustains our hope that experimental Neuroscience will guide us to deeper understanding of human function and behavior.

A number of useful books may be cited for the interested reader: texts by Poulton (115) and Welford (155), an introductory overview by Keele (86), a recent monograph by Phillips and Porter (112) and an older one by Granit (59), an older review by Paillard (109), as well as recent compendia edited by Desmedt (36), Evarts et al. (43), Stelmach (125), Rose (118), and Homma (73).

USE OF PROGRAMS IN DIFFERENT KINDS OF MOVEMENTS

Stetson (128) summarized his views on movements and posture obtained in his work spanning 30 years of studies concerning the mechanics and EMG of tapping movements of human limbs. "The excursion of the limb may be divided into three phases: (1) the sudden impulse . . . due to the momentary contraction of the unopposed driving muscles; (2) the free phase, during which the limb swings by its own momentum and has no dynamic connection with the pivot joint; (3) the arresting phase, during which an outside obstacle, or the contraction of the antagonists stops the movement. . . ."

His classification of movements, given below, is still useful today:

I. Slow, tension movements in which groups of muscles contract in opposition to each other.
   1. The "fixation." The tension movement . . . of holding still. "Posture" is a name for the system of fixations. . . .
   2. The tension movement which is a (slow) moving fixation.

II. Ballistic movements in which the muscle group contracts without opposition and the limb swings free.
   1. The loose movement, in which there is no contraction of the opposing muscles, and the movement is a pure ballistic movement.
   2. The stiff movement, in which the sudden pulse of the ballistic contraction supervenes on the fixation of the limb. There is continuous tension between the muscle groups throughout the movement cycle.

Simple Movements of Moderate Speed

The selection of strategy is critical when human or animal subjects must terminate their movements accurately without the assistance of mechanical stops in a prescribed target zone. We will restrict ourselves to consideration of "simple" movements, in the sense of Gordon Holmes (72), ". . . such as flexion of the forearm at the elbow . . . ," i.e., movements restricted as much as possible to one joint and to one plane. Woodworth (157) entered the arena of strategy and tactical control when he found that such movements depend on the speed and also on the accuracy that is required. His subjects drew with a pencil back and forth between mechanical stops or between ruled lines on kymograph paper moving past them, to the beat of a metronome. Thus he could compare how accurately the hand could move to the targets at different movement speeds, which however, never reached those for ballistic tapping against a surface (126,127). Woodworth described a continuous "initial impulse," and a subsequent period of "current control" during which movement oscillated and often also seemed to be corrected with deliberate, little adjustment steps. Current control, marked by oscillations, was lengthened in relation to the degree of difficulty of getting into the target zone (157). Since current control slowed the movement, this yielded an "inverse relation between speed and accuracy," illustrated in Fig. 1 from Woodworth's monograph (157). He noticed that the movement ". . . governed by the eye is much the most accurate at low speeds, and that the movements with eyes shut [are] less accurate than this. . . . But

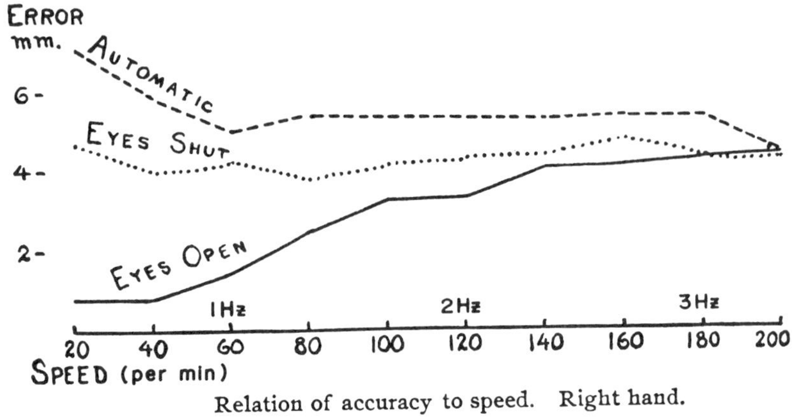

FIG. 1. Woodworth's "composite diagram" showing his "inverse relation" of accuracy to speed [reproduced from Fig. 3 of his monograph (157), unchanged except that units of "speed" have been added]. The lines are based on averaged data from 4 young subjects tested 50 to 300 times at each speed (14-cm displacement); details are given in his monograph. The subjects had to draw (to the beat of a metronome) with a pencil on paper on a kymograph rotating on a horizontal axis. The "targets" consisted either of lines drawn on the paper "at right angles to the direction of the movements" or of "stops," consisting of "straight edges of brass, exposing the moving paper, that was otherwise covered by a drum." The broken line labeled "Automatic" refers to movements when subjects had been instructed not to pay attention, but to, for instance, look out the window instead.

though this is true at low speeds, it is less and less true as speed is increased. . . . Beyond [2 to 2.5 Hz] (140 to 160 movements per min) . . . it is no longer possible to control movements separately. Much has to be left to the automatic uniformity of the hand's movements . . ." (157). This well-known phenomenon is common for movements at certain speeds of man and animals. Training apparently makes subjects estimate the "ballistic initial impulse" needed for the extent of the movement and thus to minimize final "current control." Faster movements have stronger "initial impulses," whereas slower movements have longer periods of "current control." Woodworth's "inverse relation" has since been restudied, and modified many times (cf. 52, 86, 115, 155), as will be taken up in Movement Properties.

Rapid movements, made with confidence that they will achieve their task, have offered the greatest promise to demonstrate programmed "initial impulses," because movements can be so brief as to preclude adjustments by visual, or perhaps even proprioceptive, peripheral feedback. This old problem has been well reviewed in the volumes cited at the end of the section on Hierarchies (above) and has recently been brought up to date by Evarts and Granit (49), who tested subjects' contraction or relaxation of their biceps brachii when holding a handle attached to a torque motor. They had to either supinate or pronate the wrist according to an instruction that preceded the move! command consisting of a torque-pulse perturbation exerted on the handle. They found very brief feedback times: ". . . an intended muscle discharge can occur with a latency of as little as 60 msec following a kinesthetic input whose reflex effects are inhibitory. Such discharge cannot be classified as a reflex since it is occurring in spite of, rather than because of, the reflex effects of the input. . . ." Their discussion refers back to analagous experiments by Hagbarth (62) and related ones by Hammond (64). Previous work with trained monkeys by Evarts (44) and Evarts and Tanji (47) had already demonstrated equivalent short latencies for triggering intended movements, according to a visual instruction delivered prior to the perturbation move! command. Thus, as modern methods have revealed ever-briefer reaction times, "proof" of programming on the basis of latencies has had to be reduced to "suggestion." Stratagems that rely on initiation of movements in an intended direction have indeed defined the very brief time needed for central processing. Occurrence of corrections towards such intended movements had been shown by Angel et al. (3), whose five human subjects had to perform a step-tracking task following a "target" line, that jumped from a central position to either side on an oscilloscope screen, by moving a handle whose position was translated into movement of a cursor-line on that oscilloscope. The subjects learned this well even if the central part of the oscilloscope was masked. When the translation of handle direction was made to move the cursor in the opposite direction, subjects could still correct for this, even without seeing their movements in the wrong direction, once they knew from one previous trial that their direction had been wrong (16 of 24 such errors were so corrected). The authors concluded that: "These results support the idea that subjects make use of a short latency central process in correcting motor errors."

Let us now examine movement *strategies* as used by monkeys in our original experiments (14,147) when performing tasks similar to those of Woodworth's subjects. The monkeys were trained to turn a freely moving handle (Fig. 2B) by elbow movements into learned target zones without mechanical stops, with apparatus similar to that in Fig. 4B. The arrangement resembled that used in studies of human tracking behavior by Taylor and Birmingham (136) and Angel et al. (3), in that the animals were prevented from seeing the task area through which they moved the handle (by an opaque plate, not shown in Fig. 4B). The task was to reposition the handle after the target zone had jumped in a step from one place to another (a step-tracking task, not visual pursuit tracking). The animals were cued for the direction in which to move by indicator lights to the left and right in front of them, but the target zone was not displayed for them. They learned the required positions of the target zones because a "target light" and/or "target sound" came on when the handle had been positioned correctly, and fluid reward was supplied through a tube in front of their mouths. There was no time constraint on the self-initiated and self-terminated movements.

The easier the task was made for reward, i.e., the wider the target zones, the shorter was terminal "current control," to the point of nonexistence. Some animals that never learned to work with narrow targets nevertheless performed

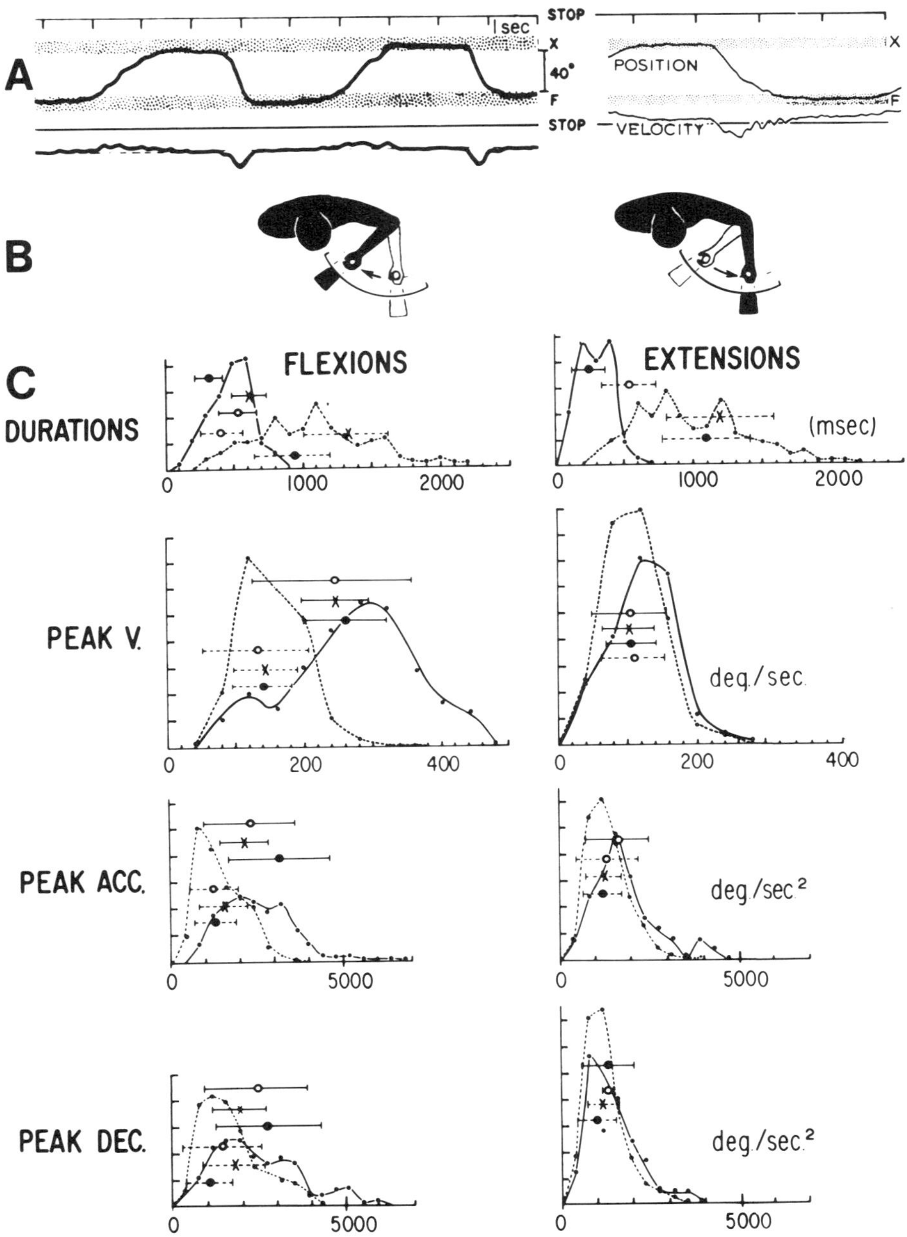

FIG. 2. Properties of two types of elbow movements, made at moderate speed by Cebus monkeys into learned target zones (*stippled bands:* 10 deg width, 50-deg center separation). "Continuous" and "Discontinuous" movements indicated by *solid* and *dotted lines* respectively.

well with wide targets, by making movements that were apparently sufficiently surely propelled by the initial impulse so as to terminate in the targets with none, or at worst only one final adjustment (e.g., flexions in Fig. 2A, left, and last adjustment in third row of Fig. 3). These movements, which we called "continuous," resemble Stetson's "stiff ballistic" ones. When the monkeys first learned to perform such movements, even with target cues, or if later they seemed uncertain about achieving the target for any reason, they used another strategy in which the initial impulse fell short of the target, and the handle was placed inside the target zones after some current control, e.g., as a terminal tremor as in Fig. 2A, right. Such movements we called "discontinuous," as well as those that oscillated throughout the movement (Fig. 2A, left) (14,16, 90,147). That type corresponds to Stetson's "slow tension" movements. Figure 2A (left and right) illustrates that oscillations (at about 3 Hz) appeared more clearly in the velocity traces. Properties of both types of movement are plotted in Fig. 2C, revealing consistent differences between flexions and extensions.

There was an inverse relation between the speed of the movement and the occurrence of obvious "discontinuities" or "oscillations" in them in many monkeys so tested. When the animals had experienced success from previous experience or when quicker movements were rewarded, then faster, "continuous" movements replaced the slower oscillating type. Later computer analysis, without the filtering imposed by the original paper records, revealed that "continuous" movements had a fine tremor of about 6 to 7 Hz (15) illustrated in Fig. 5A. Some animals always used one or the other strategy, while others mixed them. Flexions were more often "continuous" than extensions (Fig. 2A, left). These movements were regarded by us initially as alternatives (13,16,90,92). The analogy was stressed from the outset (90) to the demonstration by Navas and Stark (107, and see 11 for review) that human wrist movements can track visual targets in the "pursuit-correction" mode only up to 3 Hz; for higher frequencies they have to switch to a "predictive" strategy. Later it was recognized (15,16) that continuous and discontinuous movements probably represent net balances

(cont.) **A:** Extension (X) above (F) in sample traces, located well inside the mechanical "stops" drawn as horizontal lines. Upper stop-line carries time markers (sec) for both tracings. Velocity traces appear below position traces, made with pen recorders (linear to 60 Hz, 50% at 99 Hz). [Left sample: M4L, 1/6/69, reproduced from Fig. 5 in ref. 13. Right sample: showing infrequently occurring terminal tremor: M4L, 3/21/69, from Fig. 8 in ref. 14, V. B. Brooks, I. Kozlovskaya, A. Atkin, F. Horvath, and M. Uno, *J. Neurophysiol.,* 36:974–995 (1973).] Task conditions described in the text, and indicated diagrammatically in **B.** The set-up used resembled that in Fig. 4C, except that a less bulky handle was used. **C:** Histograms of movement parameters averaged from trials with 3 well-trained Cebus monkeys (M3L, ●; M7L, O; M4L, ×). Task conditions as in **A** and **B.** Units for histogram ordinates are 10 movements in all cases, except 4 for the rarer continuous extensions. Horizontal bars indicate standard deviations (SDs). [**B** and **C** of this figure appeared in unfinished form in a preliminary communication, ref. 91. The data for **C** consist of 870 movements made as controls in other experiments, and listed in detail as to number, type, source, date, etc., as controls ("CB") in Table 1A of ref. 14.]

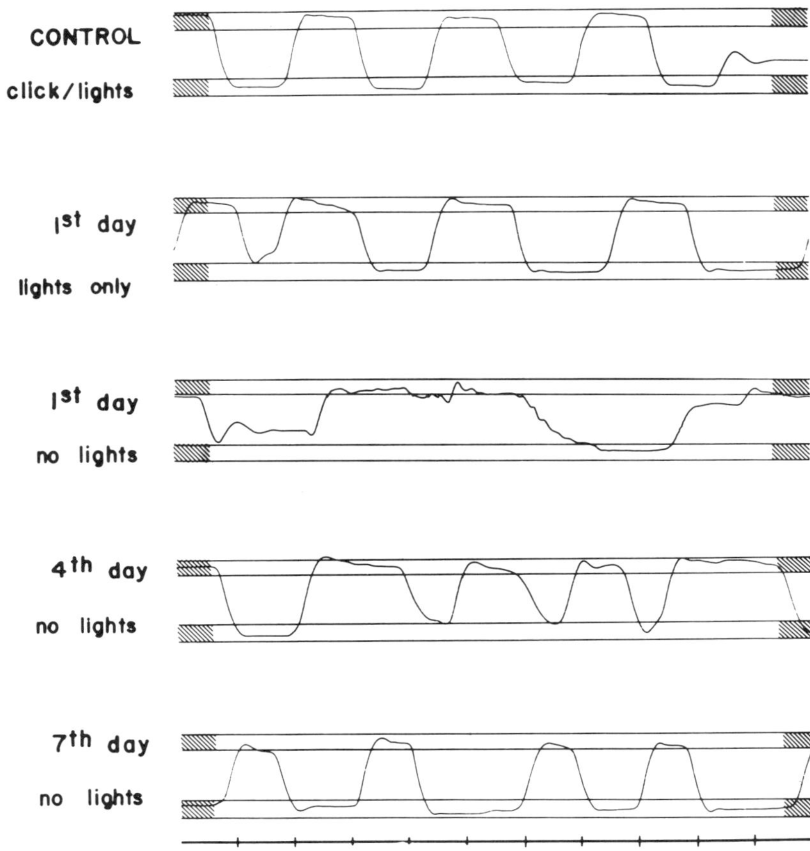

FIG. 3. Relearning of elbow movements of moderate speed made during progressive withdrawal of exteroceptive target cues. (Cebus monkey, similar arrangement to that in Fig. 2.) Extension (X) and flexion (F) target widths (15 deg) indicated as *stippled bands,* target centers separated by 50 deg. Time markers: 1 sec. Control situation in uppermost tracing ("click/lights"): "In target" indicated by target light, and "success" signaled by a "success" light as well as the "click" from the mechanical success-counter and/or by arrival of juice reward (reward ratios varied from 7:1 to 1:1). First day ("lights only"): Light cues as before but success counter no longer used. Rewards were now given after a variable delay so that juice-dispenser click and juice arrival were no longer available as position cues. 1st, 4th, and 7th days ("no lights"): Light cues were removed as well, otherwise as above. Tracings representative of 600 to 1,000 "continuous" movements made daily as observed in 2 monkeys. (M15L, 3/7–17, 1972.) [Reproduced from Fig. 2 in ref. 92, I. B. Kozlovskaya, A. Atkin, F. E. Horvath, J. S. Thomas, and V. B. Brooks: Preprogrammed and feedback-guided movements of monkeys. *Behav. Biol.,* 12:243–248 (1974).]

between two extremes in a range of cyclic force applications, since graded amplitudes of oscillations can blur rigid distinctions (e.g., Fig. 1 in ref. 16). The two main frequencies—near 3 Hz and 6 Hz—are real, however, and their coexistence has been confirmed by power spectral density analysis of (largely isotonic) elbow movements with two other monkeys (24). More recently, Neilson and

Lance (107a) have found identical peaks in spectral analysis of (largely isometric) muscle forces exerted on the elbow by humans.

After monkeys had learned the task with the aid of auditory and visual target cues, some evidence was obtained with 3 animals that fast, continuous movements could be made without tactual, visual, or auditory feedback, by unexpected withdrawal of target cues or by their unexpected insertion in unusual parts of the trajectory. Under those circumstances the movements ran their usual course. Slow, discontinuous movements, in contrast, consistently overshot when target cues were omitted, revealing their dependence on auditory or visual feedback (14,16,92). An example of successful use of "continuous" movements without visual or auditory cues, made with a fourth monkey, is illustrated in Fig. 3 (92). In such trials animals could presumably still match their memory of target position to proprioceptive information about the state of their arms.

Summary

Monkeys, like humans, can use different strategies according to their expectation of success. Relatively fast ("continuous" or "stiff ballistic") movements, but yet of moderate speed, are launched with a large "initial impulse," and can be directed correctly into learned target zones without visual, auditory, tactual feedback. In contrast, slower ("discontinuous" or "slow tension") movements, with a small "initial impulse," require some such feedback for successful target acquisition. The former type gave way to the latter when successful use of these movements was impeded in any manner. The faster type could have been guided by peripheral proprioceptive feedback or by central re-efference in ongoing manner, or through triggering central programs based on previous experience.

Ballistic Movements

The main characteristics of ballistic contractions were well described by Richer (116), who first so named them in 1895, on the basis of photographing the shape of the human quadriceps muscle during the act of kicking: ". . . It is very energetic and short lasting. It launches the limb in a set direction, and ceases long before the limb will have completed its course of action. . . ." The strategy of ballistic alternating movements requires appropriate supraspinal presetting of the final common path and use of internal re-efference, as already stated by Stetson in 1905 (126): "That such a reciprocal innervation does take place in the ballistic movement is evident, but it must be a type of coordination dependent on higher centers. . . . What furnishes the precise signal for the action of the negative muscles in catching the limb tossed to them by the positive set? [i.e., protagonists and antagonists (VB)] . . . It is probable that the cue which sets the negative set to contracting is the contraction of the positive muscle-set itself."

A demonstration of programmed "loose" ballistic arm movements (cf. 128) is illustrated in Fig. 4A [Conrad and Brooks, (20)]. Traces in the top row

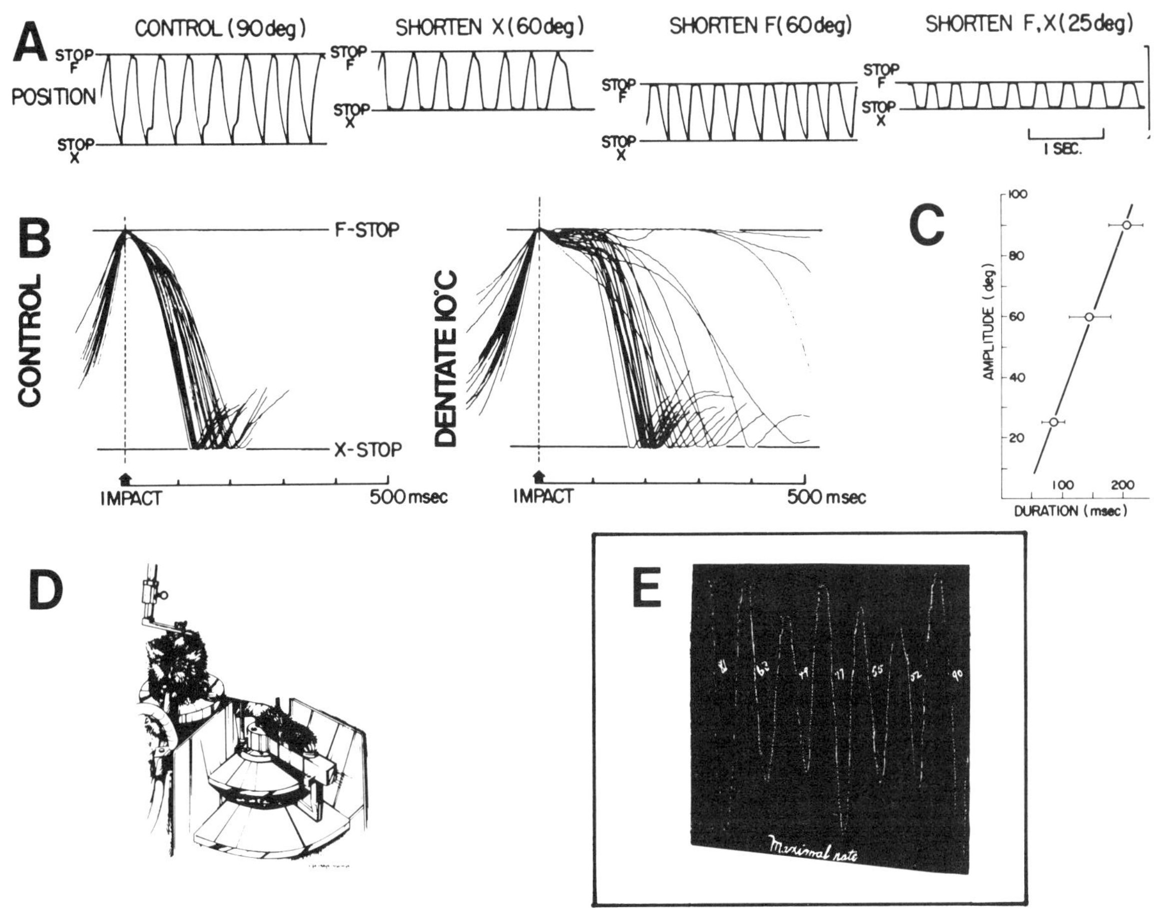
A
CONTROL (90 deg)
SHORTEN X (60 deg)
SHORTEN F (60 deg)
SHORTEN F, X (25 deg)
STOP F
POSITION
STOP X
1 SEC.
B
CONTROL
F-STOP
X-STOP
IMPACT
500 msec
DENTATE 10°C
C
AMPLITUDE (deg)
DURATION (msec)
100
80
60
40
20
200
D
E
81
63
79
77
55
52
90
Maximal rate

show position of a handle turned by a monkey by rapid elbow flexions and extensions as he slammed it back and forth between the mechanical stops (representative of 3 animals so tested), and as sketched in Fig. 4D. Traces of the sort shown in Fig. 4A (top left) appear on an expanded time-scale in Fig. 4B, left. The animal was rewarded according to the number of movements made. When the stops were moved closer together unexpectedly, by shortening extensions (X) or flexions (F) or both (Fig. 4A), the original rhythm of rapid alternations was maintained as learned for the usual "control" separation of the stops, without systemic decreases of duration errors: Fig. 4C (cf. section 4b). The animal continued to exert muscular control for the usual length of time, with the result that the handle was pressed against the stops when it would otherwise have been moved through the usual, expected arc of displacement. In the large sample, of which Fig. 4A is representative, all other parameters remained constant: accelerations, mean and peak velocities, and durations. *The rhythm of muscular effort, or some correlate of it, seems to have been programmed.* This conclusion is also supported by the fact that for the expected "control" lengths of movement, protagonist EMGs usually ceased 10 to 40 msec before impact of the handle on the mechanical stops (not illustrated), discounting peripheral feedback (also cf. 36a). The same conclusion was drawn by Hallett, Shahani, and Young (63) for ballistic elbow flexions made by 18 normal human subjects and by an essentially deafferented patient step-tracking an oscilloscope target by moving a handle. The patient had the same durations of biceps activity and previous inhibition of triceps activity. (Figure 4B, right, will be considered in relation to involvement of different brain parts (in the section on Properties of Motor Programs) and some consequences of that also. Recruitment according

←

FIG. 4. Traces of ballistic elbow movements made by a Cebus monkey between mechanical stops, demonstrating programmed movements. **A:** Movements made after unexpected changes of movement range. In the second, third and fourth tracings the mechanical stops for extension (X), flexion (F), and both were moved towards the center. Calibrations for position: 90 deg. Tracings representative of 877, 214, 306, and 392 movements so measured. (M16L, 7/30/72.) **B:** Overplots of consecutive position traces for extensions demonstrating neocerebellar effect on rhythm. Left overplots: normal brain temperature. Right plots: dentate cooled to 10°C. (M16L.) Traces are aligned at moment of impact, as indicated by *vertical broken lines* and *arrows,* by a PDP-12 computer [10-bit A-D conversion, 200-Hz sampling rate (M16L, 7/18/72)]. **C:** Regression line of amplitudes on mean durations (± SD) of movements in **A.** **D:** Arrangement: Monkey in experimental chair, holding handle of the manipulandum. Note hinged neck-restraining collar, head-holder, juice-dispensing tube, and mechanical stops at end of the traverse. [**A, B, C,** and **D** reproduced from Figs. 4, 6, and 1, from ref. 20, B. Conrad and V. B. Brooks, *J. Neurophysiol.,* 37:792–804 (1974).] **E:** Reproduction of Stetson's kymograph record of "up-and-down movements of the hand and arm," made by a human subject under task conditions similar to those of the third tracing in **A:** ". . . an obstacle against which the limb strikes does not change the character of the movement; the action of the muscles is the same as if the limb had shot down to the end of the interrupted stroke." Poor alignment of records is probably due to the fact that ". . . the subject held a baton to which a delicate rubber cord was attached. Through guides this cord passed parallel to the axis and close to the surface of the kymograph cylinder. The writing point was attached to the rubber cord itself and a slight torsion was sufficient to keep it pressed against the surface of the drum." (**E** reproduced from Fig. 1 in Stetson, ref. 126.)

to Henneman's principle (small, tonic motoneurons before large, phasic ones) obtains for ballistic contractions of the normal human first interosseous muscle (36a, but also an opposing view in ref. 60a).

We are reminded of Stetson's finding (126), illustrated in Fig. 4E, that durations of ballistic efforts are independent of the lengths of movements carried out, tested with human subjects who made similar movements to those described in Fig. 4A (also cf. 155). His description is classic:

> In rapidly repeated strokes of this kind, the member is returned to the original position by a similar sudden contraction of the negative muscle-set, which relaxes in the middle of the blow, and the stroke is in turn checked by the positive muscle-set. Thus the limb is thrown back and forth, and caught in turn at the limits of its movement by the positive muscle-sets. Such a battledore-and-shuttlecock type of movement is very aptly termed "ballistic.". . . An obstacle against which the limb strikes does not affect the character of the movement; at the end of the normal interval the negative muscle-set contracts and withdraws the limb, as if the limb had shot to the end of its course unimpeded. It is simply as if the lower part of the oscillation had been cut off by the obstacle, and its place taken by a pause at the obstacle . . . ."

Terzuolo and co-workers have sought evidence against the concept of complete preprogramming of ballistic movements in the sense of Keele's definition (85) as quoted in the Introduction. The question at issue now is not, however, whether ballistic movements are programmed in their entirety, but which of their aspects are so specified. Terzuolo's group trained human subjects and squirrel monkeys to make ballistically initiated, forearm upward flexions when they could overcome a restraining force (122,137,138,139). [This "release-and-let-go" paradigm differed in one aspect as used for humans and animals. The human subjects had to overcome just about the maximal tension that each was able to exert, and then to continue "as fast as possible" (139), whereas the tension for the monkeys was varied at random. This may account for the observation that the 5 monkeys cocontracted biceps and triceps prior to movement onset (122) whereas the 7 humans did not (139)]. Following movement onset, EMG of biceps and triceps were reciprocally organized in both humans and monkeys, with several alternate bursts during each movement. End of the first biceps burst was followed by its silent period, during which time triceps discharged and the movement decelerated unintentionally, until propelled faster once more during subsequent biceps bursts. The program for ballistic take-off was examined by instructing the human subjects to "arrest intentionally the ongoing movement as soon as possible after its onset and to maintain the position attained." Intentional deceleration commenced exactly at the same time as the unintentional one described above: after the initial biceps burst of activity, i.e., when the antagonist triceps began its discharge. The duration of the initial burst, and hence the peak acceleration reached by the movement, depended on the inertial load that had to be overcome, which was adjusted without the subjects' knowledge by shifting a mass along the lever which measured the displacement of the arm. In analagous studies with monkeys (122,137,138), the loads that the animals had to overcome were presented in random order. The monkeys had been trained to initiate the movements but not to carry them to a definite endpoint, because they were "rewarded in practically all trials—the target zone encompassing virtually the entire physiological range of motion" (122). Durations of biceps EMG bursts and the subsequent pause were prolonged when inertial mass increased, whereas peak acceleration increased according to the resistive force that the animals had to overcome. The authors concluded that ". . . all the discrete features of the motor output pattern . . . are dependent upon the magnitude of the acceleration and velocity parameters." The authors further suggested that mechanical factors, such as muscle stiffness and other anatomical restraints, probably limited the amplitudes of ballistically initiated movements, since deceleration occurred before maximum peak tension.

The foregoing relation between kinematic variables and muscle activity could be interpreted as an influence of one on the other or vice versa. This was resolved by studying the same paradigm in 3 animals before and after deafferentation of the forelimb by cutting dorsal roots C3 through T1 (138). The animal continued to make ballistic flexions, but biceps activity was ". . . no longer interrupted by the presence of pauses which bear a fixed time relation to the parameters of motion. Instead the activity either ceases more or less abruptly . . . or subsides gradually. The activity of the antagonist [triceps] after deafferentation is random. . . ." Terzuolo et al. (138) concluded that the results left "no doubt that the normal patterning of the motor output to the agonist and antagonist is strictly dependent on sensory input data, at least during fast, ballistically initiated movements. In particular, they exclude that even for a learned motor task the output to the agonist and antagonist is preprogrammed centrally."

I would agree with the above conclusion if the word "entirely" were added. The data of Terzuolo's group lead me to conclude that at least two hierarchical levels of programs were involved: first the learned strategy to gain reward, and second the application of force to initiate the movement by overcoming the restraint. Movements of deafferented animals were successful, albeit clumsy, presumably because central re-efference had to substitute for detailed peripheral afference.

These are essentially also the conclusions drawn by Bizzi et al. (6) from their own related study of monkeys' head-turning towards a visual target. Three vestibulectomized monkeys were trained to press a lever when they had detected a target light 30 or 40 degrees from the starting center gaze, which they learned to accomplish by making coordinated eye-head movements even after both center and target lights were turned off following movement onset. Thus the animals chose their own strategies to achieve the task: to reach the remembered position and then to press a lever for reward. When head-turning was impeded by constant loads applied through a torque motor, movements undershot, but were then corrected to achieve correct final head position. The authors ". . . assumed that final position is an equilibrium point resulting from the interactions of a number of factors: 1) the centrally patterned commands to alpha and gamma motoneurons, 2) the length-tension curves of agonist and antagonist muscles, 3) the passive elastic forces, and 4) other external loads. Accepting this assumption, . . . a constant-torque load is bound to change the equilibrium . . . head position." From this they concluded ". . . that the central, preprogrammed pattern of neural activity calling for a given final position is not reset by the unexpected proprioceptive feedback arising from the loaded muscles and tendons." When movements were made to overshoot by application of a sudden and unexpected increase in inertia during the centrally initiated movements, corrections returned the head to the visual target. The authors concluded: "This observation, together with those on the effect of constant-torque loads, suggests that the central program establishing final head position is not dependent on a readout of proprioceptive afferents generated during the movement but, instead is preprogrammed." To further test this hypothesis, neck proprioceptive feedback was eliminated by cutting dorsal roots C1 to T3, in addition to vestibular and visual input as before. Once more the monkeys achieved the same final head positions, although movements were less smoothly controlled. In this open-loop condition, corrections after inertial loading were presumably brought about ". . . by an increase in antagonist tension generated not by an increase in alpha motor outflow to antagonist muscles, but by the intrinsic length-tension properties of the muscle tissue. The effectiveness of this mechanism is, of course, dependent on the central programming of alpha activity to both agonists and antagonists which determines the stiffness of these muscles" (60). One-half of the monkey's precentral neurons participating in ballistic movements can also participate in small, controlled movements (48,54). The purely 'ballistic' group does not respond to kinesthetic input (torque pulses).

Summary

Monkeys, like humans, can make ballistic movements, whose initial propulsion is preprogrammed by the central nervous system and whose termination seems to follow as a consequence of the mechanical properties of the moving limb.

PROPERTIES OF MOTOR PROGRAMS

An overview has been presented of evidence for operation of motor programs in different kinds of movements. Now we need to consider (a) what parts of the brain are involved, and (b) what the controlled movement properties are. The nature of the research to answer these questions makes the two topics really one, but they are taken up as separately as possible for ease of presentation. We will begin with relatively firm knowledge and end with unanswered questions that may point to future directions of research. This has been posed to us as the task of this volume.

How Separate Are the Neural Substrates for Hierarchical Levels?

The effects of lesions and timing of neural discharge indicate some differences. Loss of higher levels of association between sensory input, its perception in the context of the situation, and appropriate motor output occurs in patients and monkeys after damage of association cortex, as has been reviewed particularly for the parietal cortex by Mountcastle (105). Conversely, it has been demonstrated by Jones and Powell (78), using histological degeneration methods in monkeys, that the combinations of sensory and central information (which would be needed for interactive hierarchies of motor programs) converge in wide areas of association cortex that project to motor areas through subcortical structures. Association cortex becomes active many seconds before actual intended movement onset as indicated by potentials recorded from the scalp of humans (33). Such anticipatory discharge *has* been substantiated for many single units in prefrontal (55,93) and parietal cortex (104) of monkeys, during the fore-period in a delayed-action task, as well as for some neurons in supplementary (133) and a few in primary (120,131,132) motor cortex. Cortico-cortical connections may be particularly important for visual-motor coordination (61), but the most important integration for motor output depends on the re-entrant circuits through the basal ganglia and cerebellum (cf. 87). Precentral neurons, including pyramidal tract neurons, can be "set" for the correct movement in less than half a second after movement instructions have been given, which may anticipate movement onset by several seconds (131). No equivalent information is available about the systems descending in the lateral and ventromedial brainstem (94,95), let alone the propriospinal connections to the segmental apparatus (97).

Different timing of overt discharge under similar circumstances applies to the basal ganglia and the cerebellum. About one-quarter of a sample of 84 caudate, 17 pallidal, and 9 putamen neurons in monkeys' basal ganglia neurons *did* discharge about as early as association cortex during the fore-period of a task (123), which prompted Buchwald's suggestion (8) that the basal ganglia are important in high-level "cognitive" programming, i.e., ". . . where information has to be transferred across a period of time before a response is initiated . . ." (8). A role in programming monkeys' elbow movements has also been considered by DeLong and Strick (34) based on unit records from these structures; and has been suggested by Hore et al., (75,76), with emphasis on the proprioceptive system, following the cues of Denny-Brown (35) and Purdon Martin (101), using cooling of the globus pallidus in 3 monkeys (75) and the putamen (76). Cerebellar neurons do *not* discharge in *pronounced* fashion during the fore-period of a planned movement in a delayed task (129) or in progressively simpler tasks studied in great detail by Thach (140,141,142,143,144). Overt neural discharges have been the most useful guide to the sequence of neural activation. On that basis one would therefore assign to the cerebellum a relatively "downstream" place in the motor hierarchy, which may well be correct, since cerebellar patients suffer neither losses of memory nor of ability to learn motor tasks within the limits of their impaired execution (71,72).

But how could the cerebellum pre-adjust movements according to previous experience, even on a low hierarchical level? Evidence is growing that subtler changes of neurons do occur in time-locked fashion (cf. 123,144), and that they can be grasped experimentally by methods such as trend-analysis even on-line (cf. 106). These new capabilities may bring about tests of Marr's theory (98) that learning occurs by lasting adjustment of synaptic efficacies of relevant inputs to central neurons. Synaptic change, suggested by Brindley (102) as a learning mechanism [as predicted by Hebb (67,68)] was envisaged to be brought about by nearly coincident discharge of climbing fibers and other inputs by Marr (98) and also independently by Albus (1a). This was spelled out specifically for cerebellar-cerebral circuits (7) but was also stated as a general proposition for other structures, such as neocortex (99) or archicortex (100). The theory, or its modifications (cf. 56), was incorporated by Ito (79,80,81, cf. 84) in an explicit formulation and experimental test as to how the cerebellum could model and substitute for brain-world interaction, for the case of rabbits' vestibulo-ocular reflex. The essential advantage of Ito's scheme is that the cerebellum could produce quick, coordinated reactions by substituting stored memorized routines for feedforward control in place of time-consuming peripheral feedback loops (cf. 117). Ito et al. have now shown that stimulation (83) or destruction (82) of the cells of origin of the climbing fibers in the inferior olive can alter the adaptive plasticity of the vestibulo-ocular reflex. More to our immediate subject, Gilbert and Thach (57) have now shown that Purkinje cells do indeed respond to climbing fiber input in time-locked manner during adaptation of monkeys' wrist movements to a new situation, until the animal learns to make smooth, skilled, movements in that situation. Purkinje cell responses to parallel (mossy) fiber input, moreover, remain altered thereafter. If that turns out to be a causal consequence of "newness," then we could imagine constellations of "Hebb" synapses encoding for long-term use appropriate courses of actions according to previous experience. It follows that any neural structure containing "Hebb" synapses in a scheme such as that described above could edit its output by *filtering* incoming convergent messages, to emit only responses appropriate to the context of the situation. Multiple "filtered" events could have escaped detection up to now, because their neural correlates are less easily measured than precisely timed, overt bursts of discharge. The cerebellum could trigger off immediately usable movement *tactics* by adjusting, for instance, thresholds for peripheral events or limb compliances, in view of desired emphasis on timing, or force. Alpha and gamma routes to spinal motoneurons are sure to be used (19,59,112). In addition, the system is bound to be re-entrant, so that many brain parts come into action repeatedly during preparation, initiation, and execution of movements (see below, What Movement Properties are Programmed). Participation of the neocerebellum in successive stages of limb control is examined in more detail elsewhere (18).

Experiments with reversible lesions of central nervous system structures, produced by brief local cooling, have enabled us to compare functions before, during, and after some reasonably well-localized effects (cf. 17). Reducing the outflow of the neocerebellum by cooling the lateral aspect of the dentate nucleus produced effects that resembled those of permanent lesions, and that were more marked in 3 monkeys if the medial aspect was cooled as well (152). Hence one could not exclude possible involvement of the outflow of the spinocerebellum: n. interpositus. It should be added though that cooling n. interpositus alone through a medially placed probe had hardly any effects at all on movements of 3 monkeys tested (147), probably because spinocerebellar information routes are sufficiently parallelled by spinocerebral ones. Cooling deeply enough also to involve the dentate in those animals brought in dentate symptoms (147). The implication of this method is not that only the "lesioned" structure itself participates in the particular function under study, but instead that all the neural circuits connected to that structure may so participate.

Figure 4B (right) shows that monkeys can continue to use *ballistic* movements even during dysfunction of some cerebellar outflow, as long as rewards can still be obtained by the animals within the task context (cf. What Movement Properties Are Programmed). Ito's model suggests that cerebellar filtering initiates stored, feedforward control for such movements. Evidence for participation of appropriate neurons has been found by Fromm and Evarts (54), who reported that two-thirds of a sample of 589 precentral neurons discharged just prior to ballistic wrist movements made by well-trained monkeys, and that one-third were related exclusively to this type of movement and not to others. This cell group contained many large PTNs that did not respond to wrist perturbations injected at the start of the ballistic movement (48). The authors concluded that, "This lack of sensory control is not simply because ballistic movements are too fast, however, since sensory inputs reach supraspinal centers within 20 msec of movement onset. Instead the relative lack of sensory input seems to depend on the nature of the central program which underlies the ballistic movement. . . ." Since the monkeys had been trained to perform these actions, more long-lasting presetting is likely to have been brought to bear than would be evident by overt increases of discharge of most precentral neurons (43) or cerebellar neurons (143) that begin only 100 to 200 msec before movement onset.

Returning from ballistic movements to those of *moderate speed* we recall that guidance of faster ones was less dependent on external cues than that of slower ones. This observation also finds some grounding in neural properties: Evarts and Fromm (48) observed that some precentral neurons participating in quick, small controlled movements did, and others did not, respond to a perturbation of the wrist. The percentages of neurons responding to such proprioceptive input varied according to the type of movement performed. The authors ". . . stressed that the depression of cortical reflexes during rapid movement must be relative, and a matter of degree, rather than 'all-or-none', with magnitude of depression depending on the velocity and type of movement. . . ."

Much emphasis has been placed on study of input-output properties of monkey pyramidal tract neurons (PTNs) and other precentral neurons in area 4 (e.g., 48,54), particularly with regard to their participation in long-loop *"cortical reflex" responses* to unexpected perturbations of the fingers (113), wrist (11,47,

48,50,54,131), or elbow (21,22,102,150). The impetus had been that PTNs, or at least generally corticofugal neurons (22), are likely to be the supraspinal instigators of muscular responses to perturbations, following Phillip's original suggestion for such a loop through motor cortex (110).

Wiesendanger et al. (156) have pointed out that an adaptive gain control at the cortical level ". . . would be most useful for the complex functions of motor cortex." The point is germane to our subject of motor programs. One of the successive three precentral response groups (first and second "early" responses, and subsequent "late" responses), namely the second "early" precentral response, is particularly dependent on the monkey's directional intent or "set" for the movement (47,131). "*Set*" is regarded here as intermediate level programming, between strategy "above" and parametric detail "below." Amplitudes of late precentral responses are also more under central than peripheral control (102). Correspondingly, the muscle responses to a perturbation that occur after the spinal stretch reflex, i.e., starting at about 20 msec after the perturbation, as measured in monkeys (25,47,134,151) are also much under the sway of "set" in man (49,62,64,65,135) as well as in monkey (44,47,131). In fact, the third muscular response, beginning about 100 msec after the perturbation in man (135,145) (75 to 100 msec on monkey) (47) merges with onset of "voluntary activity." The second (set-dependent) precentral response to elbow perturbations is under cerebellar control, as was demonstrated by its reduction for 25 out of 44 precentral neurons during cooling of cerebellar nuclei in 2 monkeys (150). "Directional set" is also expressed in pre-movement discharges of many dentate neurons (144).

One possible supraspinal relay for the second and third muscular responses could be in area 3a, i.e., the part of the buried sensory cortex that receives the greatest proprioceptive input (66,74,96,111). This stands in contrast to area 4, from which most movement-related precentral unit records have been obtained and that receives only weak facilitation from 3a (2,146,159). Yumima et al. (158) found that 13 out of 55 task-related neurons in area 3a, studied in 4 monkeys trained to move a handle, responded to passive extension at the wrist joint. Of these 6 also discharged most vigorously during active wrist extension, whereas 5 discharged during active flexion. Both types of neurons gave stronger responses if the wrist had to overcome a steady load. Tanji (130) has described responses to perturbations of 15 neurons said to be located in area 3a that were related to active maintenance, rather than movement, of wrist position of monkeys trained to hold a handle stationary. Since these perturbation-sensitive neurons were not related to intense muscle activity, he suggested that they might activate fusimotor neurons. A cortico-spinal outflow from area 3a in monkeys has now been established histologically by Coulter and Jones (27), projecting primarily to the lateral intermediate zone where it overlaps with that from area 4. [Electrical microstimulation of area 3a in 20 anesthetized and paralyzed monkeys by Coulter et al. (26) has also revealed projections to the dorsal horn, where excitability of spinothalamic neurons was modulated as a result]. No microstimulation experiments of area 3a in nonparalyzed monkeys have been reported to date, so that outflow to alpha-motoneurons remains to be tested.

It is possible that the set-dependent muscular responses to perturbations are facilitated by corticospinal projections from area 3a that converge in the spinal cord with set-dependent output from area 4, itself under neocerebellar control. Such convergent modulation of the outflow from area 4 would be in line with the anatomical and behavioral results obtained with monkeys by Lawrence and Kuypers (94,95) who showed that wrist movements of the sort used by Evarts' group, and elbow movements used by ours, can be carried out by the brainstem-descending systems, in the absence of the pyramidal tract, although the latter could improve speed and dexterity. Some aspects of elbow and wrist movements might thus be programmed through the pyramidal tract and other corticofugal tracts descending in the lateral brainstem such as the cortico-rubrospinal path, as well as through the reticulo- and vestibulospinal pathways, i.e., the medial brainstem-descending system (cf. 114). [The new monograph by Phillips and Porter (112) came to hand after this was written. They arrive at a general scheme of organization that could accommodate the above suggestions (their Fig. 8.23)].

Summary

Cognitive programming seems to depend critically on association cortex, with possible participation of supplementary and primary motor cortex, as well as the basal ganglia. Preparatory set may be transmitted to the spinal cord through cerebellum as well as motor and sensory cortex, and thus may be fed back through multiple channels to various supraspinal structures. Other parts of the sensorimotor apparatus are likely to participate in high- and low-level programs, because they are thought to interact with each other as well as with ongoing spinal controls. Implications of the Marr-Ito hypothesis for learning are considered for the method of unit-recording in the study of motor control.

What Movement Properties Are Programmed?

We must now return to Woodworth's "*inverse relation*" between speed and accuracy of movements, that was introduced earlier (Fig. 1). The topic was restudied intensively in World War II with regard to gun-laying, principally by Craik, and later Vince, who showed that large movements took longer to execute than small movements if both had to be made equally accurately, but if instead they were made in the same time, then movement accuracy (and thus movement errors) was roughly proportional to movement size (29,30,31,32). This is referred to as *Craik's ratio rule.* The matter was developed further by Fitts (53) into a more precise, semilogarithmic relation, known as *Fitts' ratio rule,* that applies to prepared, i.e., planned, or intended, movements that are fast and fairly large.

Thus, human movement *strategies* depend on task requirements: If, on the one hand, *absolute accuracy* is required, then movement time will vary with movement amplitude. If on the other hand, rapid movements are required, then absolute accuracy will suffer, although *relative accuracy* (between sizes of movements and their errors) will be preserved, something like a Weber fraction in the sensory field. This story is set out splendidly in the texts by Welford (155), Poulton (115), and Keele (86).

Fitts' rule applies to monkeys as it does to humans, both in tasks involving visual control and some without (cf. 86). Performance of movements of moderate speed made by human subjects and by monkeys under the constraints of either speed or accuracy have been compared earlier. When the animals worked within, but not at, the limits of their ability for target accuracy, the reward incentive for that level of precision caused them to make large movements at the same mean velocity as small ones (15). This is borne out by the constant slope of the plot in Fig. 5B. Since there was some leeway, however, where to terminate movements within the target zones, the *allowable* error of the movements (to still gain rewards) increased monotonically with movement amplitudes, as expressed by the SDs in Fig. 5B. This stands in contrast to small and large ballistic movements with the same mean velocity (Fig. 4A), but without systematic increase of duration error (Fig. 4C). Welford (155, p. 60) had also noted that . . . "it seems as if the accuracy of ballistic movements depends on some absolute appreciation of end position rather than of distance moved."

A similar result was obtained also in another way: when movements were prolonged by inducing experimental hypermetria, through dentate cooling (14,92) as explained in the previous section, some monkeys could still succeed to gain rewards when making continuous movements. This enables us to ask whether cerebellar dysfunction necessarily influences the choice of movement strategy. The answer depends on the task. Monkeys that were working well within their ability for accuracy could continue to obtain rewards with continuous movements, even when movement execution was changed during neocerebellar dysfunction. Figure 6 (14) provides a conspicuous example of successful use of continuous movements during dentate dysfunction produced by local cooling (14). The animal managed to terminate most hypermetric continuous movements just within the wide targets (15 degrees, whose centers were separated by only 20 degrees) although the error increased to the just allowable limit for success (compare SDs between "controls" and "cooled"). This monkey hardly decreased its use of continuous movements during many such trials, presumably because this form of rapid movement served it just well enough in the *task context:* to obtain fluid reward. It thus appears that the neocerebellum is not essential for selection of these *strategies* for movements of moderate speed. It has already been seen in *Section 4a* that monkeys continued use of *ballistic* movements during dentate cooling to obtain their desired fluid rewards despite a slowing of their normal sharp movement cycle (20).

Other monkeys that were working with narrow targets (10 degrees) at the

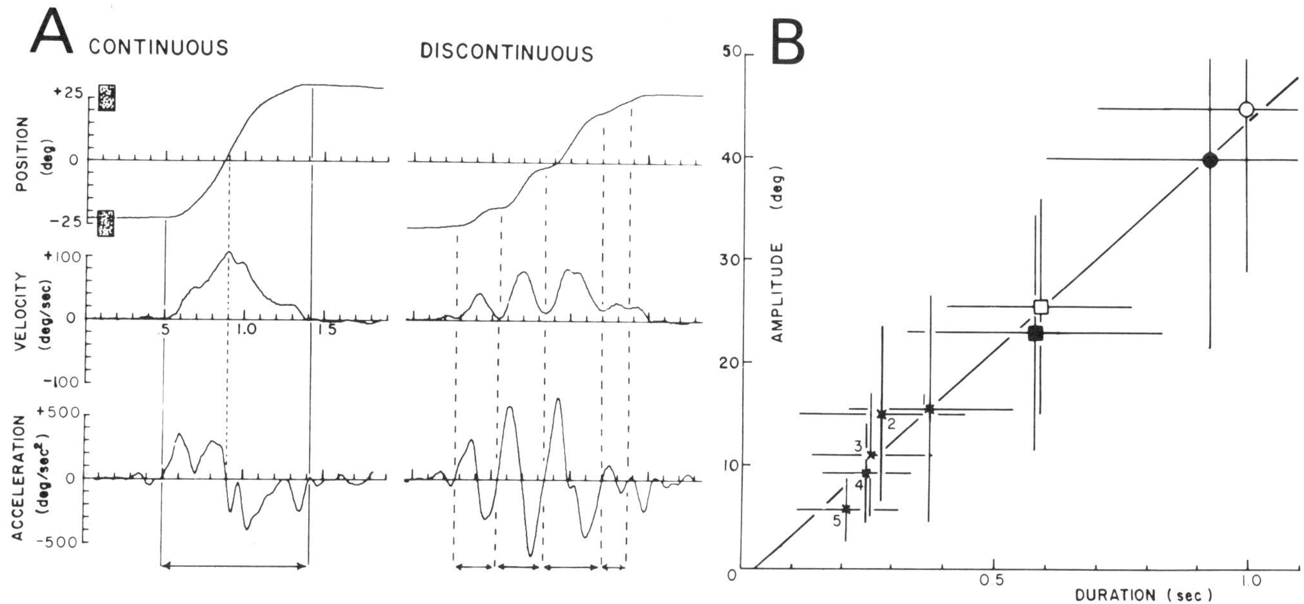

FIG. 5. Elbow movements of moderate speed made by a Cebus monkey through two expected distances, demonstrating Fitts' rule. Same experimental arrangement as in Fig. 2. **A:** "Continuous" and "discontinuous" movements in left and right columns. Position *(upper traces);* velocity *(center traces);* acceleration *(lower traces)* plotted by computer, sampling rate 100 Hz, triggered at position threshold, indicated by *first solid line at left* for the continuous movement. *Dashed vertical line* indicates maximum velocity and corresponding zero acceleration. Intervening maxima and minima of acceleration do not cross the zero acceleration line. In the discontinuous movement, points of zero acceleration bounding and defining "steps" are indicated by *broken vertical lines.* Targets width 10 deg, indicated by *boxes* on calibration of position trace. Scale of time (sec) and of position velocity and acceleration apply to all traces. (M18L, 12/2/73.) **B:** Mean amplitudes *(ordinate)* and durations *(abscissa)* of continuous movements *(circles and squares)* and of steps in discontinuous movements *(crosses)* are plotted, together with SDs. Data from two experiments; one in which target centers were separated by 50 deg (*circles* and *crosses.* M18L) and one in which target centers were separated by 25 deg (*squares:* M12L). *Open symbols:* flexions; *closed symbols:* extensions. *Numbers beside crosses* indicate sequential positions of steps in the discontinuous flexions and 183 continuous extensions for 25 deg. (M12L.) [**A** and **B** reproduced from Figs. 1 and 3 in ref. 15, V. B. Brooks, J. D. Cooke, and J. S. Thomas (1973): The continuity of movements. In: *Control of Posture and Locomotion,* edited by R. B. Stein, K. G. Pearson, R. S. Smith, and J. B. Redford, pp. 257–272. Plenum Press, New York.]

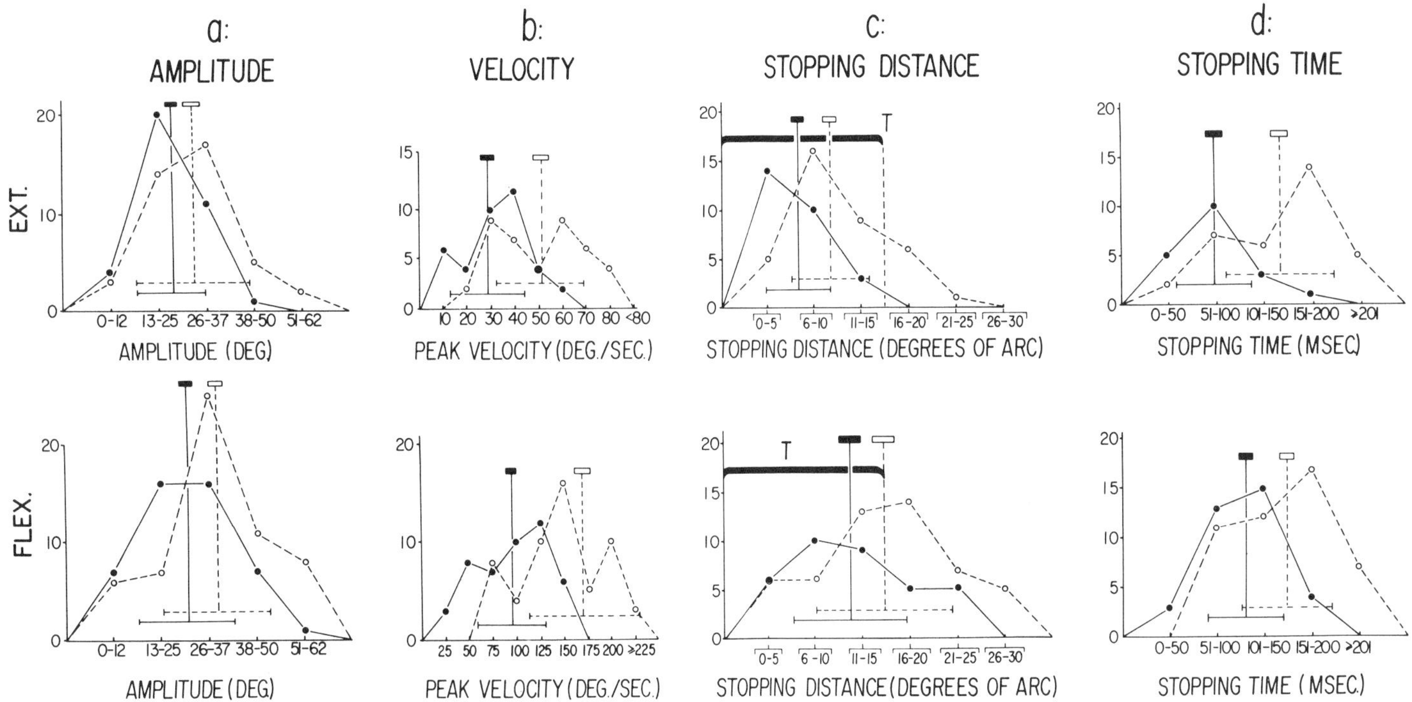
a:
AMPLITUDE
b:
VELOCITY
c:
STOPPING DISTANCE
d:
STOPPING TIME
EXT.
FLEX.
0-12 13-25 26-37 38-50 51-62
AMPLITUDE (DEG)
10 20 30 40 50 60 70 80 <80
25 50 75 100 125 150 175 200 >225
PEAK VELOCITY (DEG./SEC.)
0-5 6-10 11-15 16-20 21-25 26-30
STOPPING DISTANCE (DEGREES OF ARC)
0-50 51-100 101-150 151-200 >201
STOPPING TIME (MSEC.)

edge of their ability for accuracy can also be considered in that task context. When their rewards of fruit juice declined during dysmetria induced by dentate cooling, they seemed to "abandon" use of continuous movements in favor of discontinuous movements, which had much briefer initial impulses, and marked oscillations throughout most of the movements, as for instance in Fig. 5A, right (14). Thus, the monkeys seemed to compensate for their cerebellar dysmetria by switching to the more successful strategy of gradual target approach from hypometric initial impulses. This recalls Gordon Holmes' equivalent report about deliberate behavior of ". . . patients [with cerebellar lesions] who have experienced the risk of overshooting the mark . . ." (72).

Fitts' rule applies to the context of expected force as much as it does to that of expected amplitude: Movements of moderate speed are made faster when greater loads have to be moved. Fitts (53) described rapid movements made by human subjects to "dot" a stylus into targets, with a modern version of Woodworth's "dotting" method (157). Movement times were roughly the same whether the stylus weighed 1 oz or 1 lb. A similar experiment with monkeys is illustrated in Fig. 7. The animal was trained to move the handle against expected steady loads that were changed only between blocks of at least 20 trials, i.e., never unexpectedly or randomly. The averaged movement parameters in Fig. 7 show the limited variability of movement times that was adequate to maintain the reward-rate, even though loads against flexion (F) or against extension (X) were made heavier (23). The same result had been obtained earlier with another 2 monkeys (12). Histograms in Fig. 7 of the averaged discharge rates from a representative precentral neuron during those movements revealed corresponding changes in tonic rates during holding against various loads, as well as in onset and rise times of dynamic rates prior to moving such loads (23). This could have been the result of programming in the sense of central

FIG. 6. Use of "continuous" elbow movements made by a Cebus monkey even during cerebellar hypermetria, demonstrating that selection of a successful strategy does not depend on cerebellum. "Continuous" movements were just barely successful with an easy task, even during hypermetria induced by cooling. Same experimental arrangement as that in Fig. 2. Distributions of average parameters for controls *(filled circles* and *solid lines)* and of data obtained during cooling of dentate to about 10°C *(open circles* and *broken lines).* In **C,** *black bar* labeled "T" indicates 15-deg width of target zones, whose centers were separated by only 20 deg. **D** indicates time from entry into target zone to movement cessation within or beyond target zone. SD about means indicated by *horizontal lines* at bottom of vertical means, and one standard error (SE) by *small horizontal rectangles* at top of vertical means. Only those movements which were in the appropriate direction after preceding "success" were included in the analyses, i.e., flexions when last reinforcement had been obtained in extension target, so that the next could be achieved only by finding and remaining with flexion target; and vice versa for extensions. The plots bear comparison to those drawn with solid lines in Fig. 2. (M3L: pooled data from 3 successive runs performed on 8/16/68 with dentate temp. range 22–25°C.) Number of movements plotted *(ordinates)* representative of many trials (cf. Fig. 2: ●, and Table 1A in ref. 14). [Reproduced from Fig. 10 in ref. 14, V. B. Brooks, I. Kozlovskaya, A. Atkin, F. Horvath, and M. Uno, *J. Neurophysiol.,* 36:974–995 (1973).]

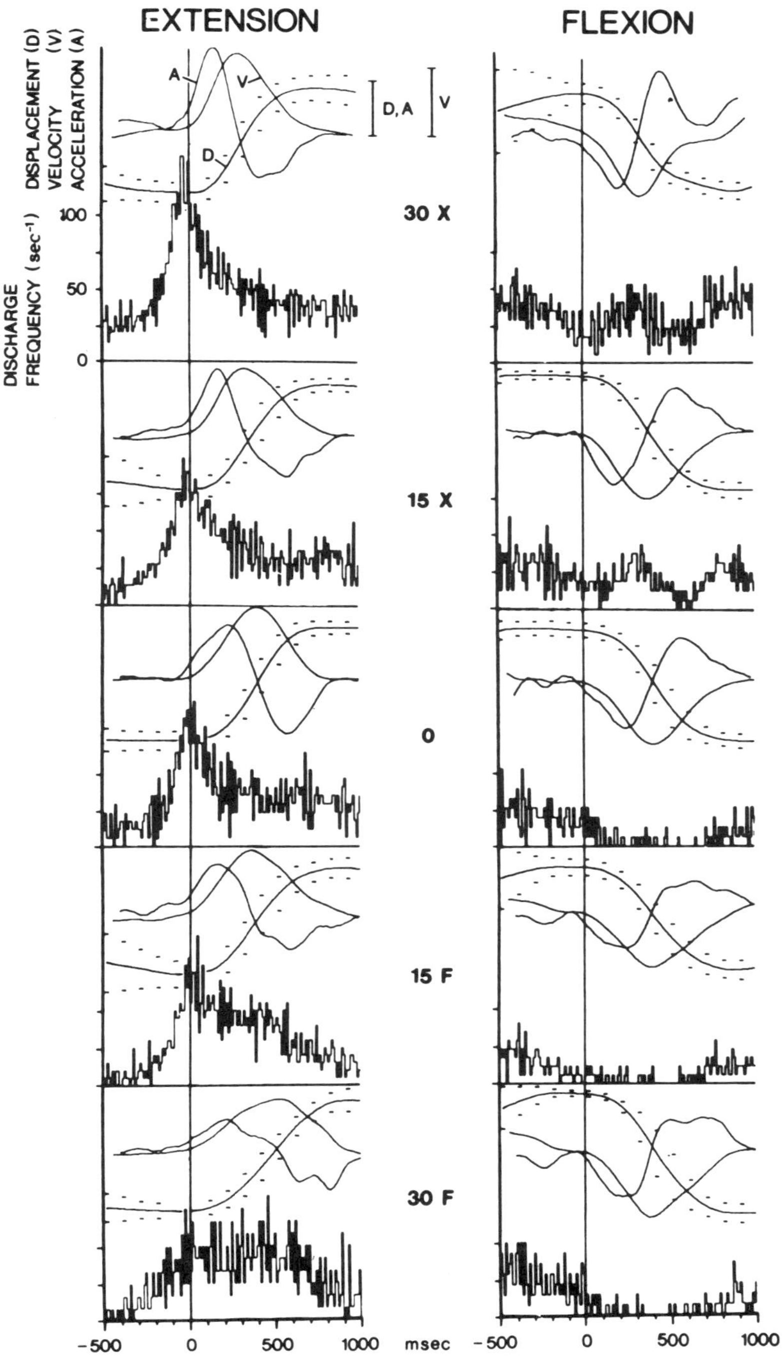
EXTENSION
FLEXION
DISPLACEMENT (D)
VELOCITY (V)
ACCELERATION (A)
DISCHARGE FREQUENCY (sec-1)
A
V
D
D, A
V
100
50
0
30 X
15 X
0
15 F
30 F
-500
0
500
1000
msec

feedback from pre-existing effort during holding against such loads (148,149) or from their cortical projection neurons in area 3a (66,74,96).

A level of programming "lower" than "set" concerns details of movement execution. Much debate has accumulated around the question of programming of *movement parameters* dating back from at least the turn of the century (cf. 126,157). At each stage of technical development for experimental analysis the question has been re-examined: Are there separate control programs for parameters such as force, speed, or displacement? (cf. 11). Even allowing that we now consider these within the context of planned movements, it is still a difficult question without final answers, although peripheral input devices exist to effect such controls (cf. 124). Even the method of correlating discharges of individual precentral neurons and single-motor-unit discharges cannot prove causality (51). The conclusions reached in the section above on ballistic movements may turn out to apply to the parametric detail of slower movements as well, except that feedback would have time to participate in guidance for termination.

For instance, Fig. 4B (right) could be interpreted as support for the idea of programming of some correlate of movement rhythm or timing by the cerebellum. Initiation of return ballistic movements away from the mechanical stop that normally occurred immediately after rebound (Fig. 4B, left) was replaced during dentate cooling by prolonged agonist activity, generating a strong tendency for the handle to be returned toward the mechanical stop (Fig. 4B, right, representative of 3 aminals so tested). The delayed termination of the preceding muscular effort thus deferred initiation of the following movement and hence prolonged the movement cycle (20). Gordon Holmes (91) noted in equivalent observations of cerebellar patients that ". . . the slowness is due directly to delay at the turn and not in time lost in the movements themselves. . . ." We are reminded of Kornhuber's elaboration (88,89) of Braitenberg's suggestion (9,10) that durations of movements and their timing in relation to each other may depend on cerebellar function (cf. 34). There is also other evidence suggesting cerebellar influence on timing: Simple reaction times of 3 Cebus monkeys

FIG. 7. Elbow movements of moderate speed made by a Cebus monkey against various expected weights, and precentral unit discharge, demonstrating Fitts' rule. Same experimental arrangement as in Fig. 2. Note in upper two rows that loads opposing extension (X) cause earlier onset and steeper rise of precentral unit discharge. Computer-averaged movement parameters and histograms of discharge of an X-related precentral neuron plotted on ordinates against time on abscissas (msec). "0" time denotes computer-trigger (50 deg/sec^2). V and A curves obtained by differentiation from D curves, which also show SDs about means. Calibrations: D = 20 deg, V = 100 deg/sec, A = 500 deg/sec^2. Loads are indicated between left and right columns. Target zone widths: 10 deg, target centers separated by 50 deg (cell M15L 36/01, representative of 8 such neurons with dynamic discharges and 13 neurons with static discharge changes during holding, among 52 so tested). Only movements within a defined range (duration 500 to 900 msec, amplitude 35 to 55 deg) were selected and sequentially sampled at 100 Hz, and analyzed. Twenty movements were averaged per load as plotted, and are representative of 400 to 500 so tested at each load. [From Fig. 1 in ref. 23, B. Conrad, M. Wiesendanger, K. Matsunami, and V. B. Brooks, *Exp. Brain Res.*, 29:85–95 (1977).]

moving a handle after a Go! command were prolonged during dentate cooling (103), once more just as has been noted for cerebellar patients (91). Furthermore, discharge of 43 out of 48 movement-related precentral neurons was equally delayed, suggesting neocerebellar control of area 4 (103). This fits well with the demonstration by Thach of the statistical likelihood that dentate neurons discharge before precentral ones (143). Yet, none of the above proves that the cerebellum programs "timing" as such. Dentate neurons are not only correlated to timing; discharge of some relates to joint position, whereas that of others relates to the direction, or for others to the force, of the intended movement (144).

A similar tale can be told about precentral neurons. The control of force for wrist movements has been stressed by Evarts (41,42,45), a conclusion that could also be reached for elbow movements from Fig. 7 (23): The heavier the load, the more force was applied. But was it preprogrammed at the cortical level? Since the pyramidal tract is not essential for these movements (94,95), it is possible that the transformations between its discharges and those of other brain parts involved in recruiting spinal motoneurons may be sufficiently linear to permit such correlations. Smith et al. (121) have confirmed the correlation of cell discharge to force, however, for 51 out of 76 appropriate precentral neurons in a typical "pyramidal" task with 3 monkeys that executed graded, isometric, precision grips between thumb and fingers. [Yet, 2 other animals could be trained to the task after pyramidal section, although movements were slower than before (69)]. Force is not the only parameter however, to which precentral neurons relate (cf. 11,14). Relation to joint position in our usual elbow task has been reported for a few (4 out of 57) precentral neurons in 1 animal (23) and relations to directional change of force for 62 out of 103 task-related precentral cells when monkeys worked against a handle whose unpredictable spring load switched at center position to oppose movements away from center (119).

The list of candidates for various forms of parametric control could be expanded, but instead it may be more useful to consider that the length-tension relationships of muscle (6) and those of force-velocity as well (77) apply under all circumstances (Houk, Bizzi, and Partridge, *this volume*). Houk (28,77) has pointed out that the spinal reflexes controlling length and tension of muscles acting on a joint, named collectively by him as the "Motor Servo," regulate the immediate needs of the "evolving movement" as Eccles called it (38,39). Furthermore, on the basis of studies with human subjects reacting to perturbations on conditional instructions such as "correct" or "do not intervene," i.e., Hammond's paradigm (64,65), Houk has argued that higher control processes send motor commands by changing the setpoints of the reflexes constituting the Motor Servo. This is envisaged to occur by triggering preprogrammed motor commands from a supraspinal "Stimulus-Response Processor," whose repertoire and trigger levels are under the sway of a separate "Supervisory Process." [This proposal ignores the effects of alpha-gamma co-activation on reflex gains (cf.

59), but it would provide a role for peripheral sensory input in control of take-off commands for preprogrammed "ballistic" movements, permitting adjustments also in view of previous experience, cf. section on Ballistic Movements, above.] Since, according to Houk, the spinal Motor Servo guides the limb through simultaneous adjustments of muscle force, length, and velocity, supraspinal motor commands would be able to change any one of these parameters only at the expense of others, even though at least two muscles are involved [but see Bizzi (6) and *this volume*]. If, just to make up one example, the neocerebello-cerebral circuits were cast in the role of the Supervisor Process, and the pyramidal system as well as the brainstem descending pathways (94,95) in that of the Stimulus-Response Processor, functions could result according to the task and its setting (cf. section on Neural Substrates).

The accent of Houk's scheme is on the "triggered" nature of the emitted "reaction-time" (28,77) or "intended" (110) movements, at which point we arrive again at the old conundrum of how *intermittent movement corrections* are made (cf. Use of Programs, above). "Operator discontinuity" for the instance of minimal aim-correction time of 0.2 sec at an approaching target was demonstrated in 1947 by Bates (4) by showing that during that period the reaction cannot be cancelled or disturbed by any further sensory input [psychological refractory period]. Craik pointed out in 1948 (29,30) that intermittent corrective movements may have a predetermined time-pattern of about 0.2 sec and could be triggered off as a whole, i.e., that they would be "ballistic" (31,32; also cf. 153,154). There was a spate of papers during World War II on this subject, e.g., by Taylor and Birmingham (136), Hick (70), and many others (cf. 115,155). Our original view of monkeys' movements in a step-tracking task, that deliberately excluded visual pursuit tracking [as in (136)] and was made to depend on proprioceptive and internal feedback (cf. section on Simple Movements, above), followed that line of reasoning which was reviewed in 1970 (11). We assumed then that oscillations in "discontinuous" movements were stepwise corrections, of about 0.3-sec duration, operated by supraspinal influences on ongoing spinal reflexes. "Continuous" movements were regarded then by us as preprogrammed, large versions of such steps, and not modifiable by peripheral feedback (13,14,92,147). I drew that extreme view into a diagram (Fig. 3 in ref. 16) that neglects servo-controlled tremor, but, in its emphasis on preprogrammed steps, approximates one aspect of the efferent limb of Houk's (Fig. 6 in ref. 77), but without the adaptive control device. We regarded oscillations (at about 3 Hz) during cerebellar dysfunction as inappropriately sized corrections (14,16). No firm distinction could be made between degraded use of ongoing spinal control and of stored supraspinal models. The occasion was a previous visit to the realm of motor programs in 1973, when "Motor aspects of behaviour and programmed nervous activities" were discussed (108).

Craik recognized that there was no physiological proof whether visual misalignment was translated into kinesthetic misalignment giving continuous, rather than intermittent, input from brain to limb [also cf. Navas and Stark, (107)].

Eccles (37,38,39,40) has vigorously affirmed that dynamic control of all "evolving movements" occurs through continuous, ongoing operation of feedback loops up to the integrating mechanisms of the cerebellar cortex that give ongoing comment to the cerebrum. The dilemma in 1973 was twofold: First, there was no apparent "clutch" between the supposed descending programmed commands for intermittent emission of a movement entity, and the everpresent smooth, ongoing, control; second, there was no accord between neurophysiology and the well-fitting models for tracking movements that described their intermittency on a "sampled data" basis (53). The full evolution of that "sampling" view, employing the concepts of information theory begun by Fitts (53), is well introduced in the text by Fitts and Posner (52). Without going through the mathematical formalisms explained so well by them (52), we can summarize: the probability of a correct response increases with more frequent sampling, and anticipation of a predictable course can greatly reduce delay of responses to a continuously changing target position.

I would like to suggest that perhaps we can regard the oscillations during normal movements as described in the figures and text not only as a byproduct of transport-lag of information around the servo-loop controlling the limb, but teleologically also as a scanning for the need of corrections of the intended limb function within the context of the situation. This is no more than a restatement of Phillips' proposal (110) that proprioceptive feedback to motor cortex could be used to align the actual state of, for instance, the finger, to the "intended" movement, by activating precentral alpha-gamma co-activation. [In a more distantly related way Allum's proposal (1) is relevant: that "medium latency" responses observed near 100 msec after perturbation of the shoulder of human subjects represent "a test signal designed to inform the central nervous system of the current loading on the muscle and thus permit the CNS to select an appropriate preprogrammed response . . ." (cf. previous section)]. The burden of the previous sections has been to show that the intent of the movement is embodied in interactive programs in the central nervous system. Normal movement oscillations are here envisaged to test and adjust this, using not only motor cortex but other relevant brain structures. The degree of anticipation represents what we called expectation of success, in other words degree of programming, marked in faster movements by more frequent oscillations than occur in slower, more feedback-dependent, movements (16) (section on Simple Movements, above). Slowing and exaggerated amplitude of oscillations such as during cerebellar dysfunction (14,102,151) [caused at least partly by reduced fusimotor drive (58)], therefore degrades the amount of information available for smooth, goal-directed movement control.

How does Fitts' Rule relate to neural mechanisms? Speed of all movements (other than the very fastest ballistic ones (61a) might be graded by the proportion of 'ballistic' speed commands as opposed to 'accurate' lower speed commands from motor cortex to spinal cord (c.f. Section 3). [The very fastest ballistic movements might permit preferential recruitment of fast, large motoneurons

(61a), whereas all others follow Henneman's rule (36a)]. Since half of the precentral PTNs participating in ballistic movements can be switched from a 'ballistic' to a 'controlled' discharge mode according to the needs of the task (48,54), it is conceivable that this capability results in graded recruitment of spinal motoneurons, plus other threshold changes (36a). Hence strategy programs can variably engage motor cortex to generate continuous gradation of fast movements (that are relatively insensitive to external inputs) to slow movements (that are strongly dependent on peripheral guidance: c.f. Use of Programs, and 16). The matter is still undecided.

We can ask another, related, question: Do "reaction-time" movements (by Houk's definition) fit into the rhythm of ongoing spinal "Motor-Servo" control? There is some indirect evidence that this may be so: Onset-timing of self-initiated arm movements of monkeys appears to fit into the ongoing oscillations, since the rhythm of oscillations cohered across at least 8 such movements in a study of 2 Cebus monkeys by Cooke and Thomas (24). This does not contradict the finding by Vilis and Hore (152) that under similar experimental circumstances perturbations reset the rhythm. The neural command chains, and their intensities, may well be different in the two cases. Terzuolo and Viviani (139) also considered this point for cerebellar lesions: ". . . The prolonged bursting activity of the antagonist, when the motion is not temporarily arrested, is suggestive of a sampling process initiated centrally by the control system which comprises the cerebellum and could represent preferential times at which intentional actions may eventually intervene to modify the ongoing movement." Furthermore, "set"-dependent cortical (46) and muscular responses to perturbations are modulated in amplitude according to intent relative to the needs of the situation, but their latencies remain unchanged in monkeys (44,47) and in humans (62,64,65,91). Essentially we are still on the horns of this dilemma (cf. preceding Section): The neurophysiological basis of adaptive controls is still beyond our grasp, and the scope of "reaction-time" processes is still obscure. The introduction of adaptive feedforward control (79,80) has given a new impetus, however; and its applications may sweep away some of the problems outlined in this presentation.

Summary

Humans and monkeys alike make movements according to Fitts' rule. This may reflect the capability of some precentral neurons to function either with ballistic or with finely controlled movements. Successful strategies, and adherence to Fitts' rule, can be preserved despite degradation of motor performance by some neural dysfunction. The implications of Houk's proposal that all movements initiated by supraspinal structures ("reaction-time movements") use programs are discussed. It is suggested that onset of "reaction-time movements," such as intermittent corrections according to intent of the subject, or other self-initiated movements might fit into the ongoing rhythm of oscillations. These oscillations might be regarded as sampling the need for corrections of intent and set for goal-achievement, embodied as interactive programs in the central nervous system.

OVERALL SUMMARY

Motor performance of human and subhuman primate is compared. Monkeys, like humans, can use different strategies according to their expectation of success. Relatively fast movements, but yet of moderate speed, are launched with a large 'initial impulse', and can be directed correctly into learned target zones without visual, auditory or tactual feedback. In contrast, slower movements, with a small 'initial impulse', require some such feedback for successful target acquisition. The fast type gives way to the slower one when successful use of these movements is impeded in any manner. The faster type could have been guided by peripheral proprioceptive feedback or by central re-efference in ongoing manner, or through triggering central programs based on previous experience. Monkeys, like humans, can make ballistic movements, whose initial propulsion is preprogrammed by the central nervous system and whose termination seems to follow as a consequence of the mechanical properties of the moving limb.

Cognitive programming seems to depend critically on association cortex, with possible participation of supplementary and primary motor cortex, as well as the basal ganglia. Preparatory set may be transmitted to the spinal cord through cerebellum as well as motor and sensory cortex, and thus may be fed back through multiple channels to various supraspinal structures. Other parts of the sensorimotor apparatus are likely to participate in high- and low-level programs, because they are thought to interact with each other as well as with ongoing spinal controls. Implications of the Marr-Ito hypothesis for learning are considered for the method of unit-recording in the study of motor control.

Humans and monkeys alike make movements according to Fitts' Rule. This may reflect the capability of some precentral neurons to function either with ballistic or with finely controlled movements. Successful strategies, and adherence to Fitts' Rule, can be preserved despite degradation of motor performance by some neural dysfunction. The implications of Houk's proposal that all movements initiated by supraspinal structures ("reaction-time movements") use programs, are discussed. It is suggested that onset of "reaction-time movements," such as intermittent corrections according to intent of the subject, or other self-initiated movements, might fit into the ongoing rhythm of oscillations. These oscillations might be regarded as sampling the need for corrections of intent and set for goal-achievement, embodied as interactive programs in the central nervous system.

ACKNOWLEDGMENTS

I wish to express grateful appreciation for research support at various times over the past years by the National Institute for Neurological and Communicative

Disorders and Stroke, USPHS; the Richard and Jean Ivey Fund, London, Canada; and the Medical Research Council of Canada. Permission to reproduce figures by the American Physiological Society, Academic Press, Elsevier Publishing Co., Plenum Publishing Corp., and Springer Verlag is also acknowledged. This manuscript was read in draft form by my colleagues Jim Chivers-Wilson, David Cooke, Jon Hore, Phil Kennedy, Alan Miller, and Tutis Vilis, and I thank them for their comments. I bear the burden of all interpretations presented in this paper, which was typed with great patience by Miss Marilyn Allen.

REFERENCES

1. Allum, J. H. J. (1975): Responses to load disturbances in human shoulder muscles The hypothesis that one component is a pulse test information signal. *Exp. Brain Res.*, 22:307–326.

1a. Albus, J. S. (1971): A theory of cerebellar function. *Math. Biosci.*, 10:25–61.

2. Andersen, P., and Phillips, C. G., unpublished. Cited in ref. 111 (Phillips, Powell, and Wiesendanger).

3. Angel, R. W., Garland, H., and Fischler, M. (1971): Tracking errors amended without visual feedback. *J. Exp. Psychol.*, 89:422–424.

4. Bates, J. A. V. (1947): Some characteristics of a human operator. *Inst. Electr. Eng. (Lond.)*, 94, pt. 2:298–304.

5. Bernstein, N. (1967): *The Coordination and Regulation of Movements.* Pergamon Press Ltd., Oxford.

6. Bizzi, E., Polit, A., and Morasso, P. (1976): Mechanisms underlying achievement of final head position. *J. Neurophysiol.*, 39:435–444.

7. Bloomfield, S., and Marr, D. (1970): How the cerebellum may be used. *Nature*, 227:1224–1228.

8. Buchwald, N. A., Hull, C. D., Levine, M. S., and Villablanca, J. (1975): The basal ganglia and the regulation of response and cognitive sets. In: *Growth and Development of the Brain*, edited by M. A. B. Brazier, pp. 171–189. Raven Press, New York.

9. Braitenberg, V. (1961): Functional interpretation of cerebellar histology. *Nature*, 190:539–540.

10. Braitenberg, V. (1967): Is the cerebellar cortex a biological clock in the millisecond range? *Prog. Brain Res.*, 25:334–346.

10a. Brindley, G. S. (1964): The use made by the cerebellum of the information that it receives from sense organs. *IBRO Bull.* 3:80 (Abst.).

11. Brooks, V. B., and Stoney, S. D., Jr. (1971): Motor mechanisms: The role of the pyramidal system in motor control. *Annu. Rev. Physiol.*, 33:337–392.

12. Brooks, V. B., Adrien, J., and Dykes, R. W. (1972): Task-related discharge of neurons in motor cortex and effects of dentate cooling. *Brain Res.*, 40:85–88.

13. Brooks, V. B. (1972): Some new experiments on cerebellar motor control. *Electroencephalogr. Clin. Neurophysiol. (Suppl.)*, 31:109–116.

14. Brooks, V. B., Kozlovskaya, I., Atkin, A., Horvath, F., and Uno, M. (1973): Effects of cooling the dentate nucleus on tracking-task performance by monkeys. *J. Neurophysiol.* 36:974–995.

15. Brooks, V. B., Cooke, J. D., and Thomas, J. S. (1973): The continuity of movements. In: *Control of Posture and Locomotion*, edited by R. B. Stein, K. G. Pearson, R. S. Smith, and J. B. Redford, pp. 257–272. Plenum Press, New York.

16. Brooks, V. B. (1974): Some examples of programmed limb movements. *Brain Res.*, 71:299–308.

17. Brooks, V. B. (1975): Roles of cerebellum and basal ganglia in initiation and control of movements. *Can. J. Neurol. Sci.*, 2:265–277.

18. Brooks, V. B. (1979): Control of intended limb movements by the lateral and intermediate cerebellum. In: *Integration in the Nervous System*, edited by H. Asanuma and V. J. Wilson, pp. 321–357. Igaku-Shoin Ltd., Tokyo.

19. Clough, J. F. M., Phillips, C. G., and Sheridan, J. D. (1971): The short-latency projection

from the baboon's motor cortex to fusimotor neurones of the forearm and hand. *J. Physiol. (Lond.)*, 216:257–279.
20. Conrad, B., and Brooks, V. B. (1974): Effects of dentate cooling on rapid alternating arm movements. *J Neurophysiol.*, 37:792–804.
21. Conrad, B., Matsunami, K., Meyer-Lohmann, J., Wiesendanger, M., and Brooks, V. B. (1974): Cortical load compensation during voluntary elbow movements. *Brain Res.*, 71:507–514.
22. Conrad, B., Meyer-Lohmann, J., Matsunami, K., and Brooks, V. B. (1975): Precentral unit activity following torque pulse injections into elbow movements. *Brain Res.*, 94:219–236.
23. Conrad, B., Wiesendanger, M., Matsunami, K., and Brooks, V. B. (1977): Precentral unit activity related to control of arm movements. *Exp. Brain Res.*, 29:85–95.
24. Cooke, J. D., and Thomas, J. S. (1976): Forearm oscillation during cooling of the dentate nucleus in the monkey. *Can. J. Physiol. Pharmacol.*, 54 (4):430–436.
25. Cooke, J. D., and Eastman, M. J. (1977): Long-loop reflexes in the tranquilized monkey. *Exp. Brain Res.*, 27:491–500.
26. Coulter, J. D., Foreman, R. D., Beall, J. E., and Willis, W. D. (1976): Cerebral cortical modulation of primate spinothalamic neurons. In: *Advances in Pain Research and Therapy, Vol. 1,* edited by J. J. Bonica and D. Albe-Fessard, pp. 271–277. Raven Press, New York.
27. Coulter, J. D., and Jones, E. G. (1977): Differential distribution of corticospinal projections from individual cytoarchitectonic fields in the monkey. *Brain Res.*, 129:335–340.
28. Crago, P. E., Houk, J. C., and Hasan, Z. (1976): Regulatory actions of human stretch reflex. *J. Neurophysiol.*, 39(5):925–935.
29. Craik, K. J. W. (1947): Theory of the human operator in control systems. I. The operator as an engineering system. *Br. J. Psychol.*, 38:56–61.
30. Craik, K. J. W. (1948): Theory of the human operator in control systems. II. Man as an element in a control system. *Br. J. Psychol.*, 38:142–148.
31. Craik, K. J. W., and Vince, M. A. (1963): Psychological and physiological aspects of control mechanisms with special reference to tank gunnery. Part I. UK MRC Military Personnel Research Committee Report, London, England, 1943. *Ergonomics,* 6:1–33.
32. Craik, K. J. W., and Vince, M. A. (1963): Psychological and physiological aspects of control mechanisms. Part II. UK MRC Military Personnel Research Committee Report BPC 44/322, London, England, 1944. *Ergonomics,* 6:419–440.
33. Deecke, L., Scheid, P., and Kornhuber, H. H. (1969): Distribution of readiness potential, pre-motion positivity, and motor potential of the human cerebral cortex preceding voluntary finger movements. *Exp. Brain Res.*, 7:158–168.
34. DeLong, M. R., and Strick, P. L. (1974): Relation of basal ganglia, cerebellum, and motor cortex units to ramp and ballistic limb movements. *Brain Res.*, 71:327–335.
35. Denny-Brown, D. (1962): *The Basal Ganglia, and Their Relation to Disorders of Movement.* Oxford University Press, London.
36. Desmedt, J. E., (editor) (1978): *Progress in Clinical Neurophysiology, Vol. 4: Cerebral Motor Control in Man: Long Loop Mechanisms.* Karger, Basel.
36a. Desmedt, J. E., and Godaux, E. (1978): Ballistic skilled movements: Load compensation and patterning of the motor commands. In: *Progress in Clinical Neurophysiology, Vol. 4: Cerebral Motor Control in Man: Long Loop Mechanisms,* edited by J. E. Desmedt, pp. 21–55. Karger, Basel.
37. Eccles, J. C., Ito, M., and Szentagothai, J. (1967): *The Cerebellum as a Neuronal Machine.* Springer Verlag, New York.
38. Eccles, J. C. (1967): Circuits in the cerebellar control of movement. *Proc. Natl. Acad. Sci. USA,* 58:336–343.
39. Eccles, J. C. (1969): The dynamic loop hypothesis of movement control. In: *Information Processing in the Nervous System,* edited by K. N. Leibovic, pp. 245–269. Springer Verlag, Heidelberg.
40. Eccles, J. C. (1973): Review lecture: The cerebellum as a computer: Patterns in space and time. *J. Physiol. (Lond.)*, 229:1–32.
41. Evarts, E. V. (1968): Relation of pyramidal tract activity to force exerted during voluntary movement. *J. Neurophysiol.*, 31:14–27.
42. Evarts, E. V. (1969): Activity of pyramidal tract neurons during postural fixation. *J. Neurophysiol.*, 32:375–385.
43. Evarts, E. V., Bizzi, E., Burke, R. E., DeLong, M., and Thach, W. T., editors (1972): *Neurosciences Research Symposium Summaries, Vol. 6: Central Control of Movement,* pp. 1–170.

44. Evarts, E. V. (1973): Motor cortex reflexes associated with learned movement. *Science*, 179:501–503.
45. Evarts, E. V. (1974): Sensorimotor cortex activity associated with movements triggered by visual as compared to somesthetic inputs. In: *The Neurosciences, Third Study Program*, edited by F. O. Schmitt, and F. G. Worden, pp. 327–337. MIT Press, Cambridge, Mass.
46. Evarts, E. V., and Tanji, I. (1974): Gating of motor cortex reflexes by prior instruction. *Brain Res.*, 71:479–494.
47. Evarts, E. V., and Tanji, J. (1976): Reflex and intended responses in motor cortex pyramidal tract neurons of monkey. *J. Neurophysiol.*, 39:1069–1080.
48. Evarts, E. V., and Fromm, C. (1977): Sensory responses in motor cortex neurons during precise motor control. *Neurosci. Letters*, 5:267–272.
49. Evarts, E. V., and Granit, R. (1978): Relations of reflexes and intended movements. *Prog. Brain Res.*, 44:1–14.
50. Evarts, E. V., and Fromm, C. (1978): The pyramidal tract neuron as summing point in a closed-loop control system in the monkey. In: *Progress in Clinical Neurophysiology, Vol. 4: Cerebral Motor Control in Man: Long Loop Mechanisms*, edited by J. E. Desmedt, pp. 56–69. Karger, Basel.
51. Fetz, E. E., Cheney, P. D., and German, D. C. (1976): Corticomotoneuronal connections of precentral cells detected by post-spike averages of EMG activity in behaving monkeys. *Brain Res.*, 114:505–510.
52. Fitts, P. M., and Posner, M. I. (1967): *Human Performance*. Brooks/Cole Pub. Co., Belmont, Calif.
53. Fitts, P. F. (1954): The information capacity of the human motor system in controlling the amplitude of movement. *J. Exp. Psychol.*, 47:381–391.
54. Fromm, C., and Evarts, E. V. (1977): Relation of motor cortex neurons to precisely controlled and ballistic movements. *Neurosci. Letters*, 5:259–265.
55. Fuster, J. M. (1973): Unit activity in prefrontal cortex during delayed response performance: Neuronal correlates of transient memory. *J. Neurophysiol.*, 34:61–78.
56. Gilbert, P. F. C. (1974): A theory of memory that explains the function and structure of the cerebellum. *Brain Res.*, 70:1–18.
57. Gilbert, P. F. C., and Thach, W. T. (1977): Purkinje cell activity during motor learning. *Brain Res.*, 128:309–328.
58. Gilman, S. (1973): A cerebello-thalamo-cortical pathway controlling fusimotor activity. In: *Control of Posture and Locomotion*, edited by R. B. Stein, K. G. Pearson, R. S. Smith, and J. B. Redford, pp. 309–329. Plenum Press, New York.
59. Granit, R. (1970): *The Basis of Motor Control; Integrating the Activity of Muscles, Alpha and Gamma Motoneurons and Their Leading Control Systems*. Academic Press New York.
60. Grillner, S. (1972): The role of muscle stiffness in meeting the changing postural and locomotor requirements for force development of the ankle extensors. *Acta Physiol. Scand.*, 86:92–108.
60a. Grimby, L., and Hannerz, J. (1977): Firing rate and recruitment order of toe extensor motor units in different modes of voluntary contraction. *J. Physiol. (Lond.)*, 264:865–879.
61. Haaxma, R., and Kuypers, H. (1974): Role of occipito-frontal cortico-cortical connections in visual guidance of relatively independent hand and finger movements in rhesus monkeys. *Brain Res.*, 71:361–366.
62. Hagbarth, K. E. (1967): EMG studies of stretch reflexes in man. In: *Electroencephalogr. Clin. Neurophysiol., Suppl. 25: Recent Advances in Clinical Neurophysiology*, edited by L. Widen, pp. 74–79. Elsevier, Amsterdam.
63. Hallett, M., Shahani, B. T., and Young, R. R. (1975): EMG analysis of stereotyped voluntary movements in man. *J. Neurol. Neurosurg. Psychiatry*, 38:1154–1162.
64. Hammond, P. H. (1956): The influence of prior instruction to the subject on an apparently involuntary neuromuscular response. *J. Physiol. (Lond.)*, 132:17–18P.
65. Hammond, P. H., Merton, P. A., and Sutton, G. C. (1956): Nervous graduation of muscular control. *Br. Med. Bull.*, 12:214.
66. Heath, C. J., Hore, J., and Phillips, C. G. (1976): Inputs from low threshold muscle and cutaneous afferents of hand and forearm to areas 3a and 3b of baboon's cerebral cortex. *J. Physiol. (Lond.)*, 257:199–227.
67. Hebb, D. O. (1949): *The Organization of Behaviour: A Neuropsychological Theory*. Wiley, New York.

68. Hebb, D. O. (1961): Distinctive features of learning in the higher animal. In: *Brain Mechanisms and Learning,* edited by J. F. Delafresnaye, pp. 37–46. Blackwell Scientific Publications, Oxford.
69. Hepp-Reymond, M. C., and Wiesendanger, M. (1972): Unilateral pyramidotomy in monkeys: Effect on force and speed of a conditioned precision grip. *Brain Res.,* 36:117–131.
70. Hick, W. E. (1948): Discontinuous functioning of the human operator in pursuit tasks. *Q. J. Exp. Psychol.,* 1:36–51.
71. Holmes, G. (1917): The symptoms of acute cerebellar injuries due to gunshot injuries. *Brain,* 40:461–535.
72. Holmes, G. (1939): The cerebellum of man. *Brain,* 62:1–30.
73. Homma, S., editor (1976): *Progress in Brain Research, Vol. 44: Understanding the Stretch Reflex,* pp. 507. Elsevier, New York.
74. Hore, J., Preston, J. B., Durkovic, R. G., and Cheney, P. D. (1976): Responses of cortical neurons (areas 3a and 4) to ramp stretch of hindlimb muscles in the baboon. *J. Neurophysiol.,* 39:484–500.
75. Hore, J., Meyer-Lohmann, J., and Brooks, V. B. (1977): Basal ganglia cooling disables learned arm movements of monkeys in the absence of visual guidance. *Science,* 195:584–586.
76. Hore, J., and Vilis, T. (1977): Difficulty in initiating arm movements during cooling of the putamen in the monkey. *Soc. Neurosci. Abstr.,* 3:39.
77. Houk, J. C. (1978): Participation of reflex mechanisms and reaction-time processes in the compensatory adjustments to mechanical disturbances. In: *Progress in Clinical Neurophysiology, Vol. 4: Cerebral Motor Control in Man: Long Loop Mechanisms,* edited by J. E. Desmedt, pp. 193–215. Karger, Basel.
78. Jones, E. G., and Powell, T. P. (1970): An anatomical study of converging sensory pathways within the cerebral cortex of the monkey. *Brain,* 93:793–820.
79. Ito, M. (1970): Neurophysiological aspects of the cerebellar motor control system. *Int. J. Neurol.,* 7:162–176.
80. Ito, M. (1972): Neural design of the cerebellar motor control system. *Brain Res.,* 40:81–84.
81. Ito, M. (1974): The control mechanisms of cerebellar motor systems. In: *The Neurosciences, Third Study Program,* edited by F. O. Schmitt and F. G. Worden, pp. 293–303. MIT Press, Cambridge, Mass.
82. Ito, M., and Miyashita, Y. (1975): The effects of chronic destruction of the inferior olive upon visual modification of the horizontal vestibulo-ocular reflex of rabbits. *Proc. Japan Soc.,* 51:716–720.
83. Ito, M., Nisimaru, N., and Yamamoto, M. (1977): Specific patterns of neuronal connexions involved in the control of the rabbit's vestibulo-ocular reflexes by the cerebellar flocculus. *J. Physiol. (Lond.),* 265:833–854.
84. Ito, M. (1979): *The Cerebellum and Neural Control.* Raven Press, New York *(in press).*
85. Keele, S. W. (1968): Movement control in skilled motor performance. *Psychol. Bull.,* 70:387–403.
86. Keele, S. W. (1973): *Attention and Human Performance.* Goodyear Publishing Co., Pacific Palisades, Calif.
87. Kemp, J. M., and Powell, J. P. S. (1971): The connexions of the striatum and globus pallidus: Synthesis and speculation. *Philos. Trans. R. Soc. Lond. [Biol.],* 262:441–457.
88. Kornhuber, H. H. (1971): Motor functions of cerebellum and basal ganglia: The cerebellocortical saccadic (ballistic) clock, the cerebellonuclear hold regulator, and the basal ganglia ramp (voluntary speed smooth movement) generator. *Kybernetic,* 8:157–162.
89. Kornhuber, H. H. (1974): Cerebellar cortex, cerebellum and basal ganglia. An introduction to their motor function. In: *The Neurosciences, Third Study Program,* edited by F. O. Schmitt and F. G. Worden, pp. 267–280. MIT Press, Cambridge, Mass.
90. Kozlovskaya, I., Uno, M., Atkin, A., and Brooks, V. B. (1970): Performance of a step-tracking task by monkeys. *Comm. Behav. Biol.,* 5:153–156.
91. Kozlovskaya, I. B., Atkin, A., Horvath, F. E., Uno, M., and Brooks, V. B. (1973): Mechanisms of motor control of two types of tracking movements in monkeys. Proc., 2nd Int. Symposium on Motor Control, Varna, Bulgaria. *Agressologie,* 14A:49–57.
92. Kozlovskaya, I. B., Atkin, A., Horvath, F. E., Thomas, J. S., and Brooks, V. B. (1974): Preprogrammed and feedback-guided movements of monkeys. *Behav. Biol.,* 12:243–248.
93. Kubota, K., and Niki, H. (1971): Prefrontal cortical unit activity and delayed alternation performance in monkeys. *J. Neurophysiol.,* 34:337–347.

94. Lawrence, D. G., Kuypers, H. G. J. M. (1968): The functional organization of the motor system in the monkey. I. The effects of bilateral pyramidal lesions. *Brain,* 91:1–14.
95. Lawrence, D. G., and Kuypers, H. G. J. M. (1968): The functional organization of the motor system in the monkey. II. The effects of lesions of the descending brain-stem pathways. *Brain,* 91:15–36.
96. Lucier, G. E., Ruegg, D. C., and Wiesendanger, M. (1975): Responses of neurones in motor cortex and in area 3A to controlled stretches of forelimb muscles in Cebus monkeys. *J. Physiol. (Lond.),* 251:833–853.
97. Lundberg, A. (1975): Control of spinal mechanisms from the brain. In: *The Nervous System, Vol. 1,* edited by D. B. Tower, pp. 253–265. Raven Press, New York.
98. Marr, D. (1969): A theory of cerebellar cortex. *J. Physiol. (Lond.),* 202:437–470.
99. Marr, D. (1970): A theory of cerebral neocortex. *Proc. R. Soc. Lond. [Biol.]* 176:161–234.
100. Marr, D. (1971): Simple memory: A theory for archicortex. *Philos. Trans. R. Soc. Lond. [Biol.],* 262:23–81.
101. Martin, J. P. (1967): *The Basal Ganglia and Posture.* Pitman Medical, London.
102. Meyer-Lohmann, J., Conrad, B., Matsunami, K., and Brooks, V. B. (1975): Effects of dentate cooling on precentral unit activity following torque pulse injections into elbow movements. *Brain Res.,* 94:237–251.
103. Meyer-Lohmann, J., Hore, J., and Brooks, V. B. (1977): Cerebellar participation in generation of prompt arm movements. *J. Neurophysiol.,* 40:1038–1050.
104. Mountcastle, V. B., Lynch, J. C., Georgopoulos, A., Sakata, H., and Acuna, C. (1975): Posterior parietal association cortex of the monkey: Command functions for operations within extrapersonal space. *J. Neurophysiol.,* 38:871–908.
105. Mountcastle, V. B. (1975): The world around us: Neural command functions for selective attention. Schmitt Lecture. *Neurosci. Res. Program Bull. (Suppl.),* 14:1–41.
106. Murphy, J. T., Wong, Y. C., and Kwan, H. C. (1975): Afferent-efferent linkages in motor cortex for single forelimb muscles. *J. Neurophysiol.,* 38:990–1014.
107. Navas, F., and Stark, L. (1968): Sampling or intermittency in hand control system dynamics. *Biophys. J.,* 8:252–302.
107a. Neilson, P. D., and Lance, J. W. (1978): Reflex transmission characteristics during voluntary activity in normal man and patients with movement disorders. In: *Progress in Clinical Neurophysiology, Vol. 4: Cerebral Motor Control in Man: Long Loop Mechanisms,* edited by J. E. Desmedt, pp. 263–299. Karger, Basel.
108. Paillard, J., and Massion, J., editors (1974): Motor Aspects of Behaviour and Programmed Nervous Activities. *Brain Res.* 71:189–575.
109. Paillard, J. (1960): The patterning of skilled movements. In: *Handbook of Physiology-Neurophysiology, Vol. 3,* edited by J. Field, pp. 1679–1707. American Physiological Society, Washington, D.C.
110. Phillips, C. G. (1969): The Ferrier Lecture 1968: Motor apparatus of the baboon's hand. *Proc. R. Soc. Lond. [Biol.],* 173:141–174.
111. Phillips, C. G., Powell, T. P. S., and Wiesendanger, M. (1971): Projection from low-threshold muscle afferents of hand and forearm to area 3a of baboon's cortex. *J. Physiol. (Lond.),* 210:419–446.
112. Phillips, C. G., and Porter, R. (1977): Corticospinal neurones: Their role in movement. In: *Monographs of the Physiological Society,* edited by P. F. Baker, A. D. M. Greenfield, R. A. Gregory, D. K. Hill, and P. B. C. Matthews, pp. 1–450. Academic Press, London.
113. Porter, R., and Rack, P. M. H. (1976): Timing of the response in the motor cortex of monkeys to an unexpected disturbance of finger position. *Brain Res.,* 103:201–213.
114. Porter, R. (1976): Influences of movement detectors on pyramidal tract neurons in primates. *Annu. Rev. Physiol.,* 38:121–137.
115. Poulton, E. (1974): *Tracking Skill and Manual Control.* Academic Press, New York.
116. Richer, P. (1895): Note sur la contraction du muscle quadriceps dans l'acte de donner un coup de pied. *C. R. Soc. Biol. (Paris),* 47:204–205.
117. Robinson, D. A. (1976): Adaptive gain control of vestibuloocular reflex by the cerebellum. *J. Neurophysiol.,* 39:954–969.
118. Rose, F., editor (1977): *The Physiological Aspect of Clinical Neurology,* pp. 179–199. Blackwells, Oxford.
119. Schmidt, E. M., Jost, R. G., and Davis, K. K. (1975): Reexamination of the force relationship

of cortical cell discharge patterns with conditioned wrist movements. *Brain Res.*, 83:213–223.
120. Schmidt, E. M., Jost, R. G., and Davis, K. K. (1975): Cortical cell discharge patterns in anticipation of a trained movement. *Brain Res.*, 75:309–311.
121. Smith, A. M., Hepp-Reymond, M. C., and Wyss, U. R. (1975): Relation of activity in precentral cortical neurons to force and rate of force change during isometric contractions of finger muscles. *Exp. Brain Res.*, 23:315–332.
122. Soechting, J., Ranish, N., Palminteri, R., and Terzuolo, C. (1976): Changes in a motor pattern following cerebellar and olivary lesions in the squirrel monkey. *Brain Res.*, 105:21–44.
123. Soltysik, S., Hull, C. D., Buchwald, N. A., and Fekete, T. (1975): Single unit activity in basal ganglia of monkeys during performance of a delayed response task. *Electroencephalogr. Clin. Neurophysiol.*, 39:65–78.
124. Stein, R. B. (1974): Peripheral control of movement. *Physiol. Rev.*, 54:215–243.
125. Stelmach, G. E., editor (1976): *Motor Control: Issues and Trends.* Academic Press, New York.
126. Stetson, R. H. (1905): A motor theory of rhythm and discrete succession. I. *Psychol. Rev.*, 12:250–271.
127. Stetson, R. H., and McGill, J. A. (1923): Mechanisms of the different types of movements. *Psychol. Monogr.*, 32:18–40.
128. Stetson, R. H., and Bouman, H. D. (1935): The coordination of simple skilled movements. *Arch. Neerland. Physiol.*, 20:179–254.
129. Strick, P. L. (1978): Cerebellar involvement in "volitional" muscle responses to load changes. In: *Progress in Clinical Neurophysiology, Vol. 4: Cerebral Motor Control in Man: Long Loop Mechanisms,* edited by J. E. Desmedt, pp. 85–93. Karger, Basel.
130. Tanji, J. (1976): Selective activation of neurons in cortical area 3a associated with accurate maintenance of limb positions. *Brain Res.*, 115:328–333.
131. Tanji, J., and Evarts, E. V. (1976): Anticipatory activity of motor cortex neurons in relation to direction of an intended movement. *J. Neurophysiol.* 39(5):1062–1068.
132. Tanji, J., Taniguchi, K., and Fukushima, K. (1978): Relation of slowly conducting pyramidal tract neurons to specific aspects of forearm movement. In: *Pyramidal Micro-connexions and Motor Control,* edited by J. Massion, J. Paillard, and M. Wiesendanger. *J. Physiol. (Paris)* 74:293–296.
133. Tanji, J., and Taniguchi, J. (1978): Does the supplementary motor area play a part in modifying motor control reflexes? In: *Pyramidal Micro-connexions and Motor Control,* edited by J. Massion, J. Paillard, and M. Wiesendanger. *J. Physiol. (Paris),* 74:317–318.
134. Tatton, W. G., Forner, S. D., Gerstein, G. L., Chambers, W. W., and Liu, C. N. (1975): The effect of postcentral cortical lesions on motor responses to sudden upper limb displacements in monkeys. *Brain Res.*, 96:108–113.
135. Tatton, W. G., and Lee, R. G. (1975): Evidence for abnormal long-loop reflexes in rigid Parkinsonian patients. *Brain Res.*, 100:671–676.
136. Taylor, F. V., and Birmingham, H. P. (1948): Studies of tracking behavior. II. The acceleration pattern of quick manual corrective responses. *J. Exp. Psychol.*, 38:783–795.
137. Terzuolo, C., Soechting, J., and Palminteri, R. (1973): Studies on the control of some simple motor tasks. III. Comparison of the EMG pattern during ballistically initiated movements in man and squirrel monkey. *Brain Res.*, 62:242–246.
138. Terzuolo, C. A., Soechting, J. R., and Ranish, N. A. (1974): Studies on the control of some simple motor tasks. V. Changes in motor output following dorsal root section in squirrel monkey. *Brain Res.*, 70:521–526.
139. Terzuolo, C. A., and Viviani, P. (1974): Parameters of motion and EMG activities during some simple motor tasks in normal subjects and cerebellar patients. In: *The Cerebellum, Epilepsy, and Behavior,* edited by I. S. Cooper, M. Riklan, and R. S. Snider, pp. 173–215. Plenum Press, New York.
140. Thach, W. T. (1968): Discharge of Purkinje and cerebellar nuclear neurons during rapidly alternating arm movements in the monkey. *J. Neurophysiol.*, 31:785–797.
141. Thach, W. T. (1970): Discharge of cerebellar neurons related to two maintained postures and two prompt movements. I. Nuclear cell output. *J. Neurophysiol.*, 33:527–536.
142. Thach, W. T. (1970): Discharge of cerebellar neurons related to two maintained postures and two prompt movements. II. Purkinje cell output and input. *J. Neurophysiol.*, 33:537–547.

143. Thach, W. T. (1975): Timing of activity in cerebellar dentate nucleus and cerebral motor cortex during prompt volitional movement. *Brain Res.*, 88:233–241.
144. Thach, W. T. (1978): Correlation of neural discharge with pattern and force of muscular activity, joint position, and direction of the intended movement in motor cortex and cerebellum. *J. Neurophysiol.*, 41:654–676.
145. Thomas, J. S., Brown, J., and Lucier, G. E. (1977): Influence of task set on muscular responses to arm perturbations in normal subjects and Parkinson patients. *Exp. Neurol.*, 55:618–628.
146. Thompson, F. J. Fernandez, J., and Asanuma, H. (1973): Relationship between 3a sensory cortex and motor cortex in the cat. *Fed. Proc.*, 32:340.
147. Uno, M., Kozlovskaya, I. B., and Brooks, V. B. (1973): Effects of cooling the interposed nuclei on tracking-task performance in monkeys. *J. Neurophysiol.*, 36:996–1003.
148. Vallbo, A. B. (1970): Discharge patterns in human muscle spindle afferents during isometric voluntary contractions. *Acta Physiol. Scand.*, 80:552–566.
149. Vallbo, A. B. (1974): Human muscle spindle discharge during isometric voluntary contractures. Amplitude relations between spindle frequency and torque. *Acta Physiol. Scand.*, 90:319–336.
150. Vilis, T., Hore, J., Meyer-Lohmann, J., and Brooks, V. B. (1976): Dual nature of the precentral responses to limb perturbations revealed by cerebellar cooling. *Brain Res.*, 117:336–340.
151. Vilis, T., and Cooke, J. D. (1976): Modulation of the functional stretch reflex by the segmental reflex pathway. *Exp. Brain Res.*, 25:247–254.
152. Vilis, T., and Hore, J. (1977): Effects of changes in mechanical state of limb on cerebellar intention tremor. *J. Neurophysiol.*, 40:1214–1224.
153. Vince, M. A. (1947): The intermittency of control movements and the psychological refractory period. *Br. J. Psychol.*, 38:149–157.
154. Vince, M. A. (1948): Corrective movements in a pursuit task. *Q. J. Exp. Psychol.* 85–103.
155. Welford, A. T. (1968): *Fundamentals of Skill.* Methuen, London.
156. Wiesendanger, M., Ruegg, D., and Lucier, G. (1975): Why transcortical reflexes? *Can. J. Neurol. Sci.*, 2:295–301.
157. Woodworth, R. S. (1899): *Psychological Monograph Suppl. 13: The Accuracy of Voluntary Movement,* pp. 1–114. Macmillan, New York.
158. Yumiya, H., Kubota, K., and Asanuma, H. (1974): Activities of neurons in area 3a of the cerebral cortex during voluntary movements in the monkey. *Brain Res.*, 78:169–177
159. Zarzecki, P., Shinoda, Y., and Asanuma, H. (1976): The projection of group I afferents to the motor cortex via area 3a. *Soc. Neurosci. Abstr.*, 2:958.

Posture and Movement, edited by
R. E. Talbott and D. R. Humphrey.
Raven Press, New York © 1979.

On the Cortical Control of Visually Directed Reaching: Contributions by Nonprecentral Motor Areas

Donald R. Humphrey

Laboratory of Neurophysiology, Emory University School of Medicine, Atlanta, Georgia 30322

INTRODUCTION

As a way of introducing the major topic of this chapter, let us consider the following behavioral act. Envision for the moment a subject who first gazes at or attends to an object within his visual field, and then, prompted by some internal stimulus, reaches for it, grasps it, and manipulates it under visual and tactual guidance. Although this act clearly involves a complicated motor sequence, it is possible to view it as being composed of the following, somewhat simpler processes: First, a visual fixation upon or a foveation of the object, a process that may require coordinated movement of both head and eyes; second, an orientation of the trunk and shoulder toward the object, so that the arm may be directed toward it along an appropriate path; third, projection of the arm; fourth, assumption by the hand as it travels toward the object of the appropriate posture for grip, an orienting movement of digits and wrist that is guided by an integration of visual cues concerning the shape of the object and somatosensory cues concerning the position and orientation of the hand; fifth, manual contact, grasping and manipulation of the object; and finally, a release of the object and withdrawal of the arm.

It is the purpose of this chapter to review selected evidence concerning the involvement of particular parietal and frontal areas of the cerebral cortex of primates in the central evolution and control of voluntary movements of the eyes, arm, and hand of this type. Before listing these cortical zones, however, it may be helpful to present first an appropriate context: a brief summary of current thinking about the major directions of information flow between central structures during the generation of voluntary movements. In this way, the location of the cortical areas that are to be discussed within this parent movement-control system may be indicated clearly, as may be the scope and limitations of the present review.

A general summary of this type is shown schematically in Fig. 1, which

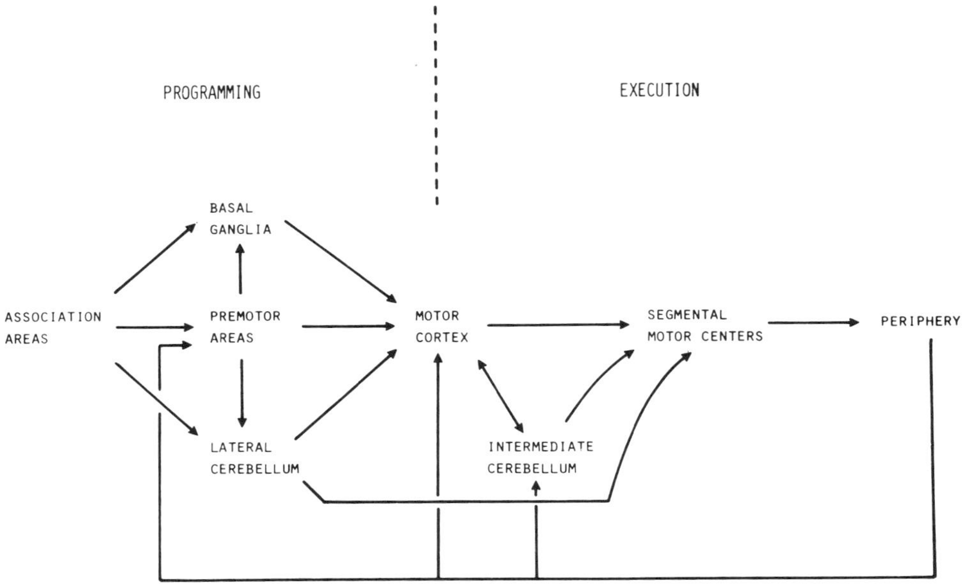

FIG. 1. Directions of activity flow between major central areas during the generation of voluntary movement. We focus in the present review upon transcortical routes which link posterior parietal association areas with frontal "premotor" zones, and both of these regions with the precentral gyrus. (After Allen and Tsukahara, ref. 4.)

has been adopted with modification from a comparatively recent review of cerebrocerebellar relations (4).

According to this view, a general decision to move toward or away from an object is transmitted from higher-order association areas of the cerebral cortex to the basal ganglia, the lateral cerebellum, and the "premotor association" areas of the cortex (for example, the posterior parietal lobe and premotor areas of the frontal lobe). These areas then act in concert to transform a general command for movement into an appropriate pattern of excitation across the somatotopically organized efferent zones of the motor cortex (7,9,164,178). The contribution of the basal ganglia to this process is not only one of reinforcing motor cortex activation (53,133,168,169), or of holding it in check when a delay in response is required (16,44,85,195,196), but also one of coordinating the emerging motor pattern with visual and other sensory data, so that the oculomotor and axial orienting responses that form the necessary base for directed attention and accurate limb-projection toward an object are generated (24,33,39, 45,53). The contribution of the lateral cerebellum is likewise complex, for it is thought to bestow upon the emerging movement appropriate temporal patterns of agonist-antagonist activation, which are based in part upon prior motor learning (4,28,81,82,191).

The outflow from the motor areas of the cortex descends as a patterned command signal not only to brainstem and spinal motor nuclei, where it is further elaborated, but also to the intermediate cerebellum and to various somatosensory nuclei, in order to modulate their responsiveness to sensory input data that will be generated by the (impending) movement (3,41,65,80,197). Recent evidence suggests that a motor outflow may also descend directly from the lateral as well as the intermediate cerebellum, by way of nuclear projections to reticulo- and rubrospinal pathways (12,13,101), and also from the basal ganglia by a nigrotectospinal route (1,5,6,64,86a). It is probable that descending outputs from these subcortical structures are used principally to control the activity of axial and proximal limb muscles during movement, whereas the highly flexible yet precise control of the distal arm and hand is accomplished primarily by the pre- and postcentral gyri (25,129,130,131,164).

Sensory inputs that are generated by the movement are fed back with varying somatotopic specificity to a number of structures in the movement-generating pathways; internal feedback loops also exist. The effectiveness of peripheral feedback in controlling the motor program varies, however, with the speed of movement, the extent to which it is highly learned or "preprogrammed," and the degree of accuracy that is required (21,29,60,72,89).

We concentrate in the present chapter upon a very limited section of this general control system: namely, upon certain areas of the cortex that project over comparatively direct, transcortical pathways—as well as by more circuitous routes through the basal ganglia and the cerebellum—to the ipsilateral precentral gyrus. These areas are: (i) the posterior parietal lobes (Brodmann's areas 5 and 7; cf. ref. 27); (ii) portions of the frontal lobe which lie, in the simian brain, in the vicinity of the arcuate sulcus (Brodmann's area 8 and posterior 9); (iii) portions of the premotor area [part of Brodmann's area 6, or area 6aβ in the classification system of the Vogts (201)]; and (iv) the supplementary motor area (SMA), as defined by Woolsey et al. (213) in the monkey, and by Penfield and Welch (160) in man. The location of these cortical areas is shown on a drawing of the hemisphere of the monkey's brain in Fig. 2.

A review of the apparent movement-control functions of these cortical areas is timely for a number of reasons. First, each appears to lie only one to a few steps back from the precentral gyrus along pathways that lead to it from higher-order association areas. Thus, an examination of the properties of these cortical areas may provide some insight into the types of input that are received by the motor cortex during voluntary movement, and perhaps also the transformations which it imposes on them. Second, each of these cortical areas projects also, either directly or by way of one cortical relay area, to subcortical structures that have been associated with control of the eyes, head, trunk, or proximal limb, either during shifts in visual attention, or during orienting-reaching movements. Thus, they may serve less as transcortical "relay" areas than as sources of parallel motor outflow which control proximal parts of the body during voluntary limb movement. Third, following a period of some 15 years, during which attention has been focused principally upon the basal ganglia and the

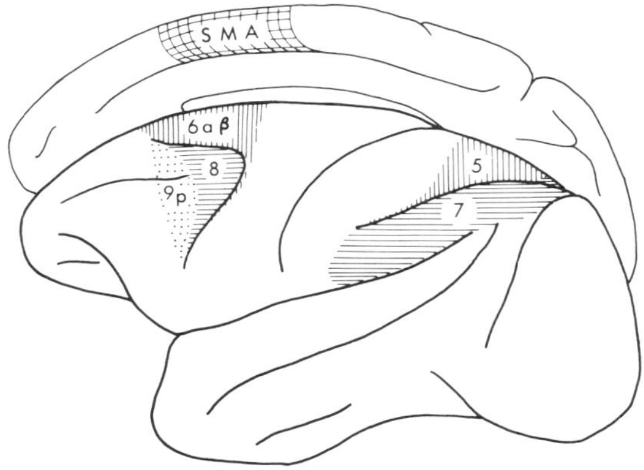

FIG. 2. Drawing of the lateral and medial surfaces of the monkey's brain, showing the locations of the cortical areas of interest. The numerical designations are those from Brodmann's (27) cytoarchitectonic classification. SMA, supplementary motor area. *In this and all subsequent, similar figures, the drawing of the medial surface of the hemisphere in the background is meant to represent the mirror image of the medial side of the hemisphere that is shown, and not the medial surface of the opposite side of the brain.*

cerebellum as the primary controllers of the motor cortex, the discovery of movement-related neuronal discharge in a number of these cortical areas has provoked renewed interest in their possible contributions to the generation and control of voluntary movement. For all of these reasons, the present review may serve as a useful adjunct to a number of comparatively recent reviews of the motor functions of the precentral gyrus, as well as others that are currently in preparation (8,30,164–166).

In discussing the motor functions of these cortical areas, the following format will be followed. First, the major afferent and efferent connections of each will be summarized. Available data from electrical stimulation, ablation, and single-unit recording studies will then be reviewed, in an attempt to provide at least partial answers, for each cortical zone, to the following questions:

1. How are goal-directed movements of the eyes and arm changed in frequency, accuracy, or dexterity by localized cooling or ablation of this region of the brain? Are observed impairments in limb movement more in the control of proximal or in that of distal muscles?
2. What are the properties of the movements that are evoked by electrical stimulation of this area? Are they muscle- or joint-specific, or do they involve more coordinated and widespread activity in several muscle groups? Are evoked responses of the arm confined to proximal muscles, or are movements of the hand also evoked?
3. What types of neuronal firing patterns have been observed in this area in relation to movements of the eye, head, or limb? Do cells within apparent limb-control zones fire in relation to the activity of specific muscles or agonist-

antagonist pairs, or are their firing patterns more general, related to the occurrence of limb movements but not to the details of muscle activity?

As is evident from these questions, some data will also be reviewed concerning the contribution of each cortical zone to the control of directed visual attention processes. These will be considered because, as will be suggested below, there is some reason to suspect that the eye movements (or the central commands for eye movement) that occur during visually directed reaching could well serve as sources of signals that are used by the limb control system for directing the arm and hand toward particular regions in space.

POSTERIOR PARIETAL LOBES

Pathways to Motor Structures

The pathways that link areas 5 and 7 with the motor areas of the frontal lobe and with various subcortical structures are summarized diagramatically in Figs. 3 and 4. These summary diagrams were constructed on the basis of a number of well-known degeneration studies (110,152,161), as well as more recent autoradiographic and retrograde cell-labeling studies (56,108,112,140,187).

Consider first area 5 (Fig. 3). This cortical zone receives a major input from areas 1 and 2 of the SI somatosensory cortex (108–110), a projection from

FIG. 3. Major anatomical connections of cortical area 5. Transcortical pathways of interest. **Subcortical connections.** *Afferents from:* thalamus (LP, pulvinar nuclei), hypothalamus, basal forebrain. *Efferents to:* putamen, caudate, thalamus, red nucleus, pons, dorsal column nuclei, spinal cord, (intermediate laminae). A similar format is followed in Figs. 4–7.

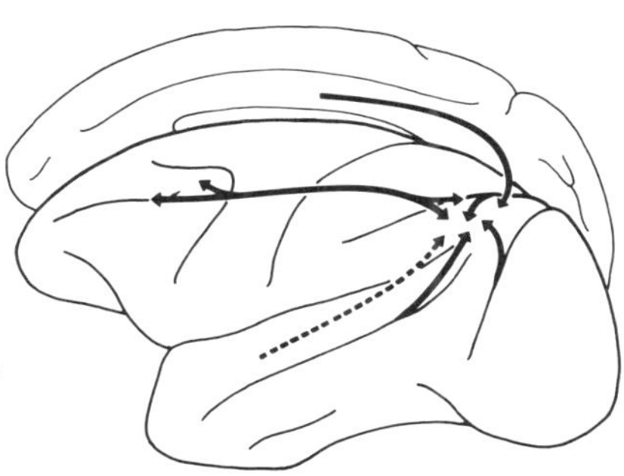

FIG. 4. Major afferent and efferent connections of cortical area 7. The *dashed line* in the drawing represents a transcortical projection to area 7 from parahippocampal regions buried within the temporal lobe. **Subcortical connections.** *Afferents from:* thalamus (AM, AV, VA, VLc, LP, pulvinar, intralaminar nuclei), hypothalamus, substantia innominata, locus ceruleus, raphe nuclei. *Efferents to:* putamen, caudate, thalamus, pretectum, superior colliculus, pons. Additional details in text.

the lateral posterior–pulvinar complex of the thalamus (106,108), and a smaller input from hypothalamic and basal forebrain structures (117). It projects to both putamen and pons (107,161), and thus may influence the motor cortex by way of basal ganglia and cerebellar circuitry. Transcortically, it projects heavily to dorsomedial area 6 (area 6aβ) and to the SMA, both of which project in turn not only to the precentral gyrus, but also *directly* to brainstem nuclei that appear to be involved in controlling axial and proximal limb muscles (the medullary reticular formation and the red nucleus; cf. refs. 122,130,131). In addition, recent autoradiographic and retrograde cell-labeling studies have shown that area 5 sends somewhat lighter projections *directly* to the arm area of the precentral gyrus (108,187), to the red nucleus (101), to the dorsal column nuclei (35), and to the dorsal and intermediate laminae of the spinal cord (40).

Thus, area 5 receives higher-order somatosensory information, and projects both directly and indirectly to structures with motor and/or somatosensory relay functions. Its anatomic connections suggest, therefore, that it may be involved in some higher-order way in the steering or guidance of limb movements on the basis of somatosensory cues.

Consider now area 7 (Fig. 4). This region also projects to the putamen and the pons and could conceivably affect the motor areas of the frontal lobe by way of the basal ganglia and cerebellum, as well as by way of transcortical pathways that are considerably less direct than those which issue from area 5. But its patterns of input and output differ from those that are known for area 5 in at least three important ways. *First,* area 7 appears to receive a much more diverse and substantial input from subcortical structures than does area 5. Inputs converge, for example, from a number of the so-called "association" nuclei of the thalamus, including the anterior (AM, AV), lateral (VA, VLc, LP, pulvinar) and midline-intralaminar groups (paracentralis, centralis lateralis) (34,56,114,140). Predominant among these thalamoparietal projections are those from the LP-pulvinar complex, nuclei which are known to receive converging inputs from somatosensory and visual pathways (17,86). Projections to area 7 have also been traced from the hypothalamus and substantia innominata (which receive inputs in turn from the limbic system), and from the locus ceruleus and raphe nuclei (56,117,140). *Second,* unlike area 5, area 7 is a site of converging transcortical input from a number of polymodal sensory association areas (56,152). Projections to the main body of area 7 arise, for example, from the intraparietal sulcus (which receives input from both somesthetic and visual parasensory association areas), the banks of the superior temporal sulcus (which receives inputs from visual and auditory association areas), and the prefrontal-arcuate area (which receives inputs from all three sensory association areas). In addition, substantial projections to area 7 arise from the cingulate gyrus and the parahippocampal region (140), areas whose activities have been associated with emotional and motivational states (138,151,208). *Third,* the projection from area 7 to the frontal lobe is directed not to the premotor (area 6) or motor areas (SMA, area 4), but instead to the prefrontal area (areas 8A, 8B, posterior 9), an area which receives inputs also from occipital and temporal association cortex, and, at its posterior region (area 8A), has eye-movement control functions (19,175,204). In addition, area 7 projects directly to the pretectal region and the mid to deeper layers of the superior colliculus (122,161), structures which are thought to be involved in the control of saccadic eye movements and directed visual attention processes (83,145,214,215), and perhaps indirectly in the coordination of eye-head movements (186,188). In contrast, the transcortical routes from area 7 to the precentral arm area are quite indirect, involving a greater number of relay sites than do those from area 5.

Thus, anatomical studies suggest that area 7 is a cortical zone that receives and processes higher-order information from a number of polymodal association areas in both thalamus and cortex, and also from structures whose activities have been correlated with motivational states, sleep-waking cycles, and levels of attention or arousal. In turn, its most direct motor projections are to structures that appear to be involved in control of the eye movements that are necessary for attending to selected aspects of the visual environment, and perhaps also for visually guided limb movement.

Movements Evoked by Stimulation of Areas 5 and 7

It has been known for several decades that both axial and limb movements may be evoked by epileptic activity or electrical stimulation within area 5 (68, 70,71,77,157,158). Stimulation studies have revealed a somatotopic organization that is less precise than but similar to that of area 4, with the leg-trunk, shoulder-arm, and neck-head musculature being represented in a mediolateral direction along the anterior border of the intraparietal sulcus (68,69,70). The requisite stimulus intensities and movements evoked by stimulation of area 5 differ significantly, however, from those associated with stimulation of the precentral gyrus (68,213). *First,* even in the very lightly anesthetized monkey, the current-voltage thresholds for evoking movement are considerably (1.5 to 3 times) higher when stimulating area 5 than when stimulating area 4. *Second,* the motor responses evoked by stimulation of area 5 are more diffuse in their distribution than the "muscle-specific" responses that are often evoked by punctate stimulation of the precentral gyrus (7,9,164). For example, though hand and finger movements may be evoked by stimulation of area 5, they are almost invariably accompanied by even more pronounced activity in the proximal limb muscles. Moreover, though contralateral responses predominate, the evoked response pattern may also involve activity in the ipsilateral trunk and proximal limb musculature. *Finally,* unlike the phasic or "twitch-like" responses that are often evoked in a single muscle or about a single joint by stimulation within area 4, the motor responses produced by stimulation within area 5 appear to be components of more complex and coordinated orienting, or limb projection-withdrawal movements, involving muscle activity at a number of joints, and persisting in time throughout a longer period of stimulation or even slightly beyond termination of the pulse train. Typically observed movements consist, for example, of a turning of the head and shoulders to the contralateral side; components of reaching or withdrawal movements of the contralateral arm, accompanied by finger flexion (grasping) or extension; or, momentary assumption of a particular limb posture. The responses appear to be mediated only in part by the primary motor areas of the cortex, for, while they are reduced in intensity and somatotopic specificity, they are not abolished by bilateral ablation of the pre- and postcentral gyri (68). It is possible that they are produced in part by projections to, and descending pathways from, the so-called "premotor" area (area 6) and the SMA (see Fig. 3). Somewhat similar movement patterns have been observed in seizure patients with epileptic foci within the parietal lobe (77,158).

It has also been known since the pioneering work of Ferrier (66) that eye movements are evoked by electrical stimuli applied within area 7 (68,70,71, 142,204). In the lightly anesthetized monkey, the evoked eye movements consist of conjugate deviations to the contralateral side, sometimes accompanied by a partial upward or downward rotation of the eyes (68,204). In the very lightly anesthetized or the unanesthetized animal, the evoked eye movements are often accompanied by a turning of the head in the same direction (134,204). The response pattern appears to be one of rotating the head and eyes so that the gaze may be directed toward some sector within the contralateral visual field. The evoked movements survive bilateral ablation of the pre-and postcentral gyri (68). The responses may be mediated by direct subcortical projections from area 7 (161), or perhaps in part by projections to the frontal eye field (152).

It is important to note that movements of the contralateral arm may also be evoked by stimulation within area 7, particularly in the unanesthetized animal (68,134). The lowest threshold regions for evocation of arm movements lie anteriorly and laterally, near the intraparietal sulcus, whereas those for evoking eye movements lie more posteriorly and medially (68).

To this author's knowledge, there have been no systematic attempts to use intracortical microstimulation techniques (cf. ref. 9) to examine in more detail the topographic organization of these regions within areas 5 and 7 from which motor activity may be evoked.

Disturbances in Centrally Directed Attention and Voluntary Movement Processes Following Posterior Parietal Lobe Lesions

The current literature contains two excellent discussions of the complex perceptual and behavioral disturbances that are produced by lesions of the posterior parietal lobe in man and in the monkey (137,149). Consequently, we will focus here only upon those lesion-induced disturbances that appear to arise primarily from deficits in the ability to attend selectively to objects within the immediate environment, to initiate appropriate voluntary movements toward or away from them, or to manipulate them skillfully under visual or proprioceptive guidance.

But before discussing these deficits further, a note of caution is perhaps in order. Clinical case studies in man have shown that lesions which involve, but are often not confined to, the posterior parietal areas, may produce profound disturbances in the perception of spatial relations, in body image, in directed visual attention processes, in the ability to initiate known movement sequences upon verbal command, and in the skill of movements once initiated (14,18,43, 51,52,78,94,148,156,182). Moreover, the pattern and severity of perceptual-motor impairment depends in part upon whether the lesion is within the dominant or the nondominant hemisphere (43,51,91). But in many of these case studies, the precise nature of the lesion—i.e., whether primarily irritative or destructive in effect—its precise location, and its anatomical boundaries have not been known with certainty.[1] Consequently, it is in general not possible to extract from these clinical reports any clear indications of the degree of functional specificity within different subregions of the parietal zone. Moreover, even in the controlled ablation studies with subhuman primates, the lesions have often encompassed a number of cytoarchitectonic areas, with significantly different patterns of afferent input and efferent outflow; for example, portions of areas 2 or 19 are often ablated, in addition to areas 5 and/or 7. Finally, in many (though not all) of the studies with experimental animals, behavioral-response indices have been used that do not allow one to determine with any certainty whether the reported post-lesion deficits were due primarily to perceptual disturbances—for example, an impaired perception of spatial relations—or to disturbances in sensorimotor coordination; i.e., a disturbance in the process of using even correctly perceived sensory data to guide voluntary movement (15,57,58,144,167,171,173).

For these reasons, it may be premature to attempt to extract from the numerous and sometimes varying descriptions of the "parietal lobe syndrome" a particular set of disturbances that appear to be primarily "attentive" or "motor" in nature, and to attempt to relate these to damage to particular intraparietal areas. But for purposes of this discussion, an attempt has been made to do so. The resulting list of disturbances may be broken down into three basic categories, which are: (a) apparent defects in visuomotor control and directed attention processes; (b) changes in the ease of initiation and frequency of arm and hand movements; and (c) defects in the sensory guidance of skilled arm and hand movements, once initated. Each of these categories will be described separately.

Deficits in Central Visuomotor Control and Directed Visual Attention Processes

For purposes of this discussion, let us consider the process of visually attending to the environment while it is scanned in search of a particular object of interest to consist of the following steps:

1. A recall from memory of the major features of the desired object.
2. The development of a central "set" or "filter," which will allow or enhance responses only to objects with the expected visual characteristics.

---

[1] An important exception to this statement are the careful descriptions by Hecaen et al. (93), where the lesions were placed neurosurgically and the anatomical boundaries were known with some certainty. As noted by Mountcastle et al. (149), the descriptions by Hecaen et al. are particularly valuable for this reason.

3. Visual exploration of the environment, a process which involves first a fixation upon a sector of the field, a breaking of fixation if the expected stimulus pattern is not encountered, and a saccade to another sector.

4. Recognition of the object when an acceptable match between visual input and expectation (memory) occurs.

5. Maintained fixation upon the object, accomplished when necessary (i.e., when it lies within the periphery of the visual field) by oculomotor pursuit and body-orienting responses.

Available descriptions of the parietal lobe syndrome suggest that an impairment may exist in any one or all of stages 2 through 5 of this hypothetical sequence. A most commonly observed symptom of unilateral, posterior parietal lobe damage is an "attention agnosia" of the contralateral body and extracorporeal space (43,49,51,57,148). There is a failure to attend to—or even a profound neglect of—objects and body parts within the contralateral hemispace, with a corresponding reduction in centrally directed movements of the contralateral arm, such as reaching and manual exploration (51,54,57). Defects have also been observed in the central control of eye movements. Following bilateral posterior parietal lobe lesions, for example, deficits have been observed in a subject's ability to fixate visually upon an object, to break fixation once it is achieved, or to track smoothly a moving visual target (37,38,92). Perhaps as a result of these impairments, or an impairment of the visual recognition process suggested above, monkeys with parietal lobe lesions may exhibit marked deficits in the ability to locate visually (or recognize) an object of motivational significance when it is surrounded by extraneous objects (125). Perceptions of the form and spatial location of an object may also be impaired, a disturbance that would seriously affect visual recognition processes (170,171,192).

It seems clear that these disturbances do *not* result from any impairment of the basic oculomotor control system. For example, spontaneous and many stimulus-evoked eye movements are apparently unaffected by parietal lobe lesion (154,155). Instead, the lesion-induced state appears to reflect a true oculomotor apraxia, a condition in which those central structures that normally use the oculomotor apparatus to "focus" visual attention upon, or to shift it from, objects of motivational interest are now either no longer intact, or are disconnected from the oculomotor system.

Although the parietal lobe lesions in many of the studies referred to above extended beyond the boundaries of area 7, it is probable in the light of more recent studies that the observed impairments in visuomotor control were due principally to destruction of this region (104,137,149).

Reduced Ease of Initiation and Frequency of Arm and Hand Movements

Though the movements are less skillful than those performed by normal subjects, subjects with unilateral lesions of the posterior parietal lobe *are* able to reach for, to grasp, and to manipulate objects with the contralateral arm and hand. However, in both man and monkey the ease of initiating these movements by verbal command or visual stimuli, their smoothness and degree of coordination, and their frequency of spontaneous occurrence are all markedly reduced (43,51,53a,90).

With dominant-hemisphere lesions in man, a peculiar condition referred to as "ideational apraxia" may exist (79,94). In such cases, the patient is unable to initiate a known motor sequence upon verbal command; for example, to use a key to open a lock when requested to do so. Yet, the request appears to be fully understood, and the patient may be able to imitate the required sequence of movements when they are performed by the examiner, or to recognize the correct sequence when it is presented in a multiple-choice situation. Thus, the basic defect appears to be in those processes which lie between comprehension of a verbal request for movement, and the encoding of the appropriate motor sequence.

In both man and monkey, a marked reduction occurs also in the frequency of spontaneous, purposive movements of the arm and hand contralateral to the lesioned side; indeed, the limb may be "neglected" in all but the simplest of bilateral motor acts (43,51,53a,54,57,148). With lesions confined to area 7 in the monkey, visual placing responses may in addition be grossly impaired or absent, whereas tactile placing may be impaired following lesions of areas 5 and 2 (51,53a). In the latter case, the remnants of spontaneous limb movement may effectively be abolished by blindfolding the animal, so that the affected limb is held catatonically in fixed, abnormal postures for surprising periods of time (51). Thus, the normal frequency of "spontaneous," exploratory-manipulative movements of the arm, and indeed even of changes in limb posture, may depend importantly upon sensory inputs which affect the motor system either by way of, or in conjunction with, the posterior parietal lobes.

It must be noted, however, that this reduced frequency of exploratory movements is in no way the result of a lesion-induced paresis of the affected limb. Indeed, at least one noted observer has reported that normal approach-exploratory movements are replaced postoperatively by avoiding movements, i.e., movements which promote an avoidance of contact with objects on the contralateral side (51). It is not known, however, to what extent this avoidance behavior is due to a released sequence of innate motor responses, normally suppressed by the posterior parietal lobe, and to reactions that are secondary to the subject's inability to attend to and perceive accurately objects within the contralateral visual world.

Defects in the Sensory Guidance of Arm and Hand Movements

The act of reaching for and manipulating an object within the visual field may be viewed as a three-stage process.

First, the limb must be launched on the correct path. According to traditional views, this would be accomplished by a continual comparison of sensory data concerning the actual position of the limb with the desired position; continual adjustments in motor output would then be used to correct any discrepancy between the actual and the desired flight path. According to the provocative new view presented by Bizzi and Polit *(this volume),* a central memory or program could be used to specify the pattern of agonist-antagonist equilibrium conditions that would exist across the muscles of the arm and trunk, *if the limb were in the correct position;* the arm would then move automatically toward the position that satisfied these requirements.

Second, the hand must assume en route the proper attitude for grip or manual contact with the object. Normally, this sequence appears to be initiated by visual cues, just prior to contact with the object.

Third, manipulation of the object, a response sequence that is set in motion by somesthetic cues when the object is touched.

Combined lesions of areas 5 and 7 appear to impair all of these subsidiary processes. In man, the errors in reaching are greatest when the arm contralateral to the lesion is used to reach for objects on that side, but smaller errors may also be detected when the ipsilateral arm is directed into the contralateral field (43,51). In monkeys, the defect is somewhat more limb-specific, being greatest for the contralateral limb when it is used to reach into its hemispace, but detectable also when the affected limb is used to cross the midline for objects on the side ipsilateral to the lesion (90). The errors in reaching observed in the lesioned monkey are on the order of 12° and are toward the lesioned side (123).

A normal monkey is able to reach with some accuracy for an object in the dark, once proprioceptive and other cues have been used to establish its position. Animals with posterior parietal lesions are totally deficient in such tasks, however, indicating an impairment in the use of proprioceptive data to estimate the relative position of the object with respect to the body, or to guide a limb accurately toward it (90). Visually guided movements are also impaired, and visual placing responses of the limb may be absent (51,90).

Defects in smoothly grasping and manipulating an object once it is reached are also seen (51). Monkeys with combined lesions of areas 5 and 7 have great difficulty in performing discrete hand and finger movements that require visual or propricoceptive cues for successful task-completion (181a). Moreover, during local cooling of area 5, monkeys have difficulty in grasping a small object once it is contacted, apparently because of some deficit in the ability of somesthetic cues to trigger or guide manipulatory responses (186a). When the cooling probe is applied to area 7, the contralateral limb becomes *grossly* clumsy only when it is used to reach for objects on the same side, suggesting that the primary defect is in the visual guidance of movements on that side of the body (186a). These results have been viewed as support for the idea that area 5 is concerned principally with the somesthetic (or proprioceptive) guidance of contralateral arm and hand movements, particularly those that require fine sensory tracking for execution, whereas area 7 is concerned principally with the visual guidance of these same movements.

Movement-Related Behavior of Posterior Parietal Lobe Neurons

The summary description to be given here was derived from the single-unit recording studies of Hyvarinen and Poranen (104), Mountcastle et al. (149), and Lynch et al. (137), each of which was performed with unanesthetized monkeys. Additional data on area 5 cells may be found in reference (179).

A summary of the various classes of behaviorally defined neurons that have been observed in areas 5 and 7 is shown in Table 1, along with the approximate percentages of the observed unit samples that were accounted for by neurons of each type. Also shown are the cortical layers within which cells of each type were observed most frequently, with the first layer shown being that of greatest observed cell density. Finally, a code is given for each movement- (or eye fixation) related class of cells, to indicate whether changes in their firing rates began and reached their peaks prior to, approximately coincident with, or after the onset and peak of the motor response. The major features of the cell populations observed within area 5 will be considered separately from those observed within area 7.

TABLE 1. *Summary of types and locations of neurons observed within the posterior parietal lobe of the unanesthetized monkey*

Cortical area and type of neuron	Percent of observed unit sample[a]	Observed most frequently in layers	Timing of discharge[b]
Area 5			
Joint rotation	64	III, IV	0, −
Muscle, other deep receptors	12	III, V	
Cutaneous	9	III, V	
Arm projection & hand manipulation	11	III, IV, VI	+
Other	4	V	
Area 7[c]			
Visual fixation	32–57	III, II, IV	+, 0
Visual tracking	7–14	III, V	+
Saccade	17	III, II	+
Arm projection & hand manipulation	15–34	III	+
Visual fixation & arm projection	4–10	II, III	

[a] The percentage estimates for area 5 were derived from Mountcastle et al. (149) and those for area 7 from Mountcastle et al., Lynch et al. (137), and Hyvarinen and Poranen (104).

[b] Estimates of the timing of unit discharge in relation to movement or to eye fixation were derived from the data of Mountcastle et al. and Lynch et al. The symbols indicate whether the observed onset and peak of change in unit discharge frequency most frequently led (+), was nearly simultaneous with (0), or lagged (−) in time the correlated behavioral event.

[c] Although data are not included in this table, Yin and Mountcastle (218) have also observed recently that cells with visual receptive fields exist in area 7. They postulate that these neurons may provide the visual drive for area 7 visuomotor neurons.

Area 5 Neurons

As might be expected on the basis of the known projections from area 2 to area 5, the latter contains a major set of neurons that respond to somatic stimuli delivered to the contralateral side. Some 9% of the observed cells responded to cutaneous stimuli, and an additional 12% to stimulation of muscle or other deep receptors. But by far the largest set of peripherally driven neurons were *joint cells,* more than 80% of which responded to passive rotation of only a single, contralateral joint (149).

These latter cells were found to behave, however, in a manner that differed significantly from the previously observed behavior of joint neurons in the primary somatosensory cortex. Although they responded to passive joint rotation, they fired much more briskly during rotations of similar amplitude and speed that were produced actively, i.e., during spontaneous and conditioned limb movements. Since the cells did not fire in this way when muscles were contracted across a fixed joint, they did not appear to be driven by tendon organs. Consequently, Mountcastle et al. suggested that the increased firing of area 5 joint cells during movement could be due to a corollary discharge from central (presumably motor) structures, which facilitated their responsiveness to peripheral input. Presumably, this facilitation would occur at some high level within the sensorimotor system, for a similar augmentation is apparently not seen within lemniscal pathways or the somatosensory cortex (149). In this regard it is interesting to note that area 5 receives transcortical projections from precentral motor areas, as well as projecting anteriorly to them (108,187); thus, a candidate pathway for such corollary input from cortical motor centers exists.

Mountcastle et al. also observed another set of cells within area 5 whose properties are of particular interest here. These cells comprised about 11% of the neurons observed within area 5 and could be subdivided into two subsets of approximately the same size: arm-projection and hand-manipulation neurons.

Arm-projection neurons appeared to be unresponsive to any form of passive sensory stimulation, firing only during reaching movements toward objects of motivational interest within the immediate extrapersonal space (e.g., food or reinforcement-related stimuli). Their discharge was thus conditional and did not occur during even similar limb movements to the same regions in space when an object of interest was not present. Slowly rising increases in their discharge frequencies were present as early as 1 sec prior to the onset of the reaching movement, with the peak change occurring well before the movement was complete (149, Figs. 6 and 7). Moreover, the details of their firing patterns were independent of the sensory channel used in the behavioral paradigm of Mountcastle et al. to trigger the movement, and, over a range of up to 32° along the horizontal meridian, also of the direction of arm projection. Thus, the discharge of these cells appeared to be motivationally dependent, to occur early enough in time to serve as a candidate source for the driving of motor structures, and to be related to the molar, behavioral act of reaching, rather than to the quantitative patterns of muscle activation.

Hand-manipulation neurons were similar, firing during tactile or manual exploration of objects of interest. They were also active during grooming behavior, and were, in this regard, less conditional in their discharge dependencies than were arm-projection neurons.

It is of interest to note that all of these categories of cells were observed with greatest frequency within cortical layer III, a layer that is known, at least within the sensorimotor areas of the cortex, to be a principal source of transcortical efferents (108,111).

Area 7 Neurons

The cells that have been observed within area 7 may be divided into five classes, three of which discharge in relation to attentional or visuomotor processes, and two in relation to arm movements.

An important property of the first three classes of cells is that their discharge in relation to visual stimuli appears to be as much centrally as peripherally determined. None appears to respond directly, for example, to light stimuli of *moderate* intensity that are sufficient to drive visual-area cells (137,149). Rather, they appear to fire only when visual attention is directed toward some object of motivational interest within the nearby visual field, when it is tracked visually, or when fixation is broken during a saccadic eye movement to a new object of interest. The three classes of cells were referred to by Lynch et al. (137) as visual fixation, visual tracking, and saccade neurons. They appear to possess the following properties.

Visual fixation neurons fire only during foveation of an object that is both of motivational interest and that is obtainable; for example, food, a reward-related manipulandum that is within reach, or visual stimuli that signal the availability of food. The discharge begins some 100 to 200 msec prior to the onset of fixation as defined by electrooculograms. The discharge declines or ceases abruptly if vision of the desired object is suddenly blocked or if it is removed from the visual field, suggesting that the cell is driven by visual input from an object with particular (known) stimulus characteristics. Yet, the *conditional* or "gated" effectiveness of such peripheral input is indicated clearly by the fact that the same fixated object may no longer cause the cells to fire if it is unobtainable, no longer signals reward, or if the motivation to obtain it is sated (e.g., the visual fixation of food by a sated animal). The discharge of some 40% of these cells (that fraction within layer V) is also suppressed during saccadic eye movements of the type described below. Finally, each visual fixation cell displays a different (though often overlapping) *preferred gaze field;* i.e., its discharge is most brisk when the fixated object lies within a particular sector of the visual field, most often contralaterally.

Visual tracking neurons are active only during smooth oculomotor pursuit of slowly moving objects of interest. Their discharge begins some 50 to 70 msec prior to the small saccade that initiates the pursuit and brings the eyes onto the track of the moving target (137, Fig. 14). The discharge of a given cell is maximal during pursuit movements *in a particular direction,* and is suppressed during saccadic eye movements of the type described below.

Saccade neurons begin to discharge some 150 msec prior to and continue throughout visually evoked saccadic eye movements; i.e., during saccades that are evoked by a rapidly moving target of interest, or by the abrupt introduction of such a target into nonfoveal regions of the visual field. Their discharge is also directionally dependent, being maximal for a given cell during eye movements in only one or two of the four major directions. The cells are inactive during visual fixation, slow pursuit, and "spontaneous" saccades such as those which occur in the dark. It was suggested by Lynch et al. that these cells may function principally to suppress or "break" visual fixation, so that the eyes may be directed rapidly to new objects within the visual field during stimulus-evoked shifts in visual attention. Since the discharge of these cells differs from that of frontal eye field neurons, it was suggested that they probably do not affect the oculomotor apparatus by way of projections to the frontal lobe; rather, they may do so by way of direct projections to the superior colliculus and pretectal region (137,161).

On the basis of these observed discharge properties, Lynch et al. suggested the following: ". . . the parietal lobe, particularly area 7, executes a matching function between the neural signals of the nature of objects and the internal drive state of the organism, and contains a neural apparatus for the direction of visual attention to objects of interest and for shifting the focus of attention from one to another" (137, p. 362).

In addition to these three classes of cells, *arm-projection* and *hand-manipulation* neurons of the type described previously were also observed within area 7. Such cells accounted for some 34% of the sample of area 7 neurons that were observed by Mountcastle et al., suggesting that they may exist with greater density within the anterior regions of area 7 than within area 5.[2]

Finally, a smaller subset of cells was observed by all investigators, which fired when an animal fixated visually upon a desired object within the visual field, or when he reached toward that object even with vision blocked. The

[2] Robinson et al. (176,177) have recently questioned the "command" function hypothesis for arm projection and other posterior parietal neurons advanced by Mountcastle et al. (149). On the basis of a more limited sample of neurons, these authors found that many of the hand-projection and saccade cells identified by their procedures did in fact respond to somatosensory, to visual, or to both types of stimuli. The discharge of the cells was *enhanced,* however, if a saccadic eye movement (saccade cells) or hand projection (projection cells) was to be made toward the effective stimulus. In the absence of the stimulus the cell did not discharge, even if the eye or hand movement occurred. On the basis of such observations, Robinson and Goldberg therefore concluded that area 7 cells may function to in some way "enhance" commands for movement, doing so in response to converging inputs from higher-order sensory and motivational structures, but do not function directly as command cells per se. Their paper may be consulted for further details, as should be the reports by Yin and Mountcastle (216,217), who also found visually responsive cells in area 7 but showed them to be clearly distinct from previous classes of neurons described by Mountcastle et al. (149) and by Lynch et al. (137). An excellent discussion of the problems encountered in attempting to define either single or populations of command neurons has also appeared recently, and may be consulted for details (121).

cells were referred to as *looking-reaching* neurons by Hyvarinen and Poranen (104), who noted that, like visual fixation neurons, they exhibited preferred gaze and reach fields. Indeed, it was strongly implied if not clearly stated by these authors that *such neurons could well serve as a source of central signals for directing motor responses toward particular objects or locations of interest within the immediate visual field,* a function that might also be postulated for visual fixation cells.

As with the cells observed within area 5, it is again of interest to note that all of these categories of neurons were observed with greatest frequency within cortical layer III.

Comment and Summary

The evidence summarized above suggests that different neuronal populations within area 5 could be involved in the *initiation* as well as the *guidance* of voluntary reaching and exploratory movements of the contralateral arm and hand. Its output pathways impinge upon the supplementary motor area, area 6 and the precentral arm area, and it sends some fibers directly to the region of the red nucleus, as well as a heavier projection into the opposite corticospinal tract. When stimulated electrically, it produces responses that appear to be partially organized fragments of purposive orienting and reaching movements, involving in particular the muscles of the arm and trunk that would normally be involved in such movements. It contains cells (projection and manipulation neurons) whose discharge leads in time and could well be involved in the central initiation of such movements in response to motivational stimuli. Yet, the detailed motor program that is necessary for the correct initiation of such movements must lie downstream within the motor system and be evoked as a composite pattern by area 5 and perhaps other sets of neurons. For, as might be expected with a hierarchically organized motor system, the discharge of area 5 cells whose activity leads movement is related only to the *behavioral act* of reaching or manipulation, and apparently not to the details of muscle activity underlying these movements. Nor are the movements eliminated from the animal's repertoire by destruction of area 5. On the other hand, it is clear that such lesions do impair an animal's ability to perform skilled movements that require a strong element of proprioceptive guidance. Such guidance could well be provided by the numerous joint neurons within area 5, whose activities are in some way brought into bold relief during active movement by unknown, but presumably central, mechanisms. It will be a major task for future research to determine the mechanisms that underlie this facilitation process, and the specific pathways over which the enhanced activity of area 5 joint cells might reach motor structures in order to more accurately guide movement.

It seems likewise clear that area 7 is involved intimately in those centrally directed visual attention processes that precede and steer limb movements toward objects of motivational interest within the visual field. The outputs from this area impinge upon both frontal lobe and subcortical structures that have been associated with visuomotor attention processes, and it receives inputs from cortical areas with known memory and motivational functions. Lesions of area 7 produce deficits in directed attention processes, and in the frequency and accuracy of limb movements toward objects in the contralateral visual field. Like area 5, area 7 contains a set of arm-projection and hand-manipulation neurons, which respond principally to central (137,149) but perhaps also to relayed sensory inputs (177), and which could serve as part of a central "command" or "enhancement" network for motivationally evoked, manual exploration of objects within the immediate extrapersonal space. Moreover, the location- and direction-specific discharge of its visual fixation, visual tracking, and "looking-reaching" cells could serve as a central signal for directing such limb movements toward specific locations and objects in space. Again, however, the movement-related neurons of area 7 must evoke or facilitate as a composite whole motor programs for reaching and manipulation that are stored (hardwired) elsewhere, for the underlying details of muscle activity to not appear to be encoded within their patterns of discharge.

PERIARCUATE AND "PREMOTOR" AREAS OF THE FRONTAL LOBE

The posterior parietal regions discussed above project heavily to frontal areas in the region of the arcuate sulcus. Area 8 and posterior 9 [posterior area 46 in the terminology of Walker (206)], the former being roughly coextensive with the frontal eye field, receive substantial transcortical inputs from area 7, whereas the adjacent "premotor" zone [area 6aβ; cf. (201)] receives a significant projection from area 5 (110,152,153). It is quite likely that each of these frontal areas contributes in a substantially different way to the control of eye and hand movements; it will be appropriate, therefore, to consider them separately.

Area 8[3]

Major Input and Output Pathways

The major anatomical connections of area 8 are summarized in Fig. 5A and 5B. In constructing these diagrams, considerable weight has been placed on recent autoradiographic studies of the frontal connections and efferent projec-

[3] Although the discussion here will refer only to area 8, it should be noted that in many of the studies of the anatomical connections and functions of pre-arcuate tissue, the lesions, observed sites of terminal degeneration and effective sites for the evocation of eye movements have included both area 8 proper and the adjacent, posterior parts of area 9 (cf. Fig. 2). It has only been recently that newer anatomical methods have been used to show that area 8 is linked with the premotor zone (area 6) only indirectly, by way of projections to posterior area 9, which projects in turn to area 6 (119,120). There is at present clear evidence to indicate that tissue within the principal sulcus region is involved in the mediation of somewhat different complex perceptual-motor behaviors than is that in the region of the arcuate sulcus (see, for example, refs. 84,162), but additional studies are needed to resolve precisely how these two adjacent areas might contribute differently to the control of directed attention processes and voluntary movement.

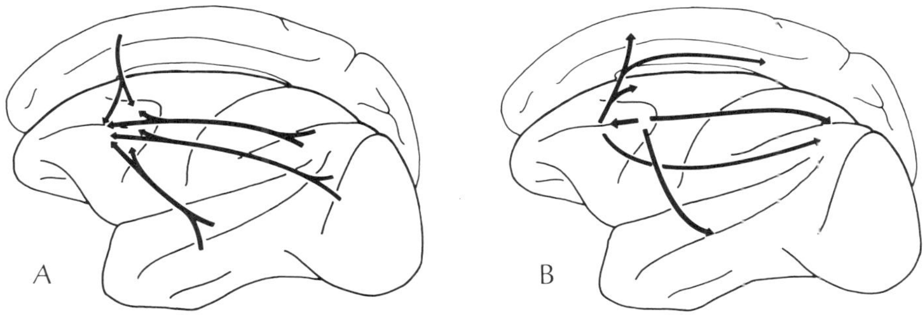

FIG. 5.A: Major afferent pathways to areas 8 (frontal eye field) and posterior 9. *Subcortical afferents from:* thalamus (MD, pulvinar nuclei), hypothalamus, basal forebrain. Additional details in text.

B: Major efferent projections from areas 8 and posterior 9. *Subcortical efferents to:* putamen, caudate nucleus, thalamus, zona incerta, pretectum, superior colliculus, mesencephalic central gray, pons.

tions of area 8 (119,120). Additional information concerning efferent projections, as well as afferent pathways to area 8 from subcortical and posterior cortical zones, was derived from earlier, degeneration studies (2,11,110,122, 152,153).

Area 8 receives transcortical inputs of substantial size from the temporal lobe, the preoccipital or peristriate area, and, according to some investigators (152), but not others (110), from area 7 (110,152).[4] A projection also arrives from the SMA (119,152). Thus, area 8 is a site of converging inputs from association areas that are linked with visual, auditory, and perhaps indirectly (via SI and SII projections to the SMA) with somatic sensory systems, and from frontal regions whose activities are associated with control of the axial muscles. This pattern of input suggests that the apparent oculomotor modulating functions of area 8 may be influenced jointly by heavily processed inputs from major sensory systems, and perhaps coordinated with the axial orienting movements that are mediated by way of bilateral outputs from the supplementary motor area. The major thalamic input to area 8 is derived from a subdivision (pars paralamellaris) of the mediodorsal nucleus (2), with a lighter projection arriving from the medial pulvinar (23).

The major transcortical projections from area 8 are directed back to the temporal lobe (superior temporal sulcus), to area 7 (intraparietal sulcus), and to adjacent area 9, which projects in turn to area 6 both dorsally and laterally, and to the cingulate gyrus (119,120,152). Subcortically, area 8 projects to the striatum (caudate and putamen), the thalamus, the pretectum, and superior colliculus, and to the pons (11,120). Thus, pathways issue from area 8 that appear to be capable of influencing both cortical and subcortical areas that

[4] Case 3C of Pandya and Kuypers (152) suggests that area 7 projects directly to area 8. However, the study by Jones and Powell (110) suggests that area 7 projects principally to posterior area 46 in the terminology of Walker (206), or dorsal posterior 9 in the cytoarchitectonic map of Brodman.

are involved in the control of eye movements, of visual attention processes, and perhaps also of the muscles of the neck and trunk.

Movements Evoked by Electrical Stimulation

It has been known since the pioneering studies of Ferrier (66) that eye movements may be evoked by electrical stimulation of regions within the frontal lobe that are now included within cytoarchitectonic area 8 in the classification of Brodmann. The response observed most often by Ferrier upon stimulation of this area was a conjugate deviation of the eyes toward the contralateral side; he therefore referred to this cortical area as the "frontal adversive (eye) field," and he noted that similar movements were evoked by stimulation of temporal and parietal areas.

These basic observations have since been confirmed and extended in both man and monkey by numerous investigators (cf. reviews 142, 210, and refs. 139,175,204). In the *anesthetized* monkey, the primary response to stimulation within the frontal eye field (FEF) is a conjugate horizontal or oblique deviation of the eyes to the contralateral side (142,175,204). The eye movements are saccadic, and their direction is independent of initial eye position (175). The amplitude of the evoked saccade depends upon the position of the stimulating electrode within area 8, but once threshold is exceeded is remarkably independent of stimulus intensity, a property that is quite unlike the intensity-dependent limb movements that are evoked by electrical stimulation within area 4 (164,175). In the *conscious* monkey, spontaneous ("voluntary") and FEF-evoked saccades in the same direction interact occlusively, and it has been suggested that both processes involve common circuitry at some subcortical site that is central to oculomotor neurons (139). Consistent with this view is the fact that the dynamic properties of spontaneous and FEF-evoked saccades are essentially identical (175).

In the conscious monkey or human, FEF-evoked saccades are often accompanied by head rotation and (less frequently) axial orienting responses in the same direction as the eye movement (159,204). Moreover, the same pattern of response may be evoked at slightly higher stimulus intensities from a wide expanse of frontal tissue adjacent to area 8 (posterior area 9), suggesting that the production of coordinated eye and body orienting responses to the contralateral side may be an important motor consequence of activity within a larger portion of the frontal lobe than is apparent from acute stimulation studies (134,204).

The subcortical routes that are involved in mediating FEF-evoked saccades are not clearly known, but a route through the superior colliculus has been rejected for two reasons. First, FEF-evoked saccades may have a shorter latency (15 msec) than those that are evoked by collicular stimulation (20 to 25 msec); and second, they are not abolished by collicular ablation (174,175,180,181). It is possible that projections from area 8 to oculomotor structures by way of the lateral pons and cerebellum are involved in FEF-evoked eye movements (cf. ref. 10 and the discussion on pp. 156–157 of ref. 120).

Sensorimotor Effects of Ablation of Area 8

There are three transient but consistent changes in behavior that have been observed in the monkey after *unilateral* lesion of area 8: (i) a contralateral visual neglect (hemiamblyopia); (ii) a marked reduction in the use of the contralateral arm and hand; and (iii) an ipsilateral deviation of the head and eyes, accompanied by bouts of continuous orienting or circling movements in the same direction (32,48,115,126,128,209). Perimetry studies have shown that the hemiamblyopia is not complete, but involves instead a decreased probability of responding to visual stimuli within the contralateral field that becomes more pronounced as they are moved toward the periphery (126). The ipsilateral deviation of the eyes is tonic, producing a shift in resting fixation of about 10° (127). Curiously, however, no obvious major defects in oculomotor control have been reported to follow unilateral lesions of area 8. While some observers have reported post-lesion changes in the fast phase of optokinetic nystagmus, others have failed to observe such effects (32,154). Moreover, spontaneous saccades appear to be normal, and a lesioned animal is able to fixate upon and to reach accurately for objects within even the affected visual field if the arm ipsilateral to the unaffected field is used (126). All of these defects are transient, with significant recovery occurring within 6 to 10 weeks postoperatively (126). If the intact FEF is subsequently removed, a similar sequence of changes occurs on the side opposite the (new) lesion.

The behavioral changes that are observed following *simultaneous bilateral ablations* of area 8 are, however, more pronounced. During the first few days postoperatively, lesioned monkeys are markedly inattentive, particularly to objects in the periphery of the field, tending to stare fixedly ahead and to exhibit little spontaneous movement (115,127,128). They may fail to reach for available food with their hands, and may instead attempt to eat in a most uncharacteristic way by bringing the mouth directly to the food. Yet, a few days later they may attempt to grasp and hold onto every surface within reach (54). A reduction in manual dexterity is also evident. For example, the time required to grasp a peanut with thumb and forefinger and lift it from a foodwell under visual guidance may be prolonged significantly (54). Within 8 to 10 weeks these motor deficits also clear, however, and a lesioned animal is able to perform accurate, visually guided movements, as well as manual tasks such as opening a simple (one-stage) latch box in order to obtain food. Curiously, throughout this period eye movements appear upon gross inspection to be normal (54,126).

There are, however, two subtle but more enduring deficits in visuomotor performance that are produced by bilateral ablations of area 8. First, monkeys with bilateral lesions exhibit a lasting deficit in the ability to perform an *ordered sequence of movements* in order to obtain reward; for example, first pressing a button and then pulling a lever in order to open a transparent box which contains food (54). The errors in performing such sequential motor acts are of two types: either (i) a perseverative repetition of one of the movements, usually that most closely associated in the sequence with obtaining the reward, or (ii) a correct performance of all movements, but in the incorrect order. Similarly, a unilateral lesion that is extended into adjacent area 6 renders an animal almost incapable of consistently reaching *around* a transparent barrier with the contralateral arm, in order to obtain a visible food reward (147). Instead, visually guided reaching movements are made accurately but directly at the food, *as if the animal is incapable of suppressing a visually evoked program for reaching, until orienting responses are first made that will allow his arm to be directed along the more indirect but correct path.* As with more restricted (but bilateral) lesions, involving area 8 and posterior 9 only, the inappropriate responses perseverate. A second enduring deficit which follows lesions of area 8 is one in locating and/or reaching toward an object within the visual field that is surrounded by similar but extraneous objects (125). A similar impairment is seen following lesions of area 7, although perhaps for different reasons.

These lesion-induced deficits in behavior in subhuman primates resemble in certain respects the various frontal lobe syndromes that have been observed in man following large frontal lesions which include area 8. For example, the inability to perform a correct sequence of movements, though each may be performed flawlessly in isolation, is a defining property of one version of apraxia, a class of disorders of voluntary movement that has been associated with both parietal and premotor lesions in man (31,46,91,163). An inability to suppress very well learned or more reflexly evoked movement patterns, so that they occur *at the appropriate time* within a movement sequence, has also been observed following frontal lesions in man and may in fact be the fundamental problem in certain forms of apraxia (46,135,143). It is of interest to note that such lack of suppression of highly learned or innate motor patterns—as required by the context—is also evident in the eye movements that are made by patients with extensive frontal lobe damage. For example, if a normal subject is asked to answer questions about a picture while he examines it visually, his eyes will follow different scan paths, dwelling longer on some aspects of the picture than others, in a manner appropriate to the question. In contrast, the scan paths followed by the eyes of a frontal lobe patient may be markedly stereotyped, regardless of the question that he is asked (and apparently understands) about the visual scene (135,136). Such perseverative responses suggest a fundamental inability to suppress certain preeminent motor patterns, so that they will not interfere with, or so that they may be called forth at more appropriate times within, a goal-directed movement sequence.

Effects of Interrupting Occipital Projections to Areas 8 and 9 on Visually Guided Hand Movements

The peristriate cortex and the inferior temporal lobe to which it projects, both project substantially to areas 8 and 9 (110,152). To test the hypothesis that these projections are important for visual guidance of fine movements of the contralateral hand and fingers, Haaxma and Kuypers (88) transected both projections in monkeys who had been trained to reach into a small foodwell to obtain a raisin, using a particular orientation of the hand and fingers that was indicated by visual stimuli placed at the periphery of the well. A number of the commissural connections were also severed. After the occipito- and temporofrontal leucotomies, the animals were still capable of projecting the arm accurately to the region of the foodwell, but were incapable of properly orienting the thumb and forefinger in relation to the visual cues so that the food reward could be extracted from the well. The residual ability for accurate limb projection was apparently due to the intact ipsilateral hemisphere, which has been shown previously to be capable of mediating visually guided movements of the trunk and proximal limb (25). The loss of fine, visually guided hand and finger movements suggests, however, that areas 8 and/or 9 are also involved in transmitting movement-guidance information from visual association areas to those regions of the precentral gyrus that mediate fine, distal control of the contralateral limb. At the moment it is not clear whether such guidance information is purely "visual" in nature, or if it is in some way relayed from area 8 in the form of signals associated with eye position (see below).

Discharge Properties of Area 8 Neurons

Table 2 summarizes much of the available data concerning the discharge properties of neurons within the frontal eye fields of the conscious monkey. These data were derived from the studies by Bizzi (19), Bizzi and Schiller (22) and Mohler et al. (146), as was much of the following summary.

TABLE 2. *Summary of discharge properties of frontal eye field neurons in the conscious monkey*

Cell type	% of observed sample		Cortical layer[d]	Timing of discharge with respect to movement onset
	[a,b]	[c]		
Saccade (Type I cells)	5	22	III–V	(−)
Fixation and slow pursuit (Type II cells)	5–7	—	III–V	(0 or +)
Head rotation	1	—	III–V	(+)
Visual cells[c,e]	—	47	—	
Other cells[f]	—	9	—	?

[a,b]Data taken from the studies by Bizzi (19) and by Bizzi and Schiller (22).
[c]Estimated from the study by Mohler et al. (146).
[d]Estimated from the depth values provided by Bizzi (19) and by Bizzi and Schiller (22).
[e]It is not known whether some of these cells might have behaved similarly to the parietal visual fixation cells described by Lynch et al. (137), for the experimental conditions in the study by Mohler et al. were not equivalent.
[f]This category includes cells which fired in relation to limb movement, in response to auditory stimuli, or during other forms of nonvisual stimulation and movement.

Three classes of neurons were identified by Bizzi and Schiller. Type I or *saccade cells* discharge in relation to *spontaneous* or *voluntary saccades.* They are thus *un*like the parietal lobe saccade cells that were discovered subsequently, which appear to fire only in relation to visually evoked saccades (137,149). Moreover, unlike these latter cells, whose discharge leads eye movement, the discharge of FEF saccade neurons was found to lag eye movement in time. Thus, these cells do not appear to be involved in the initiation of voluntary saccades, but instead appear to fire in response either to some centrally originating signal—for example, as if the cells were part of a corollary discharge pathway (cf. refs. 193,194)—or to feedback from sources at or near the oculomotor periphery. In favor of the latter possibility is the fact that the cells also fire in relation to the fast phase of optokinetic or vestibular nystagmus. The input source is not visual, however, for they continue to fire in relation to spontaneous saccades and caloric nystagmus in the dark; nor does it appear to be derived from ocular muscle stretch receptors (19). Fewer than 5% of an observed sample of FEF neurons was found by Bizzi to fall into the saccade category, and approximately one-half of these discharged in relation to saccades in the *ipsilateral* direction. Very few saccade cells responded antidromically to stimulation of the cerebral peduncle; thus, the destinations of their output signals are essentially unknown.

A second category of cells was also observed, which consisted of *fixation* and *tracking neurons* (Bizzi's Type II cells). Like the visual fixation cells of the parietal lobe, these cells fired in relation to steady positions of the eye within a certain range of movement, or during slow pursuit movements across these same position sectors; they ceased to fire during saccades. However, unlike parietal fixation cells, whose discharge appears to be in part conditional, i.e., most brisk when a conjunction of appropriate visual input and central (motivational) drive exists, FEF fixation cells fired as strongly in relation to steady eye positions and slow movements in the dark as in the light. They also fired in relation to the slow phase of nystagmus, both in the light and in the dark, and ceased to fire during either the fast phase or during spontaneous saccades. It is not clear from the available data whether the discharge of these cells most often led or lagged eye movement, for they fired only when the eye moved to within or slowly across a limited position band, in which case an ongoing eye movement was already in progress. Over its particular position band, however, the discharge of each cell appeared to be roughly proportional to eye position. The majority of the cells responded antidromically to stimulation of the cerebral peduncle, and it is possible that they are involved in the cortical production of eye movements. It is also possible, however, that they may simply exert some background modulating action on subcortical oculomotor control circuits, when driven reflexly by eye movements initiated by way of another system. In either event, the tonic deviation in eye position which occurs following unilateral FEF lesions could be due to their destruction. The possibility might also be considered that it is the activation of these cells by abruptly applied pulse trains of high frequency, rather than that of neighboring saccade (Type

I) neurons, that is in fact responsible for the saccadic eye movements that are evoked by FEF stimulation. As indicated previously, however, the pathways which mediate this effect are largely unknown, although possible routes by way of pons and cerebellum have been proposed (10,11,120).

A third class of cells was also observed by Bizzi and Schiller (22) which discharged in relation to *head turning* but not to eye movements. All of the observed cells of this type responded antidromically to stimulation of the cerebral peduncle, and their discharge clearly led in time the related head movements. Thus, they appeared to be cortical output neurons with properties similar to the pyramidal tract cells of area 4 (59,100). It is possible that this small subset of cells, which accounted for less than 1% of the sample observed by Bizzi and Schiller, is a rostral extension of a much larger subset of similar cells within adjacent areas 6 and anterior 4, which are known to produce activity in neck and other axial muscles when stimulated electrically (134,142,212).

In the study of FEF neurons by Mohler et al. (146) two additional observations of significance were made. First, visually evoked rather than spontaneous saccades were studied, and a much larger proportion of saccade neurons was observed (22% of 137 cells). Of those tested for activity in the dark, only one-half fired in relation to spontaneous saccades, indicating that the FEF as well as the parietal lobe contains neurons that fire only in relation to visually evoked saccades. It is unlikely that the FEF subset is driven directly by parietal saccade cells, however, for the latter fire well before the onset of a saccade (70 to 150 msec) and the former substantially (30 msec) after. Secondly, Mohler et al. discovered a large subset of FEF cells (47% of their sample) which responded to visual stimuli, but did not fire in relation to *saccades*. The cells had large, contralateral receptive fields, but also a discharge frequency that was graded in relation to the proximity of an effective stimulus to the center of the field. Since Mohler et al. did not study changes in fixation or slow eye movements, it is not clear to what extent their category of visual cells might have overlapped with the fixation and tracking neurons identified by Bizzi (19). However, they did note that the receptive field properties of the cells were quite similar to those of collicular neurons, and suggested that they too might be involved in the control of directed attention processes. Subsequent studies have provided support for this interpretation (214–216). It is of interest to note that a small subset of cells was observed also which fired in relation to arm movements, or to orienting responses made to auditory stimuli (9% of the sample).

Although the cells observed by Fuster (76) were in general located anterior to the FEF or along its rostral margin, his observations are nevertheless also of interest here. This investigator studied the behavior of frontal lobe cells during the performance of a delayed-response task in the monkey. In this task, the animal was required first to focus his attention upon two small foodwells while a food reward was placed in one of them. A screen was then placed in front of the foodwells for a delay period of up to 18 sec. The screen was then elevated, and the animal was allowed to reach toward the foodwells and—if the appropriate choice was made—obtain a reward. Thus, correct performance of the task required an appropriate blend of focused attention, and a memory of the foodwell toward which the reaching response should be directed. The task-related units observed most frequently by Fuster were cells that began to fire during the initial stimulus presentation and fired continuously throughout the delay period. Distracting stimuli presented during the delay period attenuated the firing of the cells and also interfered with correct task performance; either no response was made, or it was directed toward the incorrect foodwell. It was suggested by Fuster that these cells may have been part of a neuronal circuit that was responsible for *sustained attention processes,* which he viewed as being a major component of short-term memory. But is it not also possible that these or similar cells might have functioned *to hold in readiness the motor program for reaching toward a particular point in space,* until the appropriate time in the task sequence? In this connection it should be noted that pyramidal tract cells may change their discharge rates significantly during a period when an animal *waits* for a signal to perform a particular (learned) movement (189). The direction of the change in discharge rate is determined by the task instructions given to the animal and is thus related to a specific motor program. Yet, even if the premovement change is a significant increase in firing rate, there may be no detectable correlated activity in task-related muscles during the waiting period. Thus, it is clear that there are systems that are capable of sustaining a particular pattern of activity across a population of motor-cortex output cells, while at the same time suppressing the translation of that activity into movement until the appropriate stimulus occurs.

Comment and Summary

Despite the long-term designation of area 8 as the frontal eye field, available evidence suggests that it may not be a primary cortical motor area for the control of eye movements. The basic properties of spontaneous and evoked saccades, as well as slow pursuit eye movements, appear to be largely unaffected by lesions of area 8, except in the immediate postoperative period. Moreover, the eye-movement related discharge of even those area 8 cells that project subcortically appears to follow rather than to lead the movement. Although electrical stimulation of area 8 may evoke eye movements, it must be noted that such effects do not necessarily imply a *primary* motor function. Consider, for example, the postcentral gyrus.

Stimulation of the postcentral gyrus may evoke movements even in the absence of the precentral motor areas (68,212), apparently by way of direct corticobulbar and corticospinal projections (40,111,122). Corticospinal projections from this area are known to terminate, for example, within the dorsal laminae of the cord, where they may act to bias the excitability of segmental reflex arcs and ascending sensory pathways (40,41). Yet, studies in the behaving animal have shown that postcentral cells discharge primarily *after* the initiation of voluntary movement (61) suggesting that they are part of a loop which monitors and modulates a voluntary movement once it is in progress, rather than being components of the movement-initiation pathway. An analogous argument might be applied to area 8 in relation to the control of eye movements, for its neurons appear to *monitor* activity, rather than to evoke it, within more fundamental parts of the oculomotor control system.

But what specifically, then, might be the contributions of this cortical zone to the control of directed attention processes and to visually guided limb movements? Although the data are as yet too few to allow definitive answers to this question, we may nonetheless consider four possible contributions on the basis of those that are currently available.

Does area 8 participate with area 7 in the control of directed visual attention? Consistent with this view are the anatomical connections between areas 7 and 8 (or adjacent area 9; cf. ref. 152), the convergence onto area 8 of inputs from auditory, somatic, and visual association areas, the efferent projections to the superior colliculus and pontine tegmentum, and the contralateral neglect which follows unilateral lesions of area 8. The available single-unit recording data suggest, however, that area 8 and adjacent regions may operate in parallel with area 7 rather than in series with it, to control somewhat different aspects of directed attention processes. For example, the conditions for discharge of the saccade and fixation cells in the two cortical areas, as well as their temporal relations to eye movement, are fundamentally different. Area 7 cells fire only in relation to visually evoked saccades (saccade cells), or when the gaze is directed toward an object of motivational interest in a particular subregion of the visual field (fixation cells); in each case, the discharge leads the eye movement or eye fixation. In contrast, FEF cells may discharge in relation either to evoked or spontaneous saccades (Type I cells), or to changes in eye position and slow tracking movements (Type II cells), whether these movements occur in the light or in the dark. Moreover, their discharge lags the movement. FEF neurons appear, therefore, to *monitor* (and perhaps modulate) eye position and movement rather than to direct it, and it seems unlikely that they participate directly with area 7 in the *initiation* of eye movements and fixation during shifts in visual attention. Indeed, the only cell populations within areas 7 and 8 that appear to exhibit basically similar discharge properties are the visual cells described for area 8 by Mohler et al. (146) and recently for area 7 by Yin and Mountcastle (217,218). The latter authors suggest that these cells provide the afferent visual input to the visuomotor cells of area 7 which control shifts in directed attention. It is possible that a similar visual input is used in area 8, in conjunction with inputs which signal eye position and movement, to perform the complementary functions suggested next.

Does area 8 "sense" eye movements or commands for eye movement and, by way of its projections, aid in coordinating these with orienting movements of the head, body, and arm? Much of the data reviewed above would appear to be basically consistent with this view, as well as the view that area 8 may also participate in orienting responses to auditory stimuli. As indicated previously, inputs converge upon areas 8 and posterior 9 from association areas linked with visual and auditory systems, and also from frontal regions whose activities are associated with control of axial and proximal limb muscles (the SMA). The outputs from area 8 are directed not only back to area 9, which projects in turn to these frontal motor areas (119,120), but also to regions of the striatum, midbrain (pretectum and superior colliculus), and pontine tegmentum that have been associated with eye-head movements. Stimulation of area 8 (and posterior 9) evokes orienting responses in the lightly anesthetized animal, and its destruction leads to a failure not only to orient toward objects in the periphery of the contralateral visual field—though they apparently *are* seen and eye movements *are* directed toward them—but also to a disruption in orienting responses to somatic and auditory stimuli (209). Finally, the timing and properties of cell discharge in area 8 in relation to eye-head movements are consistent with this notion. Such a general function was also envisioned by Bizzi (19), who noted that: "It is possible that the discharges of type II FEF cells, which represent a specific record of eye position, are involved in those complex coordinations whereby orienting and reaching movements are adjusted to each other in such a way that shifts of gaze are followed by appropriate head and limb movements" (p. 79).

It is clear, however, that the potential for such coordination is also possessed by other structures, for visually guided orienting and reaching movements are recovered 6 to 8 weeks after bilateral ablation of area 8. In addition, as noted by Bizzi and Schiller (22), the existence of separate populations of cells which discharge in relation to eye and to head movements indicates that the "program" which coordinates these functions must be elaborated elsewhere. It is possible that such elaboration involves so-called "premotor" and/or supplementary motor areas, to which area 8 may project by way of adjacent area 9 (119,120,153).

Does area 8 participate in the visual guidance of fine contralateral hand and finger movements? The pathway transection experiments of Haaxma and Kuypers (88), coupled with the reduction in eye-hand coordination seen following area 8 ablation (54) would suggest an affirmative answer to this question. It is possible that such a steering function might be performed in part by area 8 visual cells of the type described by Mohler et al. (146). Although each of these cells has a large (contralateral) receptive field, its discharge is smoothly graded in relation to the distance of the stimulus from the center of that field. Thus, a *population* of such cells might provide rather precise information about stimulus location. It is not clear, however, how such location-specific information might reach the precentral gyrus for the guidance of hand and finger movements, for there are apparently no direct connections from area 8 to this cortical zone (120). Transmission to the precentral gyrus by way of subcortical routes, or by way of transcortical pathways from adjacent area 9 to periarcuate, premotor (area 6) and supplementary motor areas are remaining possibilities.

Is area 8 part of a system that suppresses visually evoked eye- and arm-movement patterns until the appropriate orienting responses that are required by a given context are first made? The perseverative movement patterns (143) and the inability to suppress visually evoked reaching movements which follow bilateral ablations of area 8, or of areas 6, 8, and 9, suggest such a function. The hyperactivity which follows prefrontal ablation is a well-known phenomenon, and it is of significance that it appears to occur only in the light or in the presence of visual stimuli (87). Denny-Brown observed also perseverative reaching and manual exploratory movements following ablation of the temporal areas which project heavily to areas 8 and 9, an effect which he attributed to a "release" of positive parietal lobe influences on the basic circuitries which produce such movements (51). It is likely, therefore, that areas 8 and posterior 9 are only part of a much more extensive network, involving basal ganglia and a wider expanse of temporofrontal cortex, which performs such a basic response-suppression or movement-sequencing function.[5]

Area 6aβ

Prior to the 1950s, both anatomical and cortical stimulation studies had given rise to the concept of a "premotor" zone, which occupied the anterior portions of Brodmann's area 4 and all of area 6, including the medial region now referred to as the "supplementary motor area" (73–75,105,213). Electrical stimulation within this zone often evoked complex movement synergies, involving the eye and face when delivered to most lateral portions of area 6, and the eyes, head, trunk, and limbs when delivered medially; in the latter case, the synergies often resembled complex orienting and reaching motions (97–99,134,141). Also observed was an inhibition of the grasp reflexes of the hand, and of the tone in the leg flexors (98). These effects were thought to be mediated by way of outputs to extrapyramidal nuclei, either directly or, as appeared more likely at the time, by way of synaptic relay through the primary motor cortex (area 4). The zone was also known to receive inputs from prefrontal and posterior association areas and to project to more posterior portions of the precentral gyrus. Thus, it was viewed as part of a transcortical pathway which linked association with primary motor areas, and was referred to, following the terminology of Fulton (73), as the "premotor zone."

As the result of careful mapping studies in the anesthetized monkey, however, Woolsey et al. (213) clarified the somatotopic organization of the precentral and supplementary motor areas, and redefined their boundaries to include much

[5] Although our interest here is principally in the contribution of areas 8 and posterior 9 to the control of visually directed arm movements, the "corollary discharge" hypothesis of Teuber (193) should not be overlooked. According to this hypothesis, the descending output from area 8 would act upon subcortical structures (pretectum, superior colliculus, pons) to "preset" their responsiveness to the visual inputs that are generated by centrally induced eye movements, so that—in spite of a changing retinal image—the visual world would be seen as a stationary rather than as a moving one.

of the tissue that had previously been viewed as part of the premotor zone. At present, the only cortical tissue *of interest here* that remains part of the *redefined* premotor area is that portion of Brodmann's area 6 (27) that lies on the dorsal surface of the hemisphere, medial to the upper limb of the arcuate (precentral) sulcus and lateral to the SMA (cf. Fig. 2). In the classification system of the Vogts (201) this area corresponds to the dorsolateral portion of area 6aβ; in that of Von Bonin and Bailey (203) to areas FB and FC. At present, the precise motor functions of this region are uncertain. However, much of the earlier evidence which suggests that it is in some way involved in linking eye, head, and arm movements cannot be dismissed. For this reason, and for completeness, a brief review of the properties of area 6aβ is included here.

Anatomical Connections

The major anatomical connections of area 6aβ are summarized in Figs. 6A and 6B. Major transcortical *afferent* pathways (Fig. 6A) are derived from the superior temporal gyrus, the medial portions of area 5, posterior area 9, the SMA, and the anterior and medial portions of area 4 (110,119,152). The major thalamic input is derived from the medial zone of the VL nucleus (2,113,201).

The *efferent* projections from area 6aβ are summarized in Fig. 6B. In constructing this figure, considerable weight has been placed on recent autoradiographic studies of area 6 projections (119), with somewhat less emphasis upon earlier degeneration studies (110,152,153). *Transcortical* efferents from area 6aβ are directed posteriorly to area 7, both medially (in the region of the cingulate sulcus), within the depths of the intraparietal sulcus, and on the lateral convexity of the hemisphere. Projections are also sent to posterior area 9, to more lateral portions of area 6 within the arcuate sulcus, and to the

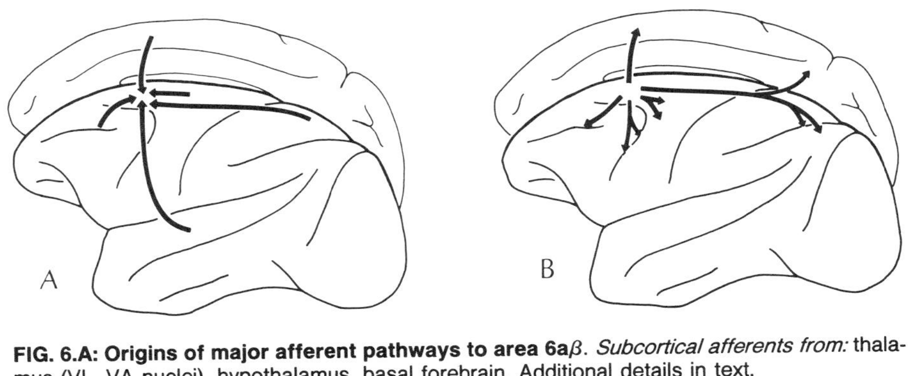

FIG. 6.A: Origins of major afferent pathways to area 6aβ. *Subcortical afferents from:* thalamus (VL, VA nuclei), hypothalamus, basal forebrain. Additional details in text.

B: Major efferent projections from area 6aβ. *Subcortical efferents to:* putamen, caudate, thalamus, red nucleus and adjacent reticular formation, pons, medial medullary reticular formation. Additional details in text.

anterior regions of area 4.[6] A substantial projection is also directed to the SMA. *Subcortically,* projections are sent to the putamen and caudate nucleus, to the pons, and to those medial brainstem and medullary cell groups that have been associated with the control of axial and proximal limb muscles (the red nucleus, and the midbrain, pontine, and medullary reticular zones; cf. refs. 101,119,122,131). Thus, on anatomical grounds one might expect this area to exert effects on the muscles which orient trunk and limbs and to perhaps coordinate these with inputs from areas 5 and 8–9_p (posterior 9) that are related, respectively, to joint and to eye position. Note that area 6aβ is the first of the cortical zones that we have considered which appears to receive substantial projections from areas that are related separately to these two position variables.[7]

Movements Evoked by Electrical Stimulation[8]

In the monkey under moderate anesthesia, stimulation of the surface of the middle region of area 6aβ with sinusoidally varying current of moderate intensity may evoke only pupillary dilatation (213). With current pulses of greater strength or with electrode positions that are nearer the periphery of this zone, however, movements may be evoked that resemble those produced by stimulation of adjacent areas: from regions near area 8–9_p (see Fig. 2) eye-head movements may be evoked; from regions nearer the border of area 4, or near the SMA, activity may be evoked in proximal leg, trunk, shoulder, or arm areas, depending upon electrode position (50,97–99,141). Although such findings suggest the possibility of a spread of stimulating current to surrounding tissue, similar movements may be evoked with considerably lower current strengths in the very lightly anesthetized or unanesthetized animal (134) or following ablation of area 4 (50). These facts, coupled with the known efferent projections from area 6aβ, suggest that movements evoked by stimulation within this zone are in fact mediated by its output pathways.

The movements that are evoked by stimulation of area 6aβ in the lightly anesthetized or unanesthetized monkey, or of a homologous region in man, exhibit a similar topography but are considerably more complex. Stimulation along the anterior to mid-region of the arcuate sulcus (adjacent to area 8–9_p) in the monkey may evoke coordinated turning movements of the eyes and head

---

[6] It is important to note that the projection to area 4 is apparently to those anterior regions which affect axial and proximal limb muscles (cf. refs. 212,213). Recent autoradiographic (119) and retrograde cell-labeling studies (P. L. Strick, *personal communication*) indicate that area 6aβ does not project directly to the precentral hand area.

[7] Area 7 may be an exception to this statement. Robinson et al. (177) have recently described a subset of hand-projection neurons that are responsive to both passive visual and somatosensory stimuli.

[8] The summary given here is based upon stimulation studies carried out prior to the development of more refined intracortical microstimulation procedures (cf. ref. 9). Additional studies of the motor responses that are evoked by stimulation of this cortical zone are needed, in which more refined stimulation methods are employed following ablation of the adjacent precentral gyrus.

to the contralateral side (98,99,134,204); stimuli delivered either more posteriorly or closer to the boundary of the SMA may evoke orienting and/or reaching movements toward the contralateral side (98,134). In the conscious human, stimuli applied to the medial regions of area 6aβ (near the SMA) may evoke: (a) "involuntary, bilaterally synergic movements, in which the contralateral hand seems to lead and the head turns to watch it" (159, p. 222); or (b) no detectable response at all, unless a voluntary movement is in progress, in which case it ceases (160).[9] Orienting (adversive) movements toward the contralateral side have also been observed upon stimulation of the posterior border region of area 6aβ, and these were not abolished by incisions between this region and adjacent area 4 (160).

In addition to these effects upon observable movement, stimulation within area 6aβ has been found to inhibit postural tone in the leg flexors of the monkey (198). These effects as well as most of those described above in subhuman primates survive pyramidotomy, indicating that they are mediated by non-pyramidal projections descending from area 6aβ and adjacent regions (98, 99,141).

Motor Effects of Ablating Area 6aβ

The results of earlier ablation studies of area 6 must be interpreted in light of the redefinition of precentral and supplementary motor areas advanced by Woolsey et al. (213) and by Penfield and Welch (160). Prior to this redefinition, ablation of all of area 6 (including area 6aα of the Vogt's and the SMA) was found to produce a release of grasp reflexes in the contralateral hand, a "plastic" rigidity of the knee, hip, shoulder, and elbow which diminished over a 6 to 10 day period, and a flexor hypertonus at these same joints (50,116,172). More recently, ablation of the entire prearcuate-premotor zone unilaterally (area 6, including the SMA, and areas 8 and 9_p as shown in Fig. 2) was found to impair a monkey's ability to first orient his body properly, and then to reach around a transparent barrier in order to obtain a visible food reward; instead, reaching movements were made directly (and accurately) toward the food, but into the barrier (147). It is not clear from these results, however, whether the primary defect was an inability to suppress visually evoked reaching movements until appropriate orienting responses of the body were first made, or an impairment in making the latter responses per se.

The specific contribution of the destruction of area 6aβ to this latter syndrome is, however, unknown. Moreover, the release of grasping reflexes and the effects on muscle tone described above are now viewed by some investigators as being due principally to destruction of the SMA (160,199: but see ref. 42). Thus,

---

[9] As noted by Penfield and Rasmussen (159), it is not clear whether such cessation of movement is due to a true inhibition of motor pathways, or is simply the subject's response to disruption of a motor program in progress. They noted that an arrest of ongoing, voluntary movement could be obtained from stimulation of many areas of the brain, including prefrontal, precentral, and supplementary motor areas.

there is at present no clear evidence concerning the effects of ablation of area 6aβ *alone* on motor performance (although the ablation experiments of Jacobsen come close to this goal; see ref. 105). Additional experiments are therefore needed in which localized cooling or destruction of this zone is combined with careful measurements of eye-head movement, proximal and distal EMG activities, and limb-movement trajectories, in order to detect the effects of its destruction on visually directed reaching. Such experiments should include bilateral lesions, for the pattern of efferent projections from area 6aβ suggests that it will exert motor effects bilaterally.

Movement-Related Behavior of "Premotor" Neurons

To this author's knowledge, there has been only one study of the behavior of neurons within the general "premotor" area of the primate cortex during the performance of conditioned arm and hand movements. The studied cells lay not within area 6aβ, however, but within the caudal banks of the arcuate fissure near its flexure (Von Bonin's and Bailey's area FB; cf. ref. 203), a region to which both area 6aβ and the SMA project (119). Kubota and Hamada (118) studied the behavior of cells within this zone and compared it with that of PT cells in caudal area 4, in monkeys that were trained in a visuomotor tracking task. The animals were trained to observe two horizontal rows of light-emitting diodes (LEDs). The position of the LED in the upper (target) row was controlled by the experimenter, whereas that of the energized LED in the lower row was determined by the position of the animal's hand as he made a wrist flexion-extension movement. The monkey was required to make stepwise changes in wrist position, in order to track similar changes in the position of the target LED. The observed unit activity was examined with respect to temporal coupling to stimulus onset, to response (movement) onset, and to direction of wrist movement.

The task-related neurons that were observed within the periarcuate zone fell into two major categories. The first and smaller group consisted of neurons that fired in close temporal relation to stimulus-light onset, but which displayed minimal coupling to movement onset, or to the parameters of wrist motion (angular velocity or acceleration). The second, larger group fired in closer temporal correlation with the movement, with changes in firing rate that led the movement by an average of 150 msec. About one-half of these cells fired reciprocally in relation to wrist flexion and extension, as if their discharge might have contributed to control of the wrist flexor and extensor muscles. The remainder fired nonreciprocally in relation to wrist movements, as if their discharges were related more to control of proximal muscles whose contraction was necessary for stabilization of the limb during wrist movement in either direction. Very few (10 to 15%) of the premotor area cells displayed a discharge pattern that correlated significantly with the parameters of wrist movement (velocity and acceleration).

The discharge of precentral PT neurons also led the movement by an amount approximately equal to that of premotor neurons. A greater percentage also fired reciprocally in relation to the direction of wrist movement and, unlike premotor neurons, tended to fire with a pattern that correlated with movement velocity or acceleration. It was concluded that the results agree with the general view that premotor cortex cells are activated just prior to or perhaps in parallel with precentral PTNs during voluntary limb movements, and it was at least implied that they may contribute in particular to control of trunk and proximal limb muscles during postural adjustments that stabilize the body and upper limb during movements of the hand. One might also infer from these results that the encoding of the parameters of motion into the signals which descend from the precentral motor area is not accomplished by way of impulses reaching it through the premotor area.

Comment and Summary

Although the data are yet unclear concerning the *specific* motor functions of dorsolateral area 6aβ, they are compatible with the view that it is part of a larger frontal complex, which includes area 8, posterior area 9, and the SMA, which functions: (i) to exert tonic position or postural effects on eye, trunk and proximal limb muscles; and (ii) to produce coordinated orienting movements of the eye, head, and trunk in response to central or peripheral commands for movement which reach it by way of cortical association areas, or via thalamic relay from cerebellar or basal ganglia structures. This area does not appear, however, to contain the basic circuitry for the generation of direct, visually evoked reaching movements, for such movements are not abolished by its bilateral ablation.

SUPPLEMENTARY MOTOR AREA

The location of the SMA in the brain of *Macaca mulatta* is shown in Fig. 2. As is well known from the mapping studies by Woolsey et al. (213), the motor outflow from the SMA appears to be organized somatotopically, provided that electrical stimuli are delivered when the precentral gyrus is intact (211). The organization is such that head and arm lie anteriorly, and the leg posteriorly. The anterior portions of the body and the distal muscles of the arm are represented dorsally, with the hand area lying near the free surface of the hemisphere, whereas the representation of the back lies partially buried within the cingulate sulcus (cf. ref. 213, Fig. 132).

Anatomical Connections

The major anatomical connections of the SMA are summarized in Fig. 7A and 7B. The major transcortical *afferents* are derived from somatosensory areas I and II (SI and SII; cf. ref. 109) and area 5 ipsilaterally, and from precentral (area 4) and premotor (area 6aβ) regions bilaterally (110,153,154). The major

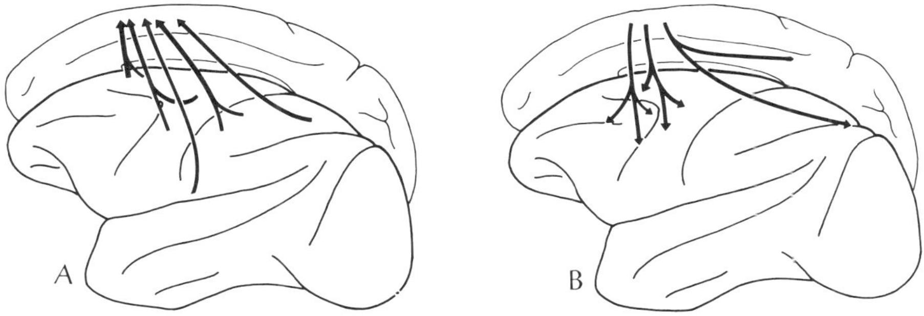

FIG. 7.A: Major pathways to the supplementary motor area. *Subcortical afferents from:* thalamus, (VL, VA nuclei), hypothalamus, basal forebrain. A summary description is given in the text.

B: Efferent projections of the supplementary motor area. Only ipsilateral projections are shown. *Subcortical efferents to:* putamen, caudate nucleus, thalamus, red nucleus and adjacent reticular formation, pons, spinal cord.

thalamic input is poorly defined but appears to come chiefly from the medial regions of the VL nucleus (2,207), and from the VA nucleus (113).

The major transcortical *efferent* pathways from the SMA are to lateral area 6 (6aβ and also regions within the depths of the ventral arcuate sulcus) and area 4 bilaterally. Ipsilateral projections are also sent to posterior area 9 (principal sulcus region), to area 8, to the cingulate gyrus, to area 5 on the medial surface of the hemisphere, to area 7 within the intraparietal sulcus (Fig. 8B; cf. refs. 119,152,211), and to areas 3a, 1, and 2 (108).

The major *subcortical projections* of the SMA are to the striatum, red nucleus, and medullary reticular formation bilaterally, and to the pontine nuclei and VL nucleus of the thalamus ipsilaterally (101,119,201). Recent cell-labeling studies have also established the existence of a light projection directly into the corticospinal tract (111,150), a fiber system not detected in earlier degeneration studies (55).

Thus, the SMA receives converging inputs from somatosensory receiving and association areas, as well as from pre- and primary motor areas. Its major outputs are directed bilaterally back to these cortical motor areas, to the opposite SMA, and to subcortical nuclei which affect primarily the motorneuron pools that control muscles of the trunk and proximal limbs. This pattern of connections suggests that the SMA may use somatosensory input data in conjunction with feedback from the precentral motor area to coordinate orienting movements of the head and trunk, which require bilateral patterns of muscle activation, with precentrally controlled movements of the opposite arm and hand.

Motor Responses to Electrical Stimulation

The nature of the responses that are evoked by stimulation within the SMA depend upon whether or not its connections with the precentral gyrus are intact. These two cases will therefore be summarized separately.

Stimulation with Intact Precentral Gyrus

When the ipsilateral area 4 is intact, stimulation within the SMA may evoke movements of the opposite arm and hand as well as activity in more proximal limb and axial muscles (160,211,213). A definite somatotopic organization is evident in the SMA of the monkey, as is shown most clearly by the maps of Woolsey et al. (213). The evoked responses differ in three ways, however, from those that are evoked by stimulation within area 4. *First,* while a somatotopic organization is clearly evident, the activity evoked by SMA stimulation is much less muscle-specific than that seen following stimulation of a discrete zone in area 4 (see, for example, ref. 211). The evoked movements may involve activity about several joints, or consist even of complex synergies. Stimulation of the arm-hand region of the SMA in the moderately anesthetized monkey, for example, may produce synergistic movements of the fingers, wrist, and elbow. Stimulation of the same zone in the very lightly anesthetized or conscious animal, however, may evoke a complex reaching-orienting movement, where the contralateral arm is projected into space, and the hand opens as if to release its grasp or to prepare for grasping (98,160). In all cases the current thresholds for evoking movement are significantly higher than those required when stimulating the precentral area. *Second,* unlike the quick, phasic movements that are typical of precentral stimulation, the movements that are evoked by stimulation of the SMA may develop more slowly—particularly those involving the proximal muscles—and may outlast for several seconds the period of stimulation. The evoked movements have been likened by some investigators to those that occur during the assumption of a new limb or body posture. *Finally,* whereas evoked movements of the arm and hand are almost exclusively on the contralateral side, those seen in the axial and occasionally the leg muscles following SMA stimulation may occur bilaterally (160,213).

Stimulation within an area in the brain of the conscious human that is presumed to be homologous to the SMA of the monkey was found by Penfield and his co-workers to produce two general classes of responses (159,160). *First,* the lowest threshold response often consisted in an inhibition or cessation of ongoing motor activity, particularly of speech or of spontaneous grasping movements of the contralateral hand. It is of interest that the ongoing movement or speech program was often simply delayed by SMA stimulation, being resumed at the correct point in the sequence upon termination of the stimulus train. Such effects suggest that the SMA may be capable of gating on and off movement sequences that are organized elsewhere, without disrupting their temporal order (a hypothesis that is consistent with recent single-unit recording experiments; see below and ref. 190). The *second* class of responses was evoked at slightly greater stimulus intensities, and consisted of observable movement patterns of two types. The first and most frequently seen movement is a raising of the arm. In the words of Penfield and Welch (160): "One gains the impression in some cases that the contralateral hand is raised in preparation for its use in a complicated act, and that meanwhile the gaze is directed by movement of head and eyes toward the hand" (p. 310); in brief, a reaching-orienting movement of the type described previously. The second set of movement patterns consisted of rhythmic, synergistic movements, such as stepping, waving of the hand, or alternate flexion-extension movements about the wrist or the elbow.

Stimulation Following Ablation of the Precentral Gyrus

After ablation of the precentral gyrus in the monkey, stimulation within the ipsilateral SMA no longer evokes movements of the opposite hand and forearm (160,211). Only activation of axial and proximal limb muscles remains. In addition, longer periods of stimulation are required for evocation of responses, and mapping of the SMA reveals less somatotopic organization than is evident in the intact (anesthetized) animal.

For these reasons, it has been suggested that the arm and hand movements that are evoked by SMA stimulation in the intact animal are due to current spread to the precentral arm area (211). This appears unlikely, however, since the hindlimb area is as close electrically to the SMA as is the precentral arm zone; yet, stimulation within the arm area of the SMA consistently evokes movements of the arm and hand rather than of the leg. It is more likely, therefore, that the hand movements that are evoked by stimulating the SMA in the intact animal are produced transynaptically, by way of projections from the SMA to the precentral gyrus.

These considerations and the known projection systems of the SMA suggest, therefore, the following hypotheses: (i) that the SMA exerts effects on the motor activity of the opposite hand only by way of its projections to the precentral gyrus; and (ii) that its bilateral effects on the proximal limb and axial muscles are exerted both by way of projections to the precentral gyri and to subcortical, motor relay nuclei.

The results of stimulating the SMA after ablation of the ipsilateral precentral gyrus in man are basically similar to those observed in the monkey: movements of the distal extremities are no longer evoked, and the degree of somatotopy is apparently less than that observed when stimulating the SMA with an intact precentral gyrus. Basic orienting responses toward the contralateral side are still evoked, however, as they are in the monkey; and inhibitory effects on speech and voluntary movement programs may still be observed (160).

Motor Effects of Lesions Within the SMA

At present there is some disagreement concerning the severity of the motor disturbances that follow ablation of the SMA in the monkey. Coxe and Landau (42) failed to observe postoperative changes in muscle tone, posture, and reflex excitability as dramatic as those described earlier by Travis (199). Despite these differences in impressions of severity, however, there does appear to be general qualitative agreement concerning the general effects of SMA lesions and their somatotopic distribution. All investigators agree, for example: (i) that the effects of bilateral lesions are much more pronounced, on both sides of the body, than are those resulting from unilateral lesions; (ii) that with one exception (release of the grasp reflex), postoperative changes are more evident in proximal than in distal muscles; and (iii) that the effects of the lesions lie more in the realm of disturbances in muscle tone and posture than in the sphere of voluntary movement.

After unilateral ablation of the SMA in the monkey, two major effects have been observed with some consistency. First, a transient release or hyperexcitability of the instinctive tactile grasp reflex of the contralateral hand occurs, an effect that lasts for 2 to 5 weeks before diminishing in intensity. The release is apparently due to removal of a tonic, net inhibitory effect by the SMA upon components of the instinctive grasp reflex arc that lie within the ipsilateral precentral gyrus, a circuit that is apparently of a basic regenerative or positive feedback configuration (see, for example, ref. 178). The enhanced reflex does not occur if the precentral gyrus is removed prior to or in conjunction with ablation of the SMA (47,48).[10]

The second major effect of ablating the SMA is a transient increase in the tone of proximal flexor muscles, so that attempts to extend or to abduct the animal's arm passively, for example, are met by a soft, "plastic" resistance that is most marked in the muscles of the shoulder (42,50,199). This effect is more pronounced on the side contralateral to the lesion and is accentuated by bilateral ablations. It appears to be due to removal of a direct inhibitory action by the SMA upon subcortical structures which facilitate tonically the proximal flexor muscles, for it is not abolished by concurrent or subsequent ablation of the precentral gyrus (50,199).

But what about the effects of such lesions on voluntary arm and hand movements, and in particular upon visually guided reaching and grasping movements of the type of general interest here? Unfortunately, a careful and detailed examination of changes in such movements following ablations confined to the SMA has not yet been reported. But there are sufficient descriptions from available lesion studies to allow some tentative conclusions.

First, it would appear by exclusion that direct, visually guided reaching is *not* abolished by lesions of the SMA. No investigator has reported, for example,

[10] It is important to distinguish between the *instinctive tactile grasping reaction* and the more primitive *grasp reflex* per se (cf. ref. 48). The former is a complex orientation of the palm toward and a grasping of any object that stimulates the back, radial, or ulnar surfaces of the hand; particularly effective are stimuli delivered to the tips or to the sides of the fingers. It is this reaction that is released or made hyperexcitable by SMA ablation, *and* that requires the integrity of the pre- and postcentral gyri for its occurrence. The more primitive grasp reflex, on the other hand, is evoked only by tactile stimuli that are delivered directly to the palm; there is no corresponding orientation of the hand or grasping in response to stimuli delivered elsewhere. The grasp reflex is released by precentral ablation, apparently because of damage to its extrapyramidal projections (47). It is often not made clear, particularly in some of the earlier studies, precisely which of these two reactions is referred to when discussing grasp reflexes.

that visually guided reaching movements for food are abolished by ablation of the SMA, although some indication of a neglect of the contralateral arm has been observed (160). Second, it seems clear that if the precentral gyri are intact, the effects of even bilateral ablation of the SMA upon voluntary movement may be difficult to detect. Thus, the SMA does not appear to be simply "in series" with the precentral gyrus during the programming and generation of voluntary arm movements, nor do any of its direct, descending contributions to the control of such movements appear to be simply additive to those of the primary motor cortex.

But is seems clear, nevertheless, that such descending contributions do exist, and that the specific functions of the SMA in the control of voluntary movement may be detectable only in experiments that are designed specifically to detect subtle changes in an animal's ability to coordinate properly basic orienting movements of the body with guided reaching and grasping motions after SMA ablation. Travis (199) noted, for example, that the contralateral paresis that followed combined SMA and precentral lesions in the monkey was invariably greater than that seen following precentral lesions alone. Moreover, following ablation of both precentral gyri and one SMA, the control of the arm opposite the intact SMA was distinctly superior to that opposite the ablated SMA. Both observations suggest that the SMA may exert some control over the contralateral arm—though perhaps not the hand—independently of the precentral gyrus. Finally, it must be recalled that the SMA was included in the so-called "premotor" lesions described by Moll and Kuypers (147), which greatly impaired a monkey's ability properly to coordinate visually evoked reaching movements of the contralateral arm with the basic orienting movements of the body that were required to obtain food reward. It is possible, therefore, that a basic function of the SMA is one of coordinating those bilateral patterns of activity in axial and proximal limb muscles that are necessary for properly orienting the body for reaching movements toward objects that are not directly ahead, or that must be directed around an obstacle. Such coordination would involve a momentary suppression of direct, visually evoked reaching movements, until the appropriate body stance or orientation was first achieved. Additional experiments are required in which tasks of this nature are combined with careful EMG measurements of sequences of proximal and distal muscle activation, in order to detect changes that might occur in these processes following ablations confined to the SMA.

Receptive Fields and Movement-Related Behavior of SMA Neurons

Although there have been few studies of SMA neurons, there are nevertheless enough data now available to allow at least a preliminary comparison of their properties with those of precentral (area 4) neurons. A comparison of this type is shown in Table 3.

As is well known, both PT and unidentified neurons within layer III-V of

TABLE 3. *Qualitative comparison of somatosensory receptive fields and movement-related discharge properties of SMA and precentral neurons*

Property	SMA neurons[a]	Precentral neurons[b]
1. Size of somatosensory receptive fields	Large, encompassing several contralateral joints, both limbs, or even both sides of the body. Approximately 15% respond to passive joint rotation; percentage affected by cutaneous inputs unknown.	Small, usually restricted to one or two contralateral joints, whose muscles are controlled by cells within the same or adjacent cortical zones. Joint movement is most effective input, but some cells also respond to cutaneous stimulation.
2. Size of muscle or movement fields	Large; discharge is usually associated with activity in several distal or proximal muscles, and often with bilateral activity.	Small; discharge related to the activity of a small set of muscles acting about one or two joints on the contralateral side, although some cells have been observed which also fire in relation to ipsilateral movements.
3. Timing of discharge in relation to movement	Most often appears to lag onset of EMG activity in opposite wrist muscles during precision grip; temporal relation to activity of proximal limb muscles unknown; additional studies are needed.	Most often leads the activity of the contralateral arm and hand muscles by 60–100 msec.
4. Relation of discharge to parameters of limb muscle activity and movement	Comparatively loose coupling; although some cells may fire in relation to gripping force (F) or dF/dt, no consistent pattern is observed across neurons.	Close correlation; cell discharge is temporally correlated with a number of parameters, and in particular with isometric muscle force or dF/dt.
5. Relation of discharge to instructional set	Cells have been observed which fire in relation to instructional set only, being uninfluenced by peripheral input.	Cells exhibit discharge patterns which reflect, to one degree or another, both instructional set and peripheral input.

[a] Summary based on refs. 26,184,185,190,211.
[b] Summary based on refs. 9,36,59,63,67,100,103,132,189.

the monkey's precentral gyrus exhibit the following properties. Their receptive fields are generally restricted to a comparatively small body region, with most cells being activated most intensely by inputs from receptors located within one or two closely spaced joints, or their related muscles (132,178). Although a small percentage of PT cells may distribute axonal branches to both cervical and thoracic or lumbar segments (183), the majority also show a restricted muscle field; i.e., their motor influence seems to be restricted to a small set of one to three synergistic muscles, operating about only one or two joints (9,36,67). The discharge of precentral cells also leads the onset of EMG activity in the muscles which they appear to control by an average of 60 to 100 msec (51,59); and their firing frequencies are correlated temporally with various measures of motor activity such as isometric muscle force *(F), dF/dt,* angular joint position, or velocity of movement (36,59,100,103). In addition, the *pre-movement* firing patterns of precentral cells are determined in part by the task instruction or movement requirements that are communicated to the animal (63,189).

The properties of SMA neurons, on the other hand, are in many ways substantially different from these. First, only a small percentage (15 to 20%) of the SMA cells that fire in relation to arm movements are responsive to somatosensory stimuli (26). Like precentral neurons, the most effective input to these cells comes from joint receptors. However, their receptive fields are large, encompassing several joints in one contralateral limb, both limbs, or even both sides of the body (26,211). Second, the muscle fields of SMA neurons are typically large. Although a percentage of the cells have been observed to fire in temporal correlation with precise gripping movements by the contralateral hand (184), a more thorough study of SMA cells using a variety of movements suggests that some 60% may fire in relation to movements at several contralateral joints, or even to movements of *either* arm (26). Third, the discharge of SMA neurons may *lag* the onset of EMG activity in the muscles of the opposite arm by up to 100 msec (185). Fourth, unlike precentral cells, the discharge of SMA neurons may correlate in no consistent manner with any particular parameter of muscle force or movement (185). On the other hand, SMA units have been observed which discharged not in relation to execution of the motor task per se, but instead to the instructional or pre-movement response set of the animal (190). These cells did not respond to peripheral stimuli, and it was suggested that they may have been members of a central programming system, which gates on and off the responsiveness of precentral neurons to peripheral inputs during somesthetically triggered movements (63,189).

Thus, while SMA neurons may fire in relation to contralateral arm movements, or respond in limited numbers to somatosensory inputs generated by such movements, they do so in a somatotopically loose fashion. Moreover, their discharge appears to often lag movement and is related in no precise way to the parameters of *limb* muscle activity or motion. It appears unlikely, therefore, that SMA neurons participate in any significant way in the precise proprioceptive or visual guidance of arm and hand movements. It is possible that their activity may prove to be more closely related to the head and axial movements which accompany oriented reaching, and perhaps also to the *release* of the grasp. Unfortunately, these classes of movement have not been carefully monitored in previous single-neuron recording experiments, and additional studies are therefore needed to evaluate these possibilities.

Summary and Comment

It is clear from the anatomical and electrical stimulation studies that have been cited above that the SMA possesses two major outflows. The first of these is directed subcortically, where it affects motor structures that contribute to maintenance of patterns of tone in axial and proximal limb muscles, and that are capable of producing basic movement patterns or synergies. These include basic orienting responses of the head and body toward the contralateral side, as well as elevation and projection of the arm in what appears to be a basic reaching motion.

The second major outflow is directed to the precentral motor area and to the premotor zone. The former possesses a degree of topographic organization, and it appears to be the route whereby the SMA exerts effects—apparently inhibitory—on somesthetically evoked orienting and grasping movements of the contralateral hand. It is perhaps a function of this projection to modulate the responsiveness of precentral hand-control cells to somesthetic inputs, so that grasping responses or release of the grasp can occur at appropriate times during movement sequences that involve also orienting movements of head and body, and arm projection or withdrawal.

But does the SMA contain any of the basic circuitry that is necessary for direct visually or proprioceptively guided reaching movements? The answer to this question would appear to be "no," for such movements are not abolished, nor does their accuracy appear to be impaired significantly, by even bilateral ablation of the SMA.

Does the SMA then control or produce basic orienting movements of the head and body? While the SMA appears to be capable of mediating such movements in the absence of the premotor-rostral precentral area, the answer to this question would appear also to be "no," or at least "not exclusively." For if the precentral-premotor zone is intact, such movements also survive bilateral ablation of the SMA, despite the production of clear changes in proximal muscle tone.

Might the SMA then serve in some way to coordinate orienting movements of the head and body that are produced by the premotor-rostral precentral zone, with movements of the distal arm and hand that are produced by the precentral gyrus? Although the evidence is far from conclusive in this regard, and the precise nature of the "coordination" process is obscure, the existing data would nonetheless appear to be consistent with a tentative "yes" to this question. The dual projection system from the SMA to proximal (subcortical) and distal (precentral) limb control areas is consistent with this view, as are the results of the "premotor" lesion experiments of Moll and Kuypers described previously (147). It may also be noted that in terms of transcortical connections, the SMA lies between and communicates reciprocally with: (i) the premotor and rostral precentral zones, which project directly to structures that control axial muscles; and (ii) the precentral gyrus, which contains the major circuitry for cortical control of the hand. In this regard the SMA is situated along a proximal-distal control gradient in a manner somewhat similar to the premotor zone, which lies between the frontal eye field and the rostral precentral zone, and which may function to coordinate movements of the eyes with those of the head, trunk, and proximal limb during orienting-reaching movements. The inhibitory effects of the SMA on grasping may also be incorporated into this proximal-distal coordination scheme, if one assumes that a system is needed for suppressing grasping movements until the reaching hand contacts an appropriate object, or for releasing an object when the hand is to be directed elsewhere. This latter function would be especially important in suppressing grasp reflexes that are initiated and maintained by tactile input from objects in the hand, a reflex whose persistence is clearly evident following SMA ablation.

It is also of interest in this regard to recall Denny-Brown's observation that the subtle avoiding movements that are released by posterior parietal ablation are themselves abolished transiently by bilateral SMA ablation (51). For this and other reasons, Denny-Brown has suggested that parietal and frontal areas may act, respectively, upon subcortical networks that produce basic approach and avoidance movements toward or away from objects in the environment (48). It is a challenge for researchers in this field to design experiments in which a more carefully selected repertoire of proximal-distal movements will be used to allow a definitive testing and sharpening of this hypothesis. Such an approach is needed in particular in combined behavioral and single-unit recording studies, where the focus has perhaps been too intense in the past upon simple flexion-extension or gripping movements of the hand. Until data from these and other experiments using a wider range of movements become available, the motor functions of the SMA are likely to remain obscure.

GENERAL SUMMARY AND SYNTHESIS

Before attempting to synthesize some of the information that has been reviewed in the present chapter, it may be appropriate first to clarify our working frame of reference by presenting the author's own perspective or point of view concerning the basic organization of cortical motor systems. This point of view is summarized in part in the diagram of Fig. 8.

In agreement with Lawrence and Kuypers (130,131), as well as others (25, 129,152), the view taken here is that the cerebral control of axial, proximal limb, and distal limb muscles is exerted by separate descending pathways, whose spheres of influence overlap only partially. The control of directed gaze and eye-head movements is exerted principally, for example, by posterior parietal

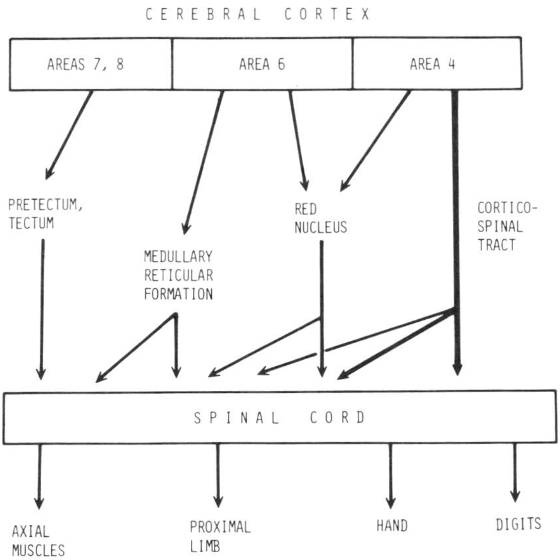

FIG. 8. Organization of corticofugal control of axial, proximal limb, and distal limb muscles. In the view summarized here, the control of eyes, head, trunk, and proximal limb is exerted, in varying degrees, by more than one cortical zone. Parietal and frontal-premotor areas exert control principally through projections to subcortical networks that contain "hardwired" circuitry for the production of primitive orienting movements of the eyes and head, as well as crude reaching movements. The basal ganglia are part of this network, and their influence is exerted in part by way of projections to subcortical structures (for example, the nigrotectal projection). The precentral gyrus also exerts control over axial and proximal limb muscles by way of corticospinal and corticorubral projections, thus bestowing flexibility and precision upon movements generated by the circuitry referred to above. The control of the distal limb muscles, on the other hand, is exerted almost exclusively by projections to the cord and red nucleus from the precentral gyrus. This general view is elaborated more fully in the text, and is consistent with three key observations: (i) that voluntary control of axial and proximal limb movement may be quite precise if the precentral gyrus is intact (consider, for example, the act of writing on a blackboard, which involves principally shoulder movement); (ii) that a degree of crude, voluntary control of these same muscles survives, however, even extensive neocortical ablations: and (iii) that voluntary control of the hand and forearm is severely impaired by precentral ablation, and it is essentially abolished if the red nucleus is also involved.

(area 7) and frontal regions (areas 8–9_p and 6) which project directly to midbrain and pontine areas; parallel routes from cortex to tectum by way of the basal ganglia may be equally involved. The cortical control of trunk and proximal limb muscles is accomplished in part, on the other hand, by projections which descend directly from area 6, the SMA, and rostral area 4 to the red nucleus and medullary reticular formation. Finally, the cortical control of the arm and hand is exerted principally by way of corticospinal and corticorubral projections which issue from caudal area 4, and to a lesser extent from areas 3, 1, 2, and 5.

Thus, the parietal and frontal areas that we have discussed above are viewed not simply as transcortical relay or motor programming regions which link the classic association areas of the cortex with a "primary motor region" in area 4;

instead, they are viewed here as additional but equally primary sources of cortical motor outflow, which affect phylogenetically older subcortical circuits that are organized for the generation of certain basic motor patterns. These patterns include: orienting movements of head and trunk, which accompany shifts in gaze or visual attention; postural adjustments of trunk and proximal limb; simple arm-projection or reaching movements; and limb withdrawal. Conceptually, these fundamental movements may be viewed as lying toward the proximal side of a gradient of proximal-to-distal muscle (or movement) control by the cerebral cortex, as is shown in the diagram of Fig. 8. These basic movements appear to be programmed largely at subcortical levels, and to be simply guided, modulated, or inhibited by cortical outputs; for this reason, no simple correlation may exist between the spatiotemporal patterns of muscle activity which they involve and the discharge patterns of the cortical cells which control or trigger them.

Lying at the distal end of this gradient of cortical control is the classic "primary" motor cortex of area 4, with its highly organized corticospinal outflow. The major functions of this cortical zone appear to be two. First, it contains the elaborate and extensive circuitry necessary for precise, flexible control of the hand and its individual digits, together with the highly organized inputs from posterior sensory and association areas that are necessary for fine sensory guidance of hand movement. But secondly, it gives rise also to less extensive projections to subcortical structures which control proximal muscles. It is probable that these projections comprise a higher-order modulating or "overlay" system, which acts to fractionate and thus bestow flexibility upon the basic movement patterns that are generated by subcortical structures, and that are accessed as well by frontal and parietal areas. It is also possible that, through developmental and motor learning processes, these overlay connections may come to assume some of the proximal muscle control that is exerted in less flexible fashion by other cortical and subcortical areas. Thus, the so-called "primary" motor cortex of area 4 is viewed not as a single cortical zone through which all cerebral control of voluntary movement must funnel, but instead as a highly specialized, phylogenetically newer zone that has developed principally for fine, sensorily guided control of the hand and fingers (as well as the muscles of speech in man), and for modulation of more primitive, proximal movement patterns that are programmed by subcortical networks.[11]

---

[11] One attractive feature of this view is that it partially resolves, or at least avoids, a difficult conceptual problem that is encountered when one adopts the area 4 "funnel" hypothesis alluded to above; namely, the problem of precisely how seemingly non-muscle-specific or holistic signals which "command" (149) or "enhance" (177) movement—and which appear to exist in the unit population discharge patterns of parietal and frontal premotor areas—are eventually transformed by the circuitry of area 4 (or by the basal ganglia or the cerebellum) into the precise, spatiotemporal patterns of muscle activity that exist in even the simplest of voluntary movements. It is also consistent with much of the electrical stimulation and lesion data reviewed in the present chapter, and with evidence which indicates that, though suppressed transiently, many organized and coordinated patterns of *proximal* muscle activity survive rather extensive cortical ablations, including, in lower primates, near-complete decortication (200).

With this frame of reference in mind, let us attempt now to synthesize some of the information that has been discussed in the present chapter. It is possible to do so with a series of flow diagrams, each of which illustrates presumed directions of activity flow between major cortical and subcortical structures during one of the basic movements described in the first paragraph of the Introduction. These diagrams will of course be inadequate, for our state of knowledge concerning the interactions between central structures during voluntary movement is incomplete; but they will serve the useful function of helping to point up more clearly some of the major deficiencies in our current data base.

1. Consider first *centrally directed fixation or shifts in visual attention.*[12] A summary flowchart for parietofrontal involvement in this motor process is shown in Fig. 9. This diagram is obviously oversimplified, but its major features are consistent with the data that were reviewed in relation to posterior parietal lobe and frontal eye field function. Area 7 is viewed here as containing two basic networks, which may *engage* the oculomotor system when activated by appropriate central and peripheral stimuli.

The first of these networks functions as a logical "AND" gate, generating an output only when it receives a conjunction of inputs from: (i) central motivational and memory circuits, whose activity constitutes a central image or memory of an object, along with a desire to obtain it; and (ii) visual areas, whose activity signals the presence of the same or a similar object within the nearby visual field. The output of this network engages the oculomotor system, so that visual fixation occurs upon the object of motivational interest. This control is presumably exerted over pathways that descend directly from the parietal area to the brainstem, although an additional (nonessential) route through the FEF may also be involved. The discharge of a population of cells within this network also encodes the direction of gaze; should this discharge be relayed to the appropriate structures, it could well provide signals which guide the motor system for orienting or reaching movements toward the foveated object. Additional experiments are therefore needed to determine the axonal projection patterns of this subclass of parietal (visual fixation) cells.

The second network interrupts or inhibits the output of the first, so that visual fixation may be broken and attention may be shifted rapidly to new objects that are introduced abruptly into the visual field. It is not clear at the moment whether such control is exerted only intracortically, or also by way of projections to saccade-generating networks within the brainstem. It is clear, however, that neither of these parietal lobe networks are those which produce

[12] In examining these flow diagrams, two points should be kept clearly in mind. First, equally and perhaps more important routes from cortical association to motor areas by way of the basal ganglia and cerebellum have been omitted, both for simplicity and because our emphasis here is on more direct, transcortical routes. Secondly, the use of numerical (cytoarchitectonic) designations is not meant to imply a unique function for each zone, nor a one-to-one correspondence between cytoarchitectonic and functional areas. Area 7, for example, may perform functions other than those suggested by a particular diagram; moreover, a given functional zone may involve only a portion of area 7, in combination with other cytoarchitectonic areas.

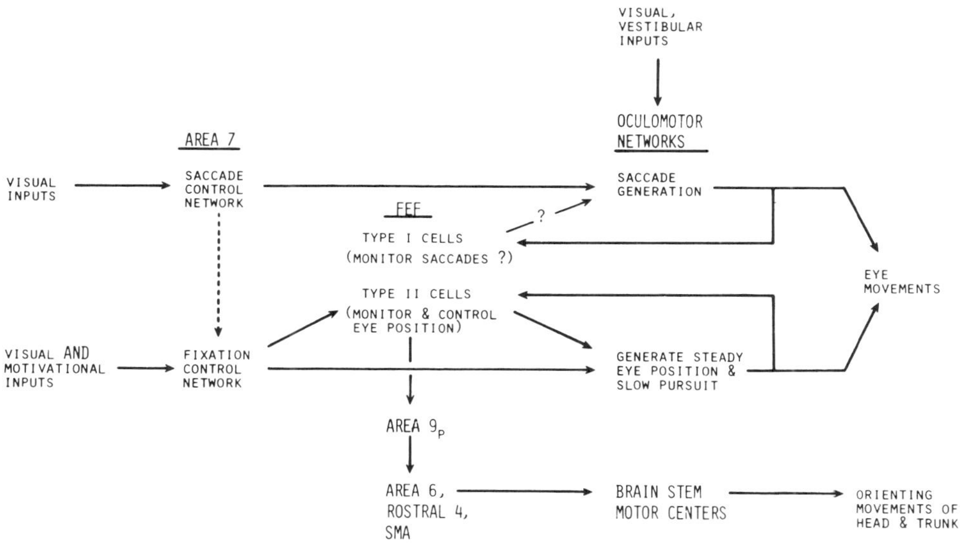

FIG. 9. Hypothetical modes of participation by parietal and frontal eye field (FEF) areas in the central control of directed attention processes, and voluntary orienting movements of eye, head, and trunk. Networks within area 7 may affect oculomotor control systems rather directly, by way of projections to the brainstem; both saccade-generating and fixation–slow pursuit networks are affected. The *dashed line* connecting area 7 networks represents an inhibitory interaction, as suggested by the data of Lynch et al. (137). Information concerning central commands for visual fixation or eye movement may reach the FEF both by way of projections from area 7, and by way of feedback pathways from oculomotor control structures at subcortical levels. Outputs from the FEF to area 7 and to the brainstem may in turn modulate such movement commands, whereas those to cortical areas 9 and (hence) 6 may serve to coordinate eye movements with voluntary orienting movements of the head and trunk. Additional description is given in the text.

spontaneous eye movements, such as those which occur in the dark, nor random visual fixation upon objects of nonmotivational significance. Thus, the destruction of these circuits produces no impairment in basic eye-movement processes; rather, it leads to a contralateral visual neglect because of an uncoupling of central memory and motivational structures from oculomotor control circuitry.

The contralateral (motor) neglect which follows FEF lesions is also viewed here as a disconnection syndrome, but one that may occur for somewhat different reasons. The studies that have been reviewed above suggest that the FEF may perform at least two functions in the cerebral control of directed attention processes. First, it may participate in the control of steady eye position and slow tracking movements, as well as modulate saccades initiated by other structures, under the influence of inputs from parietal, temporal, and occipital areas. But secondly, it appears also to "sense" or monitor commands for eye movement—whether centrally or peripherally evoked—that are generated and fed back from some point nearer the output stages of the oculomotor control system. The feedback pathways and sources of these feedback signals are of course unknown. But by relaying them to area 6 or to the brainstem, the FEF could well participate with other frontal zones in coordinating eye movements with *cortically induced* movements of the head, neck, and arm; in this way shifts in visual attention would be followed when required by appropriate orienting and reaching movements toward objects that lie, in particular, *toward the periphery of the visual field.* Thus, the contralateral neglect which follows FEF lesions may result in part from interruption of a cortical, eye-body coordination network, rather than from a primary impairment of visual attention processes. The diagram of Fig. 10 acknowledges both of these presumed FEF functions.

2. Consider now *axial orienting and arm-projection (reaching) movements.* A summary diagram for the involvement of frontoparietal structures in the cortical control of such movements is shown in Fig. 10. There are several features to note in this diagram.

First, the basic networks which contain the "programs" for these primitive movements are viewed here as lying within subcortical structures (basal ganglia, brainstem, cord), a view that is consistent with the survival of axial orienting and, to a lesser degree, crude reaching movements following extensive neocortical ablation (200). The network which generates orienting movements appears to include reticular formation, vestibular, and perhaps basal ganglia–tectospinal structures (cf. refs. 53,64), and the ventromedial spinal pathways that descend from them that have been implicated in the control of such movements (cf. refs. 122,131). Reaching movements appear to be generated, on the other hand, by a separate network, one which involves rubral as well as reticular formation, spinal, and perhaps basal ganglia structures, and that is perhaps more dependent for its activation upon cerebral input (53,53a).[13]

Second, the cortical control of these subcortical networks is exerted principally by a frontal complex consisting of parts of the SMA, rostral area 4, and perhaps area 6aβ; smaller projections from area 5 (intraparietal sulcus) to the red nucleus (101) and to the spinal cord (40,111), as well as descending pathways from the basal ganglia (e.g., the nigrotectospinal route; cf. refs. 64,86a), may serve as additional control routes. By means of its subcortical and cortico-cortical projections, this frontal complex performs at least two complementary motor functions. First, in response to inputs arising in part from areas 8 and posterior

[13] Although reference to basal ganglia structures has been largely omitted for reasons discussed previously, there is good evidence to indicate that portions of the neostriatum may be involved intimately in the control of visually evoked orienting-reaching responses. Lesions of the caudal putamen, for example, may essentially abolish visually evoked reaching, though eye movements toward the visual stimulus remain (53). Similarly, ablation of area 7 bilaterally may severely impair visually directed reaching (51,53a). Thus, routes from posterior visual association or parietal lobe structures to frontal areas of the cortex *by way of the basal ganglia* may be of more importance for the generation of visually evoked reaching than are the more direct, transcortical pathways discussed here. Recent studies have also suggested that direct pathways from the basal ganglia (substantia nigra) to the tectospinal system may be involved in the control of head turning and orienting responses (64,86a).

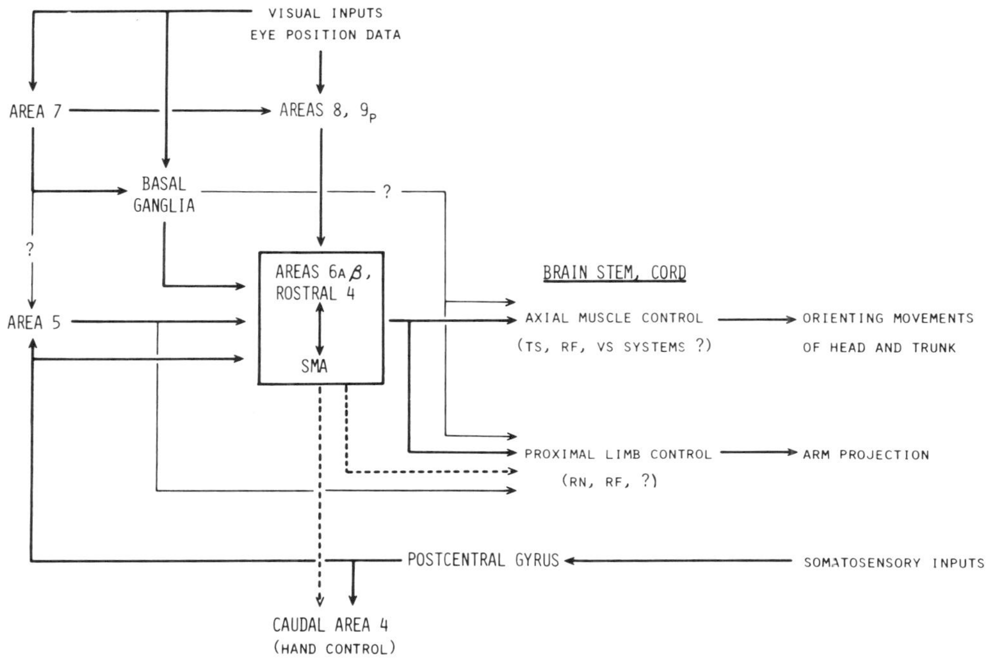

FIG. 10. Hypothetical pathways of activity flow between visuoparietal, frontal, basal ganglia, and subcortical areas during the generation of visually directed, orienting-reaching movements. A full description of this diagram is given in the text. *Dashed lines* indicate inhibitory inputs. Abbreviations: RN, red nucleus; RF, medullary reticular formation; VS, vestibulospinal; and TS, tectospinal.

9, it activates subcortical networks to produce orienting movements of the head and trunk that are coordinated with *cortically* evoked shifts in visual attention or direction of gaze.[14] Second, during such movement it acts to suppress, "gate off," or delay reaching-grasping movements toward objects of interest that would otherwise be evoked reflexly by: (a) more direct visual inputs to the subcortical structures that generate reaching (via unknown pathways through the basal ganglia, for example; cf. ref. 53 and the diagram of Fig. 10); or (b) tactile inputs to the precentral cortical areas that control grasping. In this way, reaching-

[14] There is good evidence to indicate that rapid, visually triggered eye-head movements are generated in parallel, rather than sequentially as suggested above for cortically induced movements. With such triggered movements, activation of the neck muscles may in fact precede that of the ocular muscles, and the movements of head and eyes appear to be "coordinated" rather directly, at subcortical levels, by way of afferent feedback (20). The same investigators have shown, however, that "predictive" or centrally initiated eye-head movements may be organized in a substantially different way from triggered or reflexive movements. Thus, a possible role for area 8 in the coordination of centrally initiated eye-head-arm movements does not appear to be ruled out by their observations.

grasping are delayed until they can be superimposed upon the proximal orienting movements that are necessary for launching the limb along the correct approach path toward a desired object. Thus, this complex of frontal structures performs a movement-sequencing or programming function, so that basic patterns of proximal-distal movement will occur in the temporal order that is required by a given context.[15] The apraxias which follow lesions within this frontal zone suggest such an interpretation.

Third, note that there are two major afferent input channels to this frontal complex. Inputs from visual structures arrive principally by way of area 8 and caudal area 9, although a smaller projection system by way of area 7 to area 5 in the region of the intraparietal sulcus may also exist (152). Note also that unknown pathways bypass this frontal zone to link the visual system directly with the subcortical structures that generate reaching, for direct, visually evoked reaching movements survive ablation of the FEF and the entire frontal complex considered here. Whether this pathway is separate from or involves area 7 is at the moment not entirely clear, but in favor of the latter interpretation is evidence from Denny-Brown et al. which indicates that visual placing responses and visually directed reaching are severely impaired by area 7 ablation (51,53a). Somatosensory inputs reach the frontal complex principally by way of area 5, although a direct projection also exists from the postcentral gyrus to the SMA. Thus, destruction of area 7 *or* the FEF would be expected to impair visually guided reaching, whereas that of area 5 would affect proprioceptive guidance (Denny-Brown et al. have recently drawn a similar conclusion; cf. ref. 53a). Although this interpretation may be oversimplified, it is at least consistent with interpretations that have been offered in some of the studies reviewed above, as well as with some of the functions of the frontal complex that have been outlined in the preceding paragraph.

3. Let us consider, finally, the participation of the frontoparietal areas of interest here in the *visual and proprioceptive guidance of voluntary movements of the hand and fingers.* A summary flowchart which appears to be consistent with existing data is shown in Fig. 11. Note the following features.

First, the major cortical zone for the control of hand movements is within the precentral gyrus, a concept that is supported by a large body of evidence. Movements of the hand and independent movements of the digits are produced by this cortical zone chiefly by way of its corticospinal outflow. Movements about the wrist and collective (i.e., nonindependent) movements of the digits may also be produced, however, by corticospinal projections from area 5 (40, 68,111) and/or corticorubral projections from areas 4 and (to a lesser extent) 5 (101,102,122).

Second, the precentral hand area is viewed here as being subject to control

[15] It is of interest in this connection to note that monkeys with lesions of this frontal area are apparently unable to suppress reaching-grasping movements in the presence of various visual stimuli. Instead, they reach repeatedly and compulsively for objects of interest within the visual field, even when the limb may encounter a barrier which impedes the direct reaching movement (75).

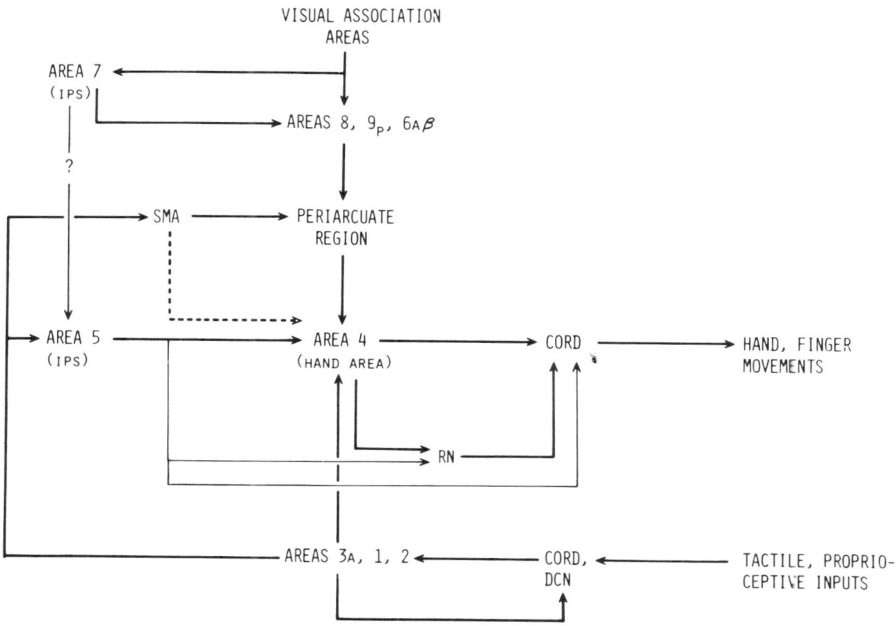

FIG. 11. Hypothetical pathways of activity flow during visually and proprioceptively guided movements of the hand. A full description of this diagram is given in the text. As before, a *dashed line* indicates an inhibitory influence; in this case, inhibition of grasping or manipulative movements of the hand by inputs from the SMA to the precentral gyrus. Abbreviations: DCN, dorsal column nuclei; IPS, intraparietal sulcus.

by three principal inputs. (a) Central commands for manipulative movements of the hand, which may originate *in part* from networks within the banks of the intraparietal sulcus and reach area 4 by pathways that are only partially known. (b) Inputs from visual and somatosensory receiving or association areas, which tend to guide hand movement. The principal route for visual input appears to be by way of area 7 and areas 8, posterior 9, and 6, and perhaps from this complex to caudal area 4 by way of connections within the depths of the arcuate sulcus. (A more direct route from area 7 to 4 by way of area 5 may also exist, through connections in the intraparietal sulcus.) Proprioceptive and tactile inputs arrive over a more direct route through the postcentral gyrus and perhaps also through area 5 (intraparietal sulcus), and are perhaps of greater importance for determining the *details* of hand and finger position than are visual inputs. The somatosensory inputs that are fed back via these channels are in turn under the control of descending projections to the dorsal column nuclei and the dorsal laminae of the cord (35,197). It is most probably the tightly coupled loop from area 4 to cord to hand, and back to area 4 by way of the postcentral gyrus (or via direct thalamocortical projections; cf. ref. 124) that is intimately involved in the oriented grasping responses that are elicited by tactile stimulation of the arm or hand. And finally (c) an input to area 4 from the SMA, which acts to gate off or modulate grasping-manipulative movements of the hand in the presence of excitatory tactile or visual inputs, until appropriate points in time during a movement sequence.

As with direct reaching movements, lesions of area 7 or of afferent pathways to the FEF would be expected to interfere principally with the visual guidance of hand movements (e.g., as in visual placing responses of the hand), whereas those within area 5 or areas 3, 1, and 2 would severely disrupt proprioceptive and tactile guidance. Again this view is perhaps an oversimplification, but it may serve as a working hypothesis until more sophisticated data are generated.

Concluding Comment and Perspective

As was indicated in the Introduction, it has been the purpose of this chapter to explore current information concerning the involvement of parietal and frontal "premotor" areas in the control of visually evoked reaching movements, either by way of their transcortical connections with area 4, or by way of their own descending, motor projections. In doing so, considerable license has been taken in attempting to integrate existing data into a series of explicit hypotheses.

But it must be emphasized that the summary diagrams that have been presented above are at the moment *only* hypotheses, and that much research remains to be done before we will know whether many of the signal paths that are shown in these flowcharts actually exist. For example, although Fig. 9 postulates a projection from Mountcastle's visual fixation cells in area 7 to some part of the oculomotor control system, it has yet to be established that this class of functionally defined neurons does in fact project to either the frontal eye field, or to some subcortical, oculomotor structure. Thus, in future single-unit recording studies of the role of cortical areas in the control of movement, an additional technical step, must be introduced. It will no longer be acceptable, for example, to simply record from neurons within a given cytoarchitectonic zone and to categorize them on the basis of their discharge patterns in relation to some behavioral measure. Instead, antidromic stimulation techniques must also be employed in order to determine the destinations of their axons; i.e., the anatomical distribution of their output signals. For it is only in this way that we will eventually arrive at a series of functional flow diagrams that are correct, and for which we have at present only indirect and rather meager evidence. In addition to these studies, experiments will also be needed to determine whether the functional strength of these connections is altered as a result of visuomotor experience, for the beautiful developmental studies of Held have shown quite clearly that the accuracy of visually guided reaching is strongly dependent upon such experience (95,96).

Finally, in addition to this methodological suggestion, there is perhaps one additional point of significance that may be drawn from the present review.

In current discussions of the central control of voluntary movement, it is sometimes implied that essentially all classes of voluntary movement may be handled by the same suprasegmental, processing systems. For example, distinctions are seldom made between voluntary movements that involve different sets of proximal-distal muscles, that are of different speed, that are only partly as opposed to very well learned, and that are triggered by different classes of sensory stimuli. Thus, the same cerebrocerebellar systems, which include in particular the primary motor cortex, are often viewed as participating equally in the generation of all classes of voluntary movement.

But evidence has accumulated to suggest that rapid, well-learned or "preprogrammed" movements may be generated in part by different neuronal subgroups than are movements that are slower and more dependent upon sensory feedback for successful execution (see, for example, refs. 45,62,72,89, and the chapter by Brooks, *this volume*). In a similar way, the studies that have been reviewed in the present chapter suggest that a distinction must be made also in future discussions between classes of movement that involve to different degrees axial, proximal limb, and hand muscles. The work of Lawrence and Kuypers (129–131) and of others (25,48) indicates quite clearly, for example, that the control of axial and distal limb muscles is most probably accomplished by largely separate, descending motor pathways. Thus, what may appear to the observer of a coordinated movement of body and hand to be the output of a single, movement-generating system, may in fact be a *multi-channel* output from a much more complex, central processor. For this reason, a clear knowledge of the anatomical output patterns of various motor areas within the CNS may prove to be as essential to an understanding of the fundamental properties of primate motor control systems, as are data concerning the basic parameters of movement control that are obtained in clever behavioral studies of the type described so well by other contributors to this volume.

SUMMARY

This chapter focuses on evidence concerning the involvement of so-called "association" and "nonprimary" motor areas of the cerebral cortex in the control of a simple behavioral act: namely, one of attending visually to, reaching for, and grasping an object within the nearby visual field. The cortical regions discussed are areas 5 and 7 of the posterior parietal lobe, the frontal eye field (area 8) and adjacent tissue (posterior area 9), the "premotor" area (6aβ), and the supplementary motor area. The major anatomical connections of each area are summarized, and data are reviewed from behavioral and physiological studies concerning its possible contributions to the central control of shifts in visual attention and voluntary reaching. An attempt is made at the end of the chapter to synthesize available data into a series of diagrams, which suggest the major directions of activity flow between these and other central areas during various components of a visually directed, reaching-grasping movement.

The review was begun with an implicitly held, but by no means unique, hypothesis. The detailed programming of patterns of muscle activity across the trunk and limb during voluntary reaching was viewed as being accomplished primarily at a cerebral, if not a cortical, level. Moreover, the programming system was viewed as a serial one, with molar "commands" for movement, generated by high-order association areas, being transformed successively into more muscle-specific and detailed commands as they are transmitted through transcortical relay areas or the basal ganglia, to and through the precentral gyrus. The review ends with a somewhat different point of view: namely, (i) that the basic "programs" for visually evoked orienting and reaching movements may in fact reside within subcortical networks, with cortical structures acting only to inhibit or modulate their expression; and (ii) that the cortical areas listed above may serve less as simple transcortical relay zones, than as sources of parallel motor outflow which may act directly upon these subcortical networks to produce primitive orienting-reaching movements.

ACKNOWLEDGMENTS

The author is indebted to his colleagues, Dr. Dwain Reed and Mr. Richard Gold, for many helpful discussions concerning the material and concepts covered in the present chapter.

This work was supported by Grant No. NS 10183 from the National Institute of Neurological and Communicative Disorders and Stroke.

REFERENCES

1. Abrahams, V. C., and Rose, P. K. (1975): Projections of extraocular, neck muscle, and retinal afferents to superior colliculus in the cat: their connections to cells of origin of the tectospinal tract. *J. Neurophysiol.,* 38:10–18.
2. Akert, K. (1964): Comparative anatomy of frontal cortex and thalamofrontal connections. In: *The Frontal Granular Cortex and Behavior,* edited by J. M. Warren and K. Akert, pp. 372–396. McGraw-Hill, New York.
3. Allen, G. I., Gilbert, P. F. C., Marini, R., Schultz, W., and Yin, T. C. T. (1977): Integration of cerebral and peripheral inputs by interpositus neurons in monkey. *Exp. Brain Res.,* 27:81–100.
4. Allen, G. I., and Tsukahara, N. (1974): Cerebrocerebellar communication systems. *Physiol. Rev.,* 54:957–1006.
5. Anderson, M. E., Yoshida, M., and Wilson, V. J. (1971): Influences of superior colliculus on cat neck motoneurons. *J. Neurophysiol.,* 34:898–907.
6. Anderson, M. E., Yoshida, M., and Wilson, V. J. (1972): Tectal and tegmental influences on cat forelimb and hindlimb motoneurons. *J. Neurophysiol.,* 35:462–470.
7. Anderson, P., Hagan, P. J., Phillips, C. G., and Powell, T. P. S. (1975): Mapping by microstimulation of overlapping projections from area 4 to motor units of the baboon's hand. *Proc. Roy. Soc. B.,* 188:31–60.
8. Asanuma, H. (1975): Recent development in the study of the columnar arrangement of neurons within the motor cortex. *Physiol. Rev.,* 55:143–156.
9. Asanuma, H., and Rosen, I. (1972): Topographical organization of cortical efferent zones projecting to distal forelimb muscles in the monkey. *Exp. Brain Res.,* 14:243–256.

10. Aschoff, J. C. (1974): Reconsideration of the oculomotor pathway. In *The Neurosciences. Third Study Program,* edited by F. O. Schmitt and F. G. Worden, pp. 305–310. M.I.T. Press, Cambridge.
11. Astruc, J. (1971): Corticofugal connections of area 8 (frontal eye field) in *Macaca mulatta. Brain Res.,* 33:241–256.
12. Bantli, H., and Bloedel, J. R. (1975): The action of dentate nucleus on the excitability of spinal motoneurons via pathways which do not involve the primary sensorimotor cortex. *Brain Res.,* 88:86–90.
13. Bantli, H., and Bloedel, J. R. (1975): Monosynaptic activation of a direct reticulo-spinal pathway by the dentate nucleus. *Pfluegers Arch.,* 357:237–242.
14. Barlow, J. S. (1970): Vestibular and non-dominant parietal lobe disorders (two aspects of spatial disorientation in man). *Dis. Nerv. Syst.,* 31:667–673.
15. Bates, J. A. V., and Ettlinger, G. (1960): Posterior biparietal ablations in the monkey. *A.M.A. Arch. Neurol.,* 3:177–192.
16. Battig, K., Rosvold, H. E., and Mishkin, M. (1960): Comparisons of the effects of frontal and caudate lesions on delayed response and alternations in monkeys. *J. Comp. Physiol. Psychol.,* 53:400–404.
17. Bayledier, C. (1977): Pulvinar-lateroposterior afferents to cortical area 7 in monkeys demonstrated by horseradish peroxidase tracing technique. *Exp. Brain Res.,* 27:501–507.
18. Benton, A. L. (1967): Constructional apraxia and the minor hemisphere. *Confinia Neurol.,* 29:1–16.
19. Bizzi, E. (1968): Discharge of frontal eye field neurons during saccadic and following eye movements in unanesthetized monkeys. *Exp. Brain Res.,* 6:69–80.
20. Bizzi, E. Kalil, R. E., Morasso, P., and Tagliasco, V. (1972): Central programming and peripheral feedback during eye-head coordination in monkeys. *Bibl. Ophthalmol.* 82 220–232.
21. Bizzi, E., and Polit, A. (1979): Characteristics of motor programs underlying visually evoked movements. *(This volume.)*
22. Bizzi, E., and Schiller, P. H. (1970): Single unit activity in the frontal eye fields of unanesthetized monkeys during eye and head movement. *Exp. Brain Res.,* 10:151–158.
23. Bos, J., and Benevento, L. A. (1975): Projections of the medial pulvinar to orbital cortex and frontal eye fields in the rhesus monkey *(Macaca mulatta). Exp. Neurol.,* 49:487–496.
24. Bossom, J. (1965): The effects of brain lesions on prism adaptation in monkey. *Psychonom. Sci.,* 2:45–46.
25. Brinkman, J., and Kuypers, H. G. J. M. (1973): Cerebral control of contralateral and ipsilateral arm, hand and finger movements in split brain rhesus monkeys. *Brain,* 69:653–674.
26. Brinkman, J., and Porter, R. (1978): Supplementary motor area of the monkey: Activity of neurones during performance of a learned motor task. *J. Physiol. (Paris),* 74:313–316.
27. Brodmann, K. (1909): *Vergleichende Lokalizationlehre der Grosshirnrinde in ihren Prinzipien dargestellt auf Grund des Zellenbaues.* J. A. Barth, Leipzig.
28. Brooks, V. B. (1975): Roles of cerebellum and basal ganglia in initiation and control of movements. *Can. J. Neurol. Sci.,* 2:265–277.
29. Brooks, V. B. (1979): Motor programs revisited. *(This volume.)*
30. Brooks, V. B., and Stoney, S. D., Jr. (1971): Motor mechanisms: The role of the pyramidal system in motor control. *Annu. Rev. Physiol.,* 33:337–392.
31. Brown, J. (1972): *Aphasia, Apraxia, Agnosia.* Charles C Thomas, Springfield, Ill.
32. Brucher, J. M. (1966): The frontal eye field of the monkey. *Int. J. Neurol.,* 5:262–281.
33. Buerger, A. A., Gross, C. G., and Rocha-Miranda, C. E. (1974): Effects of ventral putamen lesions on discrimination learning by monkeys. *J. Comp. Physiol. Psychol.,* 63:440–446.
34. Burton, H., and Jones, E. G. (1976): The posterior thalamic region and its cortical projection in new world and old world monkeys. *J. Comp. Neurol.,* 168:249–302.
35. Catsman-Berrevoets, C. E., and Kuypers, H. G. J. M. (1976): Cells of origin of cortical projections to dorsal column nuclei, spinal cord and bulbar medial reticular formation in the rhesus monkey. *Neurosci. Letters,* 3:245–252.
36. Cheney, P. D., and Fetz, E. E. (1978): Functional properties of primate corticomotoneuronal cells. *Neurosci. Abstr.,* 4:293.
37. Cogan, D. G. (1965): Ophthalmic manifestations of bilateral non-occipital cerebral lesions. *Br. J. Ophthalmol.,* 49:281–287.
38. Cogan, D. G., and Adams, R. D. (1953): A type of paralysis of conjugate gaze (ocular motor apraxia). *A.M.A. Arch. Ophthalmol.,* 50:434–442.

39. Cohen, S. M. (1972): Electrical stimulation of cortical-caudate pairs during delayed successive visual discrimination. *Acta Neurobiol. Exp.*, 32:211–233.
40. Coulter, J. D., and Jones, E. G. (1977): Differential distribution of corticospinal projections from individual cytoarchitectonic fields in the monkey. *Brain Res.*, 129:335–340.
41. Coulter, J. D., Maunz, R. A., and Willis, W. D. (1974): Effects of stimulation of sensorimotor cortex on primate spinothalamic neurons. *Brain. Res.*, 65:351–356.
42. Coxe, W. S., and Landau, W. M. (1965): Observations upon the effect of supplementary motor cortex ablation in monkeys. *Brain*, 88:763–772.
43. Critchley, J. (1969): *The Parietal Lobes.* Hafner, New York.
44. Dean, W., and Davis, G. (1959): Behavior following caudate lesions in rhesus monkeys. *J. Neurophysiol.*, 22:524–537.
45. Delong, M. R., and Strick, P. L. (1974): Relation of basal ganglia, cerebellum and motor cortex units to ramp and ballistic limb movements. *Brain Res.*, 71:327–336.
46. Denny-Brown, D. (1958): The nature of apraxia. *J. Nerv. Ment. Dis.*, 126:9–33.
47. Denny-Brown, D. (1960): Motor mechanisms—Introduction: The general principles of motor integration. In: *Handbook of Physiology*, edited by J. Field, pp. 781–796. American Physiological Society, Washington, D.C.
48. Denny-Brown, D. (1966): *The Cerebral Control of Movement.* Liverpool Univ. Press, Liverpool.
49. Denny-Brown, D., and Baker, B. (1954): Amorphosynthesis from left parietal lesion. *A.M.A. Arch. Neurol. Psychiatry*, 71:302–313.
50. Denny-Brown, D., and Botterell, E. H. (1947): The motor functions of the agranular frontal cortex. *Res. Publ. Assoc. Res. Nerv. Ment. Dis.*, 27:235–345.
51. Denny-Brown, D., and Chambers, R. A. (1958): The parietal lobe and behavior. *Res. Publ. Assoc. Nerv. Ment. Dis.*, 36:35–117.
52. Denny-Brown, D., Meyer, J. S., and Horenstein, S. (1952): The significance of perceptual rivalry resulting from parietal lesion. *Brain*, 75:433–471.
53. Denny-Brown, D., and Yanagisawa, N. (1976): The role of the basal ganglia in the initiation of movement. In: *The Basal Ganglia*, edited by M. D. Yahr, pp. 114–149. Raven Press, New York.
53a. Denny-Brown,D., Yanagisawa, N., and Kirk, E. J. (1975): The localization of hemispheric mechanisms of visually directed reaching and grasping. In: *Cerebral Localization*, edited by K. J. Zulch, O. Creutzfeldt, and G. C. Galbraith, pp. 62–75. Springer-Verlag, New York.
54. Deuel, R. K. (1977): Loss of motor habits after cortical lesions. *Neuropsychologia*, 15:205–215.
55. DeVito, J. S., and Smith, O. A. (1959): Projections from the mesial frontal cortex (supplementary motor area) to the cerebral hemispheres and brain stem of the *Macaca mulatta. J. Comp. Neurol.*, 111:261–275.
56. Divac, I., LaVail, J. H., Rakic, P., and Winston, K. R. (1977): Heterogeneous afferents to the inferior parietal lobule of the rhesus monkey revealed by the retrograde transport method. *Brain Res.*, 123:197–208.
57. Ettlinger, G., and Kalsbeck, J. E. (1962): Changes in tactile discrimination and in visual reaching after successive and simultantous bilateral posterior parietal ablations in the monkey. *J. Neurol. Neurosurg. Psychiatry*, 25:256–268.
58. Ettlinger, G., and Wegener, J. (1958): Somaesthetic alternation, discrimination and orientation after frontal and parietal lesions in monkeys. *Quart. J. Exp. Psychol.*, 10:177–178.
59. Evarts, E. V. (1966): Pyramidal tract activity associated with a conditioned hand movement in the monkey. *J. Neurophysiol.*, 29:1011–1027.
60. Evarts, E. V. (1971): Feedback and corollary discharge: A merging of the concepts. *Neurosci. Res. Progr. Bull.*, 9:86–112.
61. Evarts, E. V. (1974): Precentral and postcentral cortical activity in association with visually triggered movement. *J. Neurophysiol.*, 37:373–381.
62. Evarts, E. V., and Fromm, C. (1977): Sensory responses in motor cortex neurons during precise motor control. *Neurosci. Letters*, 5:267–272.
63. Evarts, E. V., and Tanji, J. (1976): Reflex and intended responses in motor cortex pyramidal tract neurons of monkey. *J. Neurophysiol.*, 39:1069–1080.
64. Faber, J. E., and York, D. H. (1977): The role of nigrotectal projections in relation to tectospinal tracts and head turning. *Neurosci. Abstr.*, 3:37.

65. Felix, D., and Wiesendanger, M. (1970): Cortically induced inhibition in the dorsal column nuclei of monkeys. *Pfluegers Arch.*, 320:285–288.
66. Ferrier, D. (1876): *The Functions of The Brain.* Smith, Elder and Co., London.
67. Fetz, E. E., and Cheney, P. D. (1978): Muscle fields of primate corticomotoneuronal cells. *J. Physiol. (Paris)*, 74:239–245.
68. Fleming, J. F. R., and Crosby, E. C. (1955): The parietal lobe as an additional motor area. *J. Comp. Neurol.*, 103:485–512.
69. Foerster, O. (1931): The cerebral cortex in man. *Lancet*, 2:309–312.
70. Foerster, O. (1936): Motorische felder and bahnen. In: *Handbuch der Neurologie*, edited by Bumke and Foerster. Springer, Berlin.
71. Foerster, O. (1936): The motor cortex in man in light of Hughlings Jackson's observations. *Brain*, 59:135–159.
72. Fromm, C., and Evarts, E. V. (1977): Sensory responses in motor cortex neurons during precise motor control. *Neurosci. Letters*, 5:267–272.
73. Fulton, J. F. (1934): Forced grasping in relation to the syndrome of the premotor area. *Arch. Neurol. Psychiatry*, 31:221–235.
74. Fulton, J. F. (1935): A note on the definition of the "motor" and "premotor" areas. *Brain*, 58:311–316.
75. Fulton, J. F., and Kennard, M. A. (1934): A study of flaccid and spastic paralyses produced by lesions of the cerebral cortex. *Res. Publ. Assoc. Nerv. Ment. Dis.*, 13:158–210.
76. Fuster, J. M. (1973): Unit activity in prefrontal cortex during delayed response performance: Neuronal correlates of transient memory. *J. Neurophysiol.*, 36:61–78.
77. Geier, S., Bancaud, J., Talairach, J., Bonis, A., Hossard-Bouchard, H., and Enjelvin, M. (1977): Ictal tonic postural change and automatisms of the upper limb during epileptic parietal lobe discharges. *Epilepsia*, 18:517–524.
78. Gerstmann, J. (1940): Syndrome of finger agnosia; disorientation for right and left, agraphia, and acalculia. *Arch. Neurol. Psychiatry*, 44:398–408.
79. Geschwind, N. (1965): Disconnexion syndromes in animals and man. II. *Brain*, 88:585–644.
80. Ghez, C., and Pisa, M. (1972): Inhibition of afferent transmission in cuneate nucleus during voluntary movement in cat. *Brain Res.*, 40:145–151.
81. Gilbert, P. F. C. (1974): A theory of memory that explains the structure and function of the cerebellum. *Brain Res.*, 70:1–18.
82. Gilbert, P. F. C., and Thach, W. T. (1977): Purkinje cell activity during motor learning. *Brain Res.*, 128:309–328.
83. Goldberg, M. E., and Wurtz, R. H. (1972): Activity of superior colliculus in behaving monkey: II. Effect of attention on neuronal response. *J. Neurophysiol.*, 35:560–574.
84. Goldman, P. S., and Rosvold, H. E. (1970): Localization of function within the dorsolateral prefrontal cortex of the Rhesus monkey. *Exp. Neurol.*, 27:291–304.
85. Goldman, P. S., and Rosvold, H. E. (1972): The effects of selective caudate lesions in infant and juvenile rhesus monkeys. *Brain Res.*, 43:53–66.
86. Graybiel, A. M. (1974): Studies on the anatomical organization of posterior association cortex. In: *The Neurosciences. Third Study Program*, edited by F. O. Schmitt and F. G. Worden, pp. 205–214. M.I.T. Press, Cambridge.
86a. Graybiel, A. M. (1978): Organization of the nigrotectal connection: An experimental tracer study in the cat. *Brain Res.*, 143:339–348.
87. Gross, C. G. (1963): Locomotor activity under various stimulus conditions following partial lateral frontal cortical lesions in monkeys. *J. Comp. Physiol. Psychol.*, 56:52–59.
88. Haaxma, R., and Kuypers, H. G. J. M. (1974): Role of occipito-frontal and cortico-cortical connections in visual guidance of relatively independent hand and finger movements in rhesus monkeys. *Brain Res.*, 71:361–366.
89. Hallett, M., Shahani, B. T., and Young, R. R. (1975): EMG analysis of stereotyped voluntary movements in man. *J. Neurol. Neurosurg. Psychiatry*, 38:1154–1162.
90. Hartje, W., and Ettlinger, G. (1973): Reaching in light and dark after unilateral posterior parietal ablations in monkey. *Cortex*, 9:346–354.
91. Hecaen, H. (1962): Clinical symptomatology in right and left hemispheric lesions. In: *Interhemispheric Lesions and Cerebral Dominance*, edited by V. B. Mountcastle, pp. 215–243. Johns Hopkins Press, Baltimore.

92. Hecaen, H., and Ajuriaguerra, J. (1954): Balint's syndrome (psychic paralysis of visual fixation) and its minor forms. *Brain,* 77:373–400.
93. Hecaen, H., and Penfield, W., Bertrand, C., and Malmo, R. (1956): The syndrome of apractognosia due to lesions of the minor cerebral hemisphere. *A.M.A. Arch. Neurol. Psychiatry,* 75:400–434.
94. Heilman, K. M. (1973): Ideational apraxia—A redefinition. *Brain,* 96:861–864.
95. Held, R., and Bauer, J. A., Jr. (1967): Visually guided reaching in infant monkeys after restricted rearing. *Science,* 155:718–720.
96. Held, R., and Bauer, J. A., Jr. (1974): Development of sensorially guided reaching in infant monkeys. *Brain Res.,* 71:265–271.
97. Hines, M. (1940): Movements elicited from precentral gyrus of adult chimpanzee by stimulation with sine wave currents. *J. Neurophysiol.,* 3:442–466.
98. Hines, M. (1947): The motor areas. *Fed. Proc.,* 6:441–447.
99. Hines, M. (1949): Significance of precentral motor cortex. In: *The Precentral Motor Cortex,* edited by P. Bucy, pp. 459–499. Univ. of Illinois Press, Urbana.
100. Humphrey, D. R. (1972): Relating motor cortex spike trains to measures of motor performance. *Brain Res.,* 40:7–18.
101. Humphrey, D. R., Gold, R., and Reed, D. J. (1978): Origins of cortical and cerebellar projections to the red nucleus in the monkey. *Neurosci. Abstr.,* 4:297.
102. Humphrey, D. R., and Rietz, R. (1976): Cells of origin of corticorubral projections from the arm area of primate motor cortex and their synaptic actions in the red nucleus. *Brain Res.,* 110:162–169.
103. Humphrey, D. R., Schmidt, E. M., and Thompson, W. D. (1970): Predicting measures of motor performance from multiple cortical spike trains. *Science,* 170:758–762.
104. Hyvarinen, J., and Poranen, A. (1974): Function of the parietal associative area 7 revealed from cellular discharges in alert monkeys. *Brain,* 97:673–692.
105. Jacobsen, C. F. (1932): Influence of motor and premotor areas lesions upon retention of skilled movements in monkeys and chimpanzees. *Res. Publ. Assoc. Nerv. Ment. Dis.,* 13:225–247.
106. Jones, E. G., and Burton, H. (1976): Areal differences in the laminar distribution of thalamic afferents in cortical fields of insular, parietal and temporal regions of primates. *J. Comp. Neurol.,* 168:197–247.
107. Jones, E. G., Coulter, J. D., Burton, H., and Porter, R. (1977): Cells of origin and terminal distribution of corticostriatal fibers arising in the sensory-motor cortex of monkeys. *J. Comp. Neurol.,* 173:53–80.
108. Jones, E. G., Coulter, J. D., and Hendry, S. H. C. (1978): Intracortical connectivity of architectonic fields in the somatic sensory, motor and parietal cortex of monkeys. *J. Comp. Neurol.,* 181:291–348.
109. Jones, E. G., and Powell, T. P. S. (1969): Connexions of somatic sensory cortex of the Rhesus monkey. I. Ipsilateral cortical connexions. *Brain,* 92:477–502.
110. Jones, E. G., and Powell, T. P. S. (1970): An anatomical study of converging sensory pathways within the cerebral cortex of the monkey. *Brain,* 93:793–820.
111. Jones, E. G., and Wise, S. P. (1977): Size, laminar and columnar distribution of efferent cells in the sensory-motor cortex of monkeys. *J. Comp. Neurol.,* 175:391–438.
112. Kaas, J. H., Lin, C. S., and Wagor, E. (1977): Cortical projections of posterior parietal cortex in owl monkeys. *J. Comp. Neurol.,* 171:387–408.
113. Kalil, K. (1975): Thalamo-cortical projections of the VA and VL in the rhesus monkey. *Neurosci. Abstr.,* 1:171.
114. Kasdon, D. L., and Jacobsen, S. (1978): The thalamic afferents to the inferior parietal lobule of the rhesus monkey. *J. Comp. Neurol.,* 177:685–706.
115. Kennard, M. A., and Ectors, L. (1938): Forced circling in monkeys following lesions of the frontal lobe. *J. Neurophysiol.,* 1:45–54.
116. Kennard, M. A., and Fulton, J. F. (1933): The localizing significance of spasticity, reflex grasping and the signs of Babinski and Rossolimo. *Brain,* 56:213–225.
117. Kievit, J., and Kuypers, H. G. J. M. (1975): Basal forebrain and hypothalamic connections to the frontal and parietal cortex in rhesus monkey. *Science,* 187:660–662.
118. Kubota, K., and Hamada, I. (1978): Visual tracking and neuron activity in the post-arcuate area in monkeys. *J. Physiol. (Paris),* 74:297–312.

119. Kunzle, H. (1978): An autoradiographic analysis of the efferent connections from premotor and adjacent prefrontal regions (areas 6 and 9) in *Macaca fascicularis. Brain, Behav. Evol.,* 15:185–234.
120. Kunzle, H., and Akert, K. (1977): Efferent connections of cortical area 8 (frontal eye field) in *Macaca fascicularis.* A reinvestigation using the autoradiographic technique. *J. Comp. Neurol.,* 173:147–164.
121. Kupferman, I., and Weiss, K. R. (1978): The command neuron concept. *Behavioral and Brain Sciences,* 1:3–39.
122. Kuypers, H. G. J. M., and Lawrence, D. G. (1967): Cortical projections to the red nucleus and the brain stem in the rhesus monkey. *Brain Res.,* 4:151–188.
123. Lamotte, R. H., and Acuna, C. (1975): Defects in accurate projection of the contralateral arm into extrapersonal space in monkeys after removal of the posterior parietal association cortex. *Neurosci. Abstr.,* 1:131.
124. Larsen, K. D., Yumiya, H., and Asanuma, H. (1978): Direct somatosensory inputs from the thalamus to the motor cortex. *Neurosci. Abstr.,* 4:299.
125. Latto, R. (1978): The effects of bilateral frontal eye-field, posterior parietal or superior colliculus lesions on visual search in the rhesus monkey. *Brain Res.,* 146:35–50.
126. Latto, R., and Cowey, A. (1971): Visual field defects after frontal eye-field lesions in monkeys. *Brain Res.,* 30:1–24.
127. Latto, R., and Cowey, A. (1971): Fixation changes after frontal eye-field lesions in monkeys. *Brain Res.,* 30:25–36.
128. Latto, R., and Cowey, A. (1972): Frontal eye-field lesions in monkeys. In: *Cerebral Control of Eye Movements and Motion Perception,* edited by E. Bizzi and J. Dichgans, *Bibl. Ophthalmol.,* 82:159–168.
129. Lawrence, D. G., and Hopkins, D. A. (1976): The development of motor control in the rhesus monkey: Evidence concerning the role of corticomotoneuronal connections. *Brain,* 99:235–254.
130. Lawrence, D. G., and Kuypers, H. G. J. M. (1968): The functional organization of the motor system in the monkey. I. The effects of bilateral pyramidal lesions. *Brain,* 91:1–14.
131. Lawrence, D. G., and Kuypers, H. G. J. M. (1968): The functional organization of the motor system in the monkey. II. The effects of lesions of the descending brain-stem pathways. *Brain,* 91:15–36.
132. Lemon, R., and Porter, R. (1976): Afferent input to movement related precentral neurons in conscious monkeys. *Proc. Roy. Soc. B.,* 194:313–339.
133. Liles, S., and Davis, G. D. (1969): Interaction of caudate nucleus and thalamus in alteration of cortically induced movement. *J. Neurophysiol.,* 32:564–573.
134. Lilly, J. C. (1958): Correlations between neurophysiological activity in the cortex and short-term behavior in the monkey. In: *Biological and Biochemical Bases of Behavior,* edited by H. F. Harlow and C. N. Woolsey, pp. 83–100. Univ. of Wisconsin Press, Madison.
135. Luria, A. R. (1973): The frontal lobes and the regulation of behavior. In: *Psychophysiology of the Frontal Lobes,* edited by K. H. Pribram and A. R. Luria, pp. 3–26. Academic Press, New York.
136. Luria, A. R., Karpov, B. A., and Yarbuss, A. L. (1966): Disturbances of active visual perception with lesions of the frontal lobe. *Cortex,* 2:202–212.
137. Lynch, J. C., Mountcastle, V. B., Talbot, W. H., and Yin, T. C. T. (1977): Parietal lobe mechanisms for directed visual attention. *J. Neurophysiol.,* 40:362–389.
138. MacLean, P. D. (1949): Psychosomatic disease and the "visceral brain"; Recent developments bearing on the Papez theory of emotion. *Psychosom. Med.,* 11:338–353.
139. Marrocco, R. T. (1978): Saccades induced by stimulation of the frontal eye fields: Interaction with voluntary and reflexive eye movements. *Brain Res.,* 146:23–34.
140. Mesalum, M.-M., Van Hoesen, G. W., Pandya, D. N., and Geschwind, N. (1977): Limbic and sensory connections of the inferior parietal lobule (area PG) in the rhesus monkey: A study with a new method for horseradish peroxidase histochemistry. *Brain Res.,* 136:393–414.
141. Mettler, F. A. (1947): The nonpyramidal motor projections from the frontal cerebral cortex. *Res. Publ. Assoc. Nerv. Ment. Dis.,* 27:162–199.
142. Mettler, F. A. (1964): Supratentorial mechanisms influencing the oculomotor apparatus. In: *The Oculomotor System,* edited by M. B. Bender, pp. 1–17. Harper & Row, New York.

143. Milner, B. (1964): Some effects of frontal lobectomy in man. In: *The Frontal Granular Cortex and Behavior,* edited by J. M. Warren and K. Akert, pp. 313–334. McGraw-Hill, New York.
144. Milner, A. D., Ockleford, E. M., and Dewar, N. (1977): Visuo-spatial performance following posterior parietal and lateral frontal lesions in stumptail macaques. *Cortex,* 13:350–360.
145. Mohler, C. W., and Wurtz, R. H. (1977): Role of striate cortex and superior colliculus in visual guidance of saccadic eye movements in monkeys. *J. Neurophysiol.,* 40:74–94.
146. Mohler, C. W., Goldberg, M. E., and Wurtz, R. H. (1973): Visual receptive fields of frontal eye field neurons. *Brain Res.,* 61:385–389.
147. Moll, L., and Kuypers, H. G. J. M. (1977): Premotor cortical ablations in monkeys: Contralateral changes in visually guided behavior. *Science,* 198:317–319.
148. Mountcastle, V. B. (1975): The view from within: Pathways to the study of perception. *Johns Hopkins Med. J.,* 136:109–131.
149. Mountcastle, V. B., Lynch, J. C., Georgopoulos, A., Sakata, H., and Acuna, C. (1975): Posterior parietal association cortex of the monkey: Command functions for operations within extrapersonal space. *J. Neurophysiol.,* 38:871–908.
150. Murray, E. A., and Coulter, J. D. (1977): Corticospinal projections from the medial cerebral hemisphere in the monkey. *Neurosci. Abstr.,* 3:275.
151. Nauta, W. J. H. (1964): Some efferent connections of the prefrontal cortex of the monkey. In: *The Frontal Granular Cortex and Behavior,* edited by J. M. Warrent and K. Akert, pp. 397–409. McGraw-Hill, New York.
152. Pandya, D. P., and Kuypers, H. G. J. M. (1969): Cortico-cortical connections in the rhesus monkey. *Brain Res.,* 13:13–36.
153. Pandya, D. P., and Vignolo, L. A. (1971): Intra- and interhemispheric projections of the precentral, premotor and arcuate areas in the rhesus monkey. *Brain Res.,* 26:217–233.
154. Pasik, P., and Pasik, T. (1964): Oculomotor functions in monkeys with lesions of the cerebrum and superior colliculi. In: *The Oculomotor System,* edited by M. B. Bender, pp. 40–80. Harper & Row, New York.
155. Pasik, P., Pasik, T., and Krieger, H. P. (1959): Effect of cerebral lesions upon optokinetic nystagmus in monkeys. *J. Neurophysiol.,* 22:297–304.
156. Paterson, A., and Zangwill, O. L. (1944): Disorders of visual space perception associated with lesions of the right cerebral hemisphere. *Brain,* 67:331–358.
157. Peele, T. L. (1944): Acute and chronic parietal lobe ablations in monkeys. *J. Neurophysiol.,* 7:269–286.
158. Penfield, W., and Jasper, H. (1954): *Epilepsy and the Functional Anatomy of the Human Brain,* pp. 54–102, 350–388. Little, Brown and Co., Boston.
159. Penfield, W., and Rasmussen, T. (1950): *The Cerebral Cortex of Man,* pp. 46–132, 209–220. Macmillan, New York.
160. Penfield, W., and Welch, K. (1951): The supplementary motor area of the cerebral cortex. *A.M.A. Arch. Neurol. Psychiatry,* 66:289–317.
161. Petras, J. M. (1971): Connections of the parietal lobe. *J. Psychiatr. Res.,* 8:189–201.
162. Petrides, M., and Iversen, S. D. (1975): Cross-modal matching and the primate frontal cortex. *Science,* 192:1023–1024.
163. Petrovici, I. (1968): Apraxia of gait and trunk movements. *J. Neurol. Sci.,* 7:229–243.
164. Phillips, C. G. (1969): Motor apparatus of the baboon's hand. *Proc. Roy. Soc. B.,* 173:141–174.
165. Phillips, C. G. (1975): Laying the ghost of muscles versus movements. *Can. J. Neurol. Sci.,* 2:209–218.
166. Phillips, C. G., and Porter, R. (1977): *Corticospinal Neurones: Their Role in Movement.* Academic Press, London.
167. Pohl, W. (1973): Dissociation of spatial discrimination deficits following frontal and parietal lesions in monkeys. *J. Comp. Physiol. Psychol.,* 82:227–239.
168. Purpura, D. P. (1976): Physiological organization of the basal ganglia. In: *The Basal Ganglia,* edited by M. D. Yahr, pp. 91–114. Raven Press, New York.
169. Purpura, D. P., Frigyesi, T. L., McMurtry, J. G., and Scarff, T. (1966): Synaptic mechanisms in thalamic regulation of cerebro-cortical projection activity. In: *The Thalamus,* edited by D. P. Purpura and M. D. Yahr, pp. 154–170. Columbia Univ. Press, New York.
170. Ratcliff, G., and Davies-Jones, G. A. B. (1972): Defective visual localization in focal brain wounds. *Brain,* 95:49–60.

171. Ratcliff, G., Ridley, R. M., and Ettlinger, G. (1977): Spatial disorientation in the monkey. *Cortex,* 13:62–65.
172. Richter, C. P., and Hines, M. (1932): The production of the "grasp reflex" in adult macaques by experimental frontal lobe lesions. *Proc. Assoc. Res. Nerv. Ment. Dis.,* 13:211–224.
173. Ridley, R. M., and Ettlinger, G. (1975): Tactile and visuo-spatial discrimination performance in the monkey: The effects of total and partial posterior parietal removals. *Neuropsychologia,* 13:191–196.
174. Robinson, D. A. (1972): Eye movements evoked by collicular stimulation in the alert monkey. *Vision Res.,* 12:1795–1808.
175. Robinson, D. A., and Fuchs, A. F. (1969): Eye movements evoked by stimulation of frontal eye fields. *J. Neurophysiol.,* 32:637–648.
176. Robinson, D. L., and Goldberg, M. E. (1978): Sensory and behavioral properties of neurons in posterior parietal cortex in awake, trained monkey. *Fed. Proc.,* 37:2258–2261.
177. Robinson, D. L., Goldberg, M. E., and Stanton, G. B. (1978): Parietal association cortex in the primate: Sensory mechanisms and behavioral modulations. *J. Neurophysiol.,* 41:910–932.
178. Rosen, I., and Asanuma, H. (1972): Peripheral afferent inputs to the forelimb area of the monkey motor cortex: Input-output relations. *Exp. Brain Res.,* 14:257–273.
179. Sakata, H., Takaoka, Y., Kawarasaki, A., and Shibutani, H. (1973): Somatosensory properties of neurons in the superior parietal cortex (area 5) of the rhesus monkey. *Brain Res.,* 64:85–102.
180. Schiller, P. H. (1972): Some functional characteristics of the superior colliculus of the rhesus monkey. *Bibl. Ophthalmol.,* 82:122–129.
181. Schiller, P. H. (1972): The role of the superior colliculus in eye movement and vision. *Invest. Ophthalmol.,* 11:451–460.
181a. Schwartzman, R. J., Gran, B., and Marcos, J. (1975): A functional analysis of cortical motor sensory convergence areas in the monkey. *Trans. Am. Neurol. Assoc.,* 100:114–117.
182. Semmes, J., Weinstein, S., Ghent, L., and Teuber, H. L. (1963): Correlates of impaired orientation in personal and extrapersonal space. *Brain,* 86:747–772.
183. Shinoda, Y., Arnold, A. P., and Asanuma, H. (1976): Spinal branching of corticospinal axons. *Exp. Brain. Res.,* 26:215–234.
184. Smith, A. M. (1977): The activity of neurons in the supplementary motor area during a maintained precision grip. *Neurosci. Abstr.,* 3:279.
185. Smith, A. M. (1979): The activity of neurons in the supplementary motor area during a maintained precision grip. *Brain Res., (in press).*
186. Sprague, J. M. (1972): The superior colliculus and pretectum in visual behavior. *Invest. Ophthalmol.,* 11:473–482.
186a. Stein, J. F. (1976): The effect of cooling parietal lobe areas 5 and 7 upon voluntary movement in awake rhesus monkey. *J. Physiol. (Lond.),* 258:62–63P.
187. Strick, P. L., Zarzecki, P., and Asanuma, H. (1977): Input to primate motor cortex from posterior parietal cortex. *Neurosci. Abstr.,* 3:279.
188. Stryker, M. P., and Schiller, P. H. (1975). Eye and head movements evoked by electrical stimulation of monkey superior colliculus. *Exp. Brain Res.,* 23:103–112.
189. Tanji, J., and Evarts, E. V. (1976): Anticipatory activity of motor cortex neurons in relation to direction of an intended movement. *J. Neurophysiol.,* 39:1062–1068.
190. Tanji, J., and Taniguchi, K. (1978): Does the supplementary motor area play a part in modifying motor cortex reflexes? *J. Physiol. (Paris),* 74:339–345.
191. Terzuolo, C. A., and Viviani, P. (1973): Parameters of motion and EMG activities during some simple motor tasks in normal subjects and cerebellar patients. In: *The Cerebellum, Epilepsy and Behavior,* edited by I. S. Cooper, M. Riklan, and R. S. Snider, pp. 173–215. Plenum Press, New York.
192. Teuber, H.-L. (1963): Space perception and its disturbances after brain injury in man. *Neuropsychologia,* 1:47–57.
193. Teuber, H.-L. (1964): The riddle of frontal lobe function in man. In: *The Frontal Granular Cortex and Behavior,* edited by J. M. Warren and K. Akert, pp. 410–441. McGraw-Hill, New York.
194. Teuber, H.-L. (1966): The frontal lobes and their functions: Further observations in rodents, carnivores, subhuman primates and man. *Int. J. Neurol.,* 5:282–300.

195. Teuber, H.-L. (1976): Complex functions of basal ganglia. In: *The Basal Ganglia*, edited by M. D. Yahr, pp. 151–168. Raven Press, New York.
196. Teuber, H.-L., and Proctor, F. (1964): Some effects of basal ganglia lesions in subhuman primates and man. *Neuropsychologia*, 2:85–93.
197. Towe, A. L. (1973): Somatosensory cortex: Descending influences on ascending systems. In: *Handbook of Sensory Physiology. II. Somatosensory System*, edited by A. Iggo, pp. 701–718. Springer-Verlag, New York.
198. Tower, S. S., and Hines, M. (1935): Dissociation of pyramidal and extrapyramidal functions of the frontal lobes. *Science*, 82:376–378.
199. Travis, A. M. (1955): Neurological deficiencies following supplementary motor area lesions in *Macaca mulatta. Brain*, 78:155–173.
200. Travis, A. M., and Woolsey, C. N. (1956): Motor performance of monkeys after bilateral partial and total cerebral decortications. *Am. J. Phys. Med.*, 35:273–310.
201. Vogt, C., and Vogt, O. (1919): Allegemeinere ergebnisse unserer hirnforschung. *J. Psychol. Neurol.*, 25:279–461.
202. Vogt, C., and Vogt, O. (1926): Die vergleichend-architectonische und die vergleichend-reizphysiologische felderung der grosshirnrinde unter besonderer berucksichtigung der menschlichen. *Naturwissenschaften*, 14:1190–1194.
203. Von Bonin, G., and Bailey, P. (1947): *The Neocortex of Macaca Mulatta*. Univ. of Illinois Press, Urbana.
204. Wagman, I. H. (1961): Eye movements induced by electric stimulation of cerebrum in monkeys and their relationship to bodily movements. In: *The Oculomotor System*, edited by M. B. Bender, pp. 18–39. Harper & Row, New York.
205. Wagman, I. H., Krieger, H. P., Papatheodorou, C. A., and Bender, M. B. (1961): Eye movements elicited by surface and depth stimulation of the frontal lobe of *Macaca mulatta. J. Comp. Neurol.*, 117:179–188.
206. Walker, A. E. (1940): a cytoarchitectural study of the prefrontal areas of the macaque monkey. *J. Comp. Neurol.*, 73:59–86.
207. Walker, A. E. (1949): Afferent connections. In: *The Precentral Motor Cortex*, edited by P. Bucy, pp. 111–132. Univ. of Illinois Press, Urbana.
208. Watson, R. T., Heilman, K. M., Cauthen, J. C., and King, F. A. (1973): Neglect after cingulectomy. *Neurology (Minneapolis)*, 23:1003–1007.
209. Welck, K., and Stutteville, P. (1958): Experimental production of unilateral neglect in monkeys. *Brain*, 81:341–347.
210. Whitteridge, D. (1960): Central control of eye movements. In: *Handbook of Physiology*, Sec. 1 (Neurophysiology), edited by J. Field, pp. 1089–1109. American Physiological Society, Washington, D.C.
211. Wiesendanger, M., Seguin, J. J., and Kunzle, H. (1973): The supplementary motor area—A control system for posture? In: Control of Posture and Locomotion, edited by R. B. Stein, K. G. Pearson, R. S. Smith, and J. B. Redford, pp. 331–346. Plenum Press, New York.
212. Woolsey, C. N. (1958): Organization of somatic sensory and motor areas of the cerebral cortex. In: *Biological and Biochemical Bases of Behavior*, edited by H. F. Harlow and C. N. Woolsey, pp. 63–81. Univ. of Wisconsin Press, Madison.
213. Woolsey, C. N., Settlage, P. H., Meyer, D. R., Sencer, W., Hamuy, T. P., and Travis, A. M. (1950): Patterns of localization in precentral and "supplementary" motor areas and their relation to a concept of premotor area. *Res. Publ. Assoc. Nerv. Ment. Dis.*, 30:238–264.
214. Wurtz, R., and Goldberg, M. E. (1972): Activity of superior colliculus in behaving monkey. III. Cells discharging before eye movements. *J. Neurophysiol.*, 35:575–586.
215. Wurtz, R. H., and Goldberg, M. E. (1972): Activity of superior colliculus in behaving monkey. IV. Effects of lesions on eye movements. *J. Neurophysiol.*, 35:587–598.
216. Wurtz, R. H., and Mohler, C. W. (1976): Enhancement of visual responses in monkey striate cortex and frontal eye fields. *J. Neurophysiol.*, 39:766–772.
217. Yin, T. C. T., and Mountcastle, V. B. (1977): Visual input to visuomotor mechanisms of the monkey's parietal lobe. *Science*, 197:1381–1383.
218. Yin, T. C. T., and Mountcastle, V. B. (1978): Mechanisms of neural integration in the parietal lobe for visual attention. *Fed. Proc.*, 37:2251–2257.

Posture and Movement, edited by
R. E. Talbott and D. R. Humphrey.
Raven Press, New York © 1979.

The Central Representation of Learned Motor Patterns

*Carlo A. Terzuolo and **Paolo Viviani

**Laboratory of Neurophysiology, University of Minnesota, Minneapolis, Minnesota 55455; and **Laboratoire du Physiologie du Travail du CNRS, Departement de Physiologie Neurosensorielle, 75005 Paris, France*

This chapter addresses some of the problems pertaining to learned, complex motor sequences and their central representation. We shall begin by asking first: What is a motor pattern? We shall then present and discuss some results pertaining to a particular type of skillful, learned motor pattern. Finally, we shall consider briefly a few points about the role of sensory inputs for the implementation of these patterns.

It seems to us that the term *pattern* has been most frequently used in neurophysiology to indicate a sequence of events occurring in space and/or time (most frequently both), which is taken to be typical of a whole set of such sequences. However, this meaning is listed only fourth in the Webster's dictionary, where it is explained as follows: "a representative instance; an example, a specimen offered as a sample of the whole." If this meaning were to be accepted as applicable to motor physiology, one would then expect that each instance of a given type of motor sequence should be the same as any other instance of such sequence. In particular, one would have to restrict the use of the term to those motor sequences whose distinguishable components occur, in all instances, at specific and constant times. However, several types of motor sequences commonly called "patterns" can occur at different speeds. Moreover, other attributes may also differ, for instance the amplitude of the motion. Walking is an adequate example. Although it can be accomplished by using a variety of time and space scales, nevertheless the structure of the motor *sequence* remains essentially the same. The use of the term *pattern* to indicate this type of motor sequence (with a typical and constant structure) is in keeping with other meanings attributed to the term by the Webster's dictionary. These meanings are described as follows: "a fully realized form; original or model proposed for imitation; something regarded as a normative example to be copied; archetype," and "something designed or used as a model, outline or plan." In keeping with these meanings, and on the basis of the data to be presented below, we suggest that, in the context of motor events, the term *pattern* be defined as follows: "model, or template, for a spatial and temporal distribution of motor commands." This definition poses no constraints on the absolute time as well as other characteristics of the movement. However, it implies that *some type of invariant relational characteristic among identifiable components of the motor sequence should exist.* Only if this condition is satisfied can each instance of the motor sequence qualify as belonging to the same pattern, the invariant relational characteristic itself being the pattern, or template, of the sequence. It is this template which, at a purely intuitive level, was previously called the structure of the motor sequence.

A simple example should clarify the point in question. We are all familiar with the fact that a person can write small or large, fast or slow, and even using different sets of muscles, but the writing can still be recognized as belonging to him. This phenomenon of motor equivalence (3,8,9) strongly suggests that all these different sequences derive from a common template.

The question then arises: What is the invariant relational characteristic(s) among all instances of the same pattern? By answering this question one would know of what the template consists. This of course not in terms of neuronal activities, occurring here and there in the central nervous system, but rather in terms of its logic. It should, however, be obvious that this logic is immanent to the template, that is, to the central representation of the pattern, in the sense that it results from the anatomofunctional relationships among participating neurons.

We shall now present some data which indicate that for certain skillful motor sequences, involving the fingers of both hands, the template consists of the set of ratios between the times of occurrence of successive components (or events) of the motor sequence. Indeed these ratios are the only invariant relational characteristic among segments of the motor sequence, in all instances.

The learned, skillful motor sequences we have studied are those of the fingers during typing, as done by professionals. These sequences are composed of discrete events, namely the pressing of the keys by the fingers. The time of occurrence of each of these events can be accurately and simply measured by using the electric typewriter of a digital computer. Actually, typing as such does not need to take place, since the output commands from the computer which displace the carriage and activate the ball of the typewriter can be suppressed by appropriate programming, while the computer still recognizes the letters corresponding to the keys which are depressed. These data are then retrieved together with the time (in milliseconds) between successive key pressings. The same motor skill has been the object of studies by other authors (2,5).

Figures 1 and 2 present the main findings. They came about rather unexpectedly while we were attempting to analyze the influence of linguistic information on the stated task (13). Figure 1 is divided into 3 parts: the raw data (A); the same data after some processing (B); the pattern (C). The word, *TROUBLE,* is taken from a library of about 250 English words we have studied so far.[1]

---

[1] About 200 of these words are part of meaningful sentences designed to have specific syntactic structures (4). The same words were also included in a text in which they were placed at random, to remove all syntactic constraints. The other words in our library were part of special texts designed for specific purposes, such as the study of the role of sensory inputs in the implementation of the pattern (see later) and the role of mechanical, syllabic and phonetic factors.

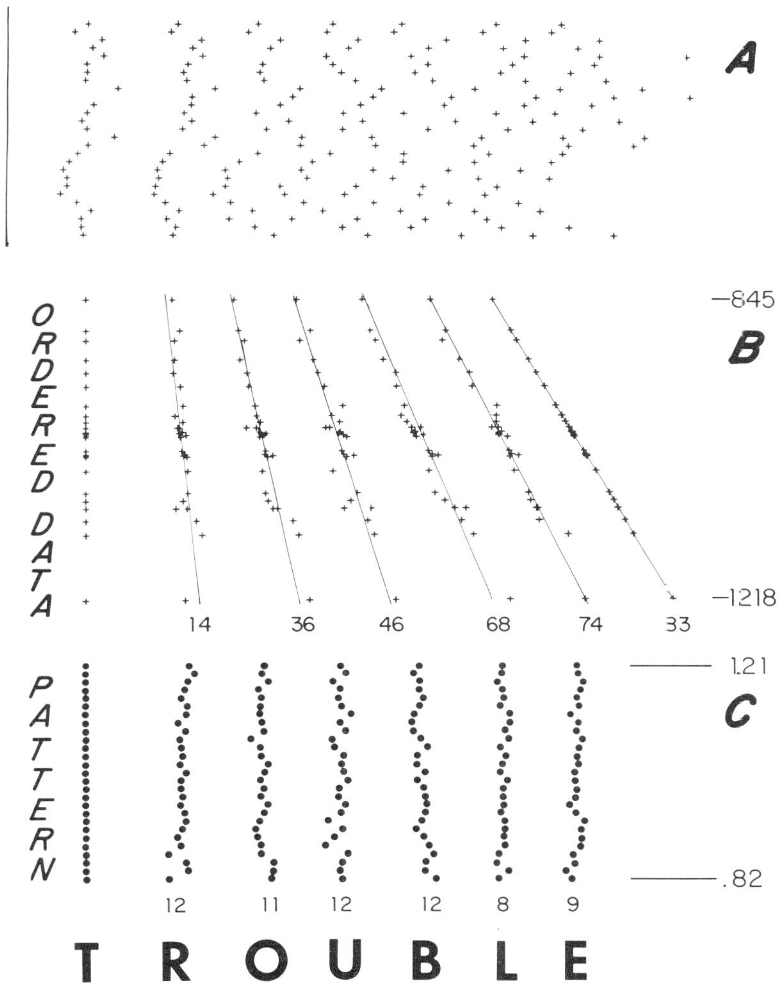

FIG. 1. In **A** and **B** the symbols indicate the time at which each of the letters in the word *TROUBLE* was typed in 27 trials. The bar in **A** (time zero) corresponds to the space preceding the word. In **B** the duration (in msec) is given for the shortest (845) and longest (1,218) instance, together with the standard deviation for the average time across instances. In **C** the numbers at the right give the amount by which the shortest instance was expanded and the longest one was contracted with respect to the average. See text for further explanation.

Four professional typists (3 female, 1 male) with typing speeds above 80 words/min participated in the experiment.

Part A of Fig. 1 presents 27 instances of the word as typed by the same subject at different times and in different contexts. The symbols mark the time at which each of the keys corresponding to the letters included in the word was depressed. The time interval space −T is also included. Hardly any sense

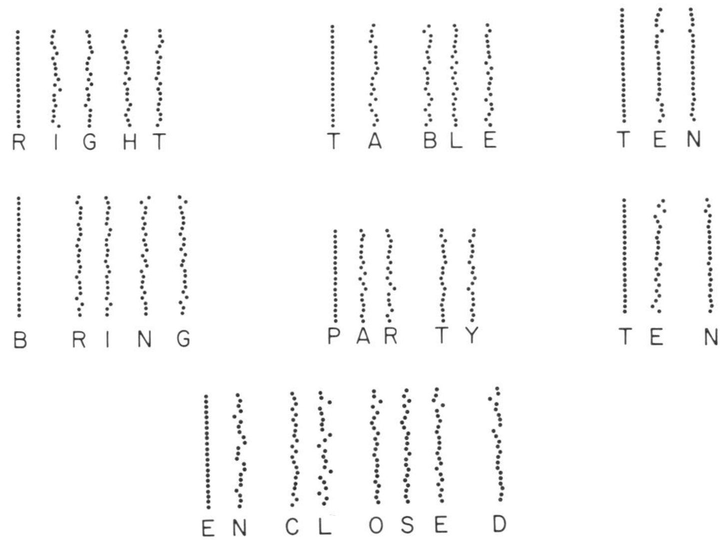

FIG. 2. Examples of patterns in two subjects.

can be made of the data as shown in A; however, after some rearrangment the same data reveal that an invariant relational characteristic is present in all instances, whereby they all belong to the same pattern. This is shown in B. Here the time intervals corresponding to space —T were removed, time zero now being that at which the key for the letter T was depressed. Moreover, each instance has been placed along the vertical axis according to the time used to type the word, the time scale being the same for both vertical and horizontal axes.

One can now see that the times at which each key is depressed fall, in all instances, approximately along lines which, if extended, converge to a single point. Thus the ratio between the times of occurrence of each pair of letters is constant irrespective of the total time used in typing the word. Differences in total time as large as 25% (and for some words up to 60%) were observed in all subjects, even for very short words.[2]

Part C of Fig. 1 shows the way we chose to identify the pattern for each of the words included in our library. For the set of data pertinent to each word, the computer was programmed to calculate first the average value (and standard deviation) of the time intervals between key pressings, and then to expand or contract each instance to approximate these average values while minimizing the variance among instances. Note that in the process of contracting or expanding each instance, the ratio between successive time intervals is kept the same as that measured experimentally.

[2] Some examples from 2 subjects are: HIS (63%); OLD (60%); NOT (50%); BOY (60%): AND (60%).

Only for the sake of convenience the time scale in C of Fig 1 is the same as that used in A and B of the same figure. Indeed, after contraction, absolute time is no longer a relevant parameter. Actually we conclude that within the template, that is, in the central representation of the motor sequence, absolute time is a free parameter and it is independently controlled.

One more point: The standard deviation across instances for the time of occurrence of each event of the pattern is very small in almost every word and in all subjects. Such values (given in milliseconds at the bottom of Fig. 1C) are typically between 0.5 and 2% of the total average time used in typing the word. Note that these values include the variability of all events occurring between the command provided by the template and the time at which the letter is recognized by the computer. Therefore, the activities and the structures leading to the motor commands must be credited with an even greater accuracy. Moreover, it was a constant finding, in all subjects, that the standard deviation for each event of the sequence does not increase with the rank order of the event. This fact excludes that successive events of the sequence are serially arranged within the template or, if so, they must be highly correlated. In all other cases the standard deviation would necessarily increase. Because of this conclusion, several plausible classes of models for the template are obviously inappropriate. Also, one may note that Bernstein (3) had long ago touched upon the same problem for other types of motor sequences. In particular he noted that during the execution of a goal-oriented motor task, the error is not larger at the most distal segment of the arm (actually at the end of a hammer striking a nail) than at the shoulder; actually the error is less. The implication of this conclusion has been that a "plan of action" exists within the central nervous system (1,11,12) for goal-oriented tasks.

The reader might have noted that the ratios belonging to the template for the word *TROUBLE* differ very little among themselves. This is the reason why this word was chosen, our purpose being to demonstrate that the "patternability" of the motor sequence (by professional typists) is not contingent upon specific and different values for the ratios. Indeed, in the examples of patterns provided in Fig. 2, these ratios vary widely within each template. This is the most common case we encountered. Even though there were differences among the four professional typists who participated in the experiments, a regular spacing between letters was the exception rather than the rule. Consequently in each subject *the pattern for a very large number of words is specific and characteristic of the word.*

It can also be noted in Fig. 2 that the template for a given word can be different in different subjects. What is important to stress is the fact that mechanical factors alone (that is, the position of the keys) cannot account for the structure of the pattern. Indeed, for many words there are large discrepancies between the pattern and the structure of the sequence predicted by collating the average time intervals between each couple of letters, as obtained from a very large sample. Actually other factors have been identified which are of a linguistic nature. This subject is considered elsewhere (13), being outside the scope of this paper. Central to it instead is the following observation: The pattern is the same whether the words are typed from a text which is read by the typist or they are typed from memory. This fact is here taken to indicate that the template and the motor engram are, in our case, but one and the same thing. If so, the simplest conclusion is that the motor sequences are engrammed in the anatomofunctional relationships which have been established as a result of learning. Thus terms such as "programmed" or "preprogrammed" motor activities are obviously inappropriate to express the operation of the template. Finally, we wish to suggest, on the basis of the literature quoted above, that no specification is present in the template as to the group of muscles to be used for its implementation. Rather the commands provided by the template can be switched to different sets of muscles, thus making the same motor engram available for different or novel motor tasks.

The last subject to be briefly considered is the role of sensory inputs in the implementation of the template. It can be contended that the decision process to initiate a learned motor task sets up an internal state encompassing both the goal to be achieved and what the subject expects and does not expect to experience while performing the task. This internal state would dictate the use to be made of different sensory inputs during the execution of the task.

If so, questions about the teleonomic significance of the availability of such inputs can be assessed only in the context of a well-defined goal. In our case this requirement is clearly met.

Several types of experiments can be performed to determine the role of exteroceptive (touch and pressure) and proprioceptive (joint and muscle receptors) inputs in the implementation of the motor patterns we are interested in. First of all, the effects of changes in load can be studied by applying weights either to all fingers, to the fingers of only one hand, or to one finger only.[3] Since these experiments cannot be described here in detail, it suffices to say that the implementation of the pattern is not affected, within reasonably large limits, although the absolute time can vary. In brief, the results indicate that the specification of the force required to implement the task is not part of the template; rather it can be independently adjusted to cope with peripheral constraints.

The respective contribution of extero- and proprioceptive inputs, in relation to the task, is further defined by the results of another series of experiments. When all the keys which are acted upon by the fingers of the left hand[4] are permanently blocked, the subjects quickly adapt to the situation. Indeed, the implementation of the template is not affected (compare Fig. 3A and 3B). Instead, if the subject is asked to perform the movement as usual but without touching the keys, the execution of the pattern is interrupted at the first letter requiring the use of the left hand (Fig. 3C). No subject was able to adapt to this situation.

[3] We used rings carrying lead weights shaped so as not to interfere with the motion of the finger. They were placed on the second phalange, their effectiveness being enhanced by this distal position. Weights up to three-fold that of the fingers were used.

[4] The choice of the left hand is dictated by the fact that spacing is done by the right thumb.

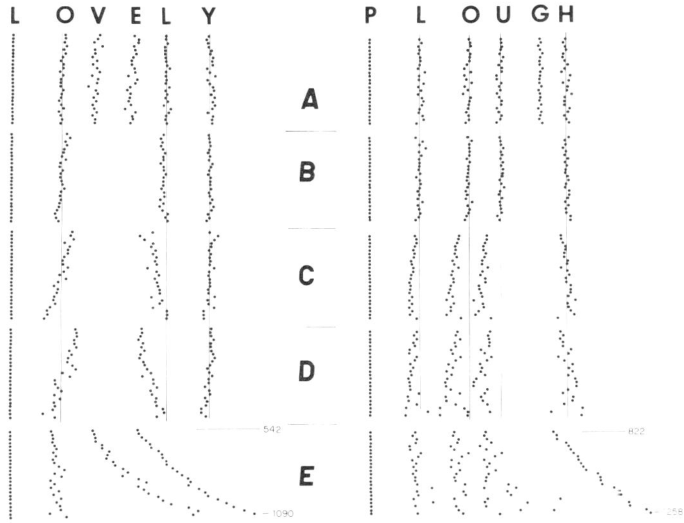

FIG. 3. The two words are from 2 subjects. In *LOVELY* the letters *V* and *E* are typed by the left hand; in *PLOUGH* only *G* is typed by this hand. In part **A** the normal patterns are shown. In **B** all keys corresponding to the left hand were blocked. However, the ratios among remaining intervals is clearly unaffected. In **C** the subjects were asked to move the fingers above the keyboard, without touching the keys. In **D** they were asked to keep the left hand closed, therefore suppressing the motor commands to the fingers. Because of data processing (see text) even those portions of the patterns correctly implemented (*LO* in *LOVELY* and *PLOU* in *PLOUGH*) appear distorted. However, the same data as in **D** when shown in real time and before processing **(E)** prove that the disruption introduced by the experimental conditions only begins with the first letter requiring the use of the left hand. Durations for the shortest and longest instances are given in msec.

This indicates that the motor commands themselves and the proprioceptive inputs generated by the resulting movements are not sufficient to ensure the implementation of the template, the exteroceptive inputs generated by touching the key being obviously necessary.[5] Instead, given the knowledge of the actual experimental conditions, the depression of the key does not contribute necessary inputs. Note that our conclusion is reinforced by the fact that if the subject is asked to suppress intentionally the commands to the left hand, no further disruption is observed (Fig. 3D). In this case all inputs are obviously eliminated.

Experiments can also be performed to assess the effect of violating expected covariances among inputs which, as suggested above, are specified as part of the internal state. For instance, a single key can be blocked unexpectedly[6] while the subject is typing a given word. This often, but *not* always, arrests the imple-

[5] One cannot rule out the presence of proprioceptive inputs resulting from the reactive forces consequent to the impact of the fingers upon the key.

[6] This is obtained by programming the computer in such a way that a key is blocked as soon as the last of a group of four specified, consecutive characters (including space) has been typed for each of the lines in which the text is subdivided. The duration of the block is automatically regulated in relation to the average typing speed, as measured over the initial 150 time intervals.

mentation of the pattern. Moreover, the modalities of the interruption are not constant.

We are currently using this approach to establish whether the templates of certain words are assembled starting from subunits shared by several words. If this can be proved, it will remain to be seen if such assembling is enacted with the assistance of sensory inputs. Also, we would like to learn whether the subunits are related to the syllabic segmentation of the word, as seen long ago in studies on Morse telegraphy (6,7). More generally we are pursuing the hypothesis that in the execution and learning of skillful motor tasks, a certain correspondence exists between the units of motor actions and those segments of sensory data which, in the context of the task, represent a self-contained perceptual event (6,7). In our case such segmentation of the relevant inputs, as provided by the text, would be determined by the linguistic constraints. Since these are relatively well understood (10), one can hope to test the above hypothesis by using the motor task described here.

Independently of the outcome of this work, the results summarized here already define some of the features of the central representation of the motor sequences we have studied. Although no specific model incorporating these features will here be considered, it suffices to say that in order to implement each ratio of the template, two neural elements are adequate. Consequently, one does not need to worry about the storage capacity by the brain of motor engrams related to language, even in the most extreme case in which each word was represented as a unit. However, we believe that it is more likely that some assembling of subunits does actually occur. This possibility does not contradict the uniqueness of the patterns shown here. In fact, interactions are likely to result as a consequence of the assembling process, much in the same way as, in handwriting, individual letters are modified to adjust to the preceding and following letters.

The points we have just made stress the difficulty of eventually obtaining direct experimental evidence about the template at the neuronal level, even in the case of other and simpler types of motor patterns not related to language and for which experimental animals can be used. This conclusion reinforces our belief that in order to define the logic used by the nervous system in organizing and controlling movements, deductions and modeling based upon measurements of the appropriate external parameters may well turn out to be the most valuable tools. Moreover, by using human subjects it is easier to define unambiguously the goal of the task and therefore the set of pertinent sensory information in their proper attuning.

ACKNOWLEDGMENTS

This work was supported by U.S. Public Health Service Grant NS-2567. Use of the IBM 1800 computer was made possible by a grant from the Air Force Office of Scientific Research, AFSC (Grant AFOSR-1221).

REFERENCES

1. Bartlett, F. C. (1932): *Remembering.* Cambridge University Press, Cambridge.
2. Bernstein, N., and Popova, T. S. (1929): Untersuchung der Biodynamik des Klavieranschlags. *Arbeitsphysiologie,* 1:5.
3. Bernstein, N. (1967): *The Coordination and Regulation of Movements.* Pergamon Press, New York, Oxford.
4. Bever, T. G., Lackner, J. R., and Kirk, R. (1969): The underlying structures of sentences are the primary units of immediate speech processing. *Percept. Psychophys.,* 5:225–234.
5. Book, W. F. (1925): *The Psychology of Skill.* Gregg Publishing Company, New York.
6. Bryan, W. L., and Harter, N. (1897): Studies in the physiology and psychology of the telegraphic language. *Psychol. Rev.,* 4:27–53.
7. Bryan, W. L., and Harter, N. (1899): Studies on the telegraphic language. The acquisition of a hierarchy of habits. *Psychol. Rev.,* 6:345–375.
8. Hebb, D. O. (1949): *The Organization of Behavior.* John Wiley & Sons, Inc., New York.
9. Lashley, K. S. (1930): Basic neuronal mechanisms in behavior. *Psychol. Rev.,* 37:1–24.
10. Lyons, J. (1968): *Introduction to Theoretical Linguistics.* Cambridge University Press, Cambridge.
11. Schmidt, R. A. (1975): A schema theory of discrete motor skill learning. *Psychol. Rev.,* 82:225–260.
12. Turvey, M. T. (1977): Preliminaries to a theory of action with reference to vision. In *Perceiving, Acting and Knowing,* edited by R. Shaw and L. Bransford, pp. 211–265. Erlbaum, Hillsdale, New Jersey.
13. Viviani, P., and Terzuolo, C. A. (1978): Engrammes moteurs dans le traitement de l'information linguistique In: *Préparation dans le Système Nerveux et Processus Psychologiques,* edited by J. Requin *(in press).*

Posture and Movement, edited by
R. E. Talbott and D. R Humphrey.
Raven Press, New York © 1979.

Immediate and Sustained Influences of Visual Olivocerebellar Activity on Eye Movement

Neal H. Barmack[1]

Department of Ophthalmology, Neurological Sciences Institute, Good Samaritan Hospital and Medical Center, Portland, Oregon 97210

INTRODUCTION

The cerebellum has provided a focal point for the study of posture and movement for over 150 years. Its role in the immediate regulation and long-term modification of postural reflexes has been documented clearly (24,49). Nevertheless, the mechanisms by which the cerebellum fulfills this regulatory and modulatory role are not well understood; nor do we understand how the anatomical organization of the cerebellum makes this structure particularly useful for the control of posture. However, one assumes that the well-documented relative homogeneity in the structure of the cerebellum places a constraint on the variety of functions which the cerebellum can perform, and that a systematic analysis of a specific part of the cerebellum could reveal general functional principles which would be descriptive of the entire structure. In this chapter, we will examine how the activity in one specific afferent pathway to the cerebellum, the visual climbing fiber pathway, may be related to the immediate and sustained modulation of a particular motor output, horizontal eye movement. We will consider how the microphysiological organization of this climbing fiber pathway renders it useful for reflex modification, and how natural sensory stimulation evokes activity in this pathway. In addition, we will consider the effects of selective artificial stimulation and disruption of this pathway on oculomotor performance. Finally, we will examine the effects of prolonged stimulation of the visual olivocerebellar pathway on oculomotor performance.

Review of Cerebellar Anatomy

There are two classes of afferent pathways to the cerebellar cortex: the mossy fibers and climbing fibers. Climbing fibers from the inferior olive cross the brainstem and enter the cerebellum primarily through the inferior cerebellar peduncle,

[1] It is a pleasure to dedicate this chapter to Professor J. M. Brookhart. His own research on motor control is but one example of his ability to employ a variety of experimental techniques in seeking answers to deceptively simple questions concerning the functions of the central nervous system at the cellular and behavioral levels.

making extensive synaptic contacts on the dendrites of Purkinje cells (22,23, 42,61,76). Mossy fibers have numerous anatomical origins and synapse upon cerebellar granule cells whose axons, the parallel fibers, comprise a second input to the dendrites of Purkinje cells (76). Physiologically, the dual afferent organization of the cerebellum is reflected by the different types of Purkinje cell responses evoked by activity in the climbing fiber and mossy fiber pathways. The characteristic multiple spiked "climbing fiber responses" (CFRs) or "complex spikes" evoked in cerebellar Purkinje cells at relatively low discharge rates (1 to 8 imp/sec) can easily be distinguished from the unitary "simple spikes" (SSs) which have a shorter duration, a higher spontaneous frequency (10 to 80 imp/sec), and are evoked by activity in the mossy fiber–granule cell afferent system (4,26,33,36,91). Another fundamental distinction between these two afferent systems is the difference in the divergence of their cerebellar projections. In the adult cat, Purkinje cells appear to receive innervation from only one climbing fiber. At present it is assumed, but without substantive anatomical or physiological proof, that every Purkinje cell is innervated by a climbing fiber. Even if this assumption is incorrect, the total number of Purkinje cells exceeds the total number of climbing fibers by almost an order of magnitude; therefore, each climbing fiber must diverge to innervate several Purkinje cells (27,73). This modest degree of divergence should be contrasted with the feline mossy fiber–granule cell system, in which a single mossy fiber may contact more than 600 granule cells (28). Since there are 900 to 1,500 granule cells for each Purkinje cell (30,60,77), and since each granule cell may be contacted by about 5 mossy fibers, it would appear that there are roughly 200 granule cells for each mossy fiber or about 5 mossy fibers for each Purkinje cell. Therefore, the mossy fiber system contains 50 times more afferent fibers than the climbing fiber system, and makes a divergent synaptic contact on granule cells within the cerebellum which is roughly 50 times greater than the divergence of climbing fibers onto Purkinje cells. This degree of divergence of the mossy fiber projection onto granule cells is mirrored in the large convergence of granule cells onto Purkinje cells.

An additional anatomic feature of the climbing fiber projection to the cerebellar cortex is its sagittal organization. The sagittal organization, which was first suggested by histochemical stains for the enzyme 5'-nucleotidase (80,81), has been subsequently confirmed by electrophysiological, anatomical, and histochemical investigations (8,10,19,29,39,68,75,79). This pattern of sagittal banding of the climbing fiber projection onto the cerebellar cortex appears to be preserved in the projection of the cerebellum onto the cerebellar nuclei (20,40).

Theories of Climbing Fiber Function

In spite of the relatively broad empirical base of anatomical and physiological information concerning the olivocerebellar system, there is only a limited understanding of its contribution to cerebellar function. Some theories concerning its function, at different levels of analysis, are considered briefly below.

Selection of Corticonuclear Strips of Activity

The sagittal CF projection onto corticonuclear strips may provide a more general framework for the control of movement. It has been hypothesized that these corticonuclear strips correspond to a proximal-distal organization of muscles. Activation of a particular strip would excite a set of muscles and reduce the excitability of neighboring corticonuclear strips (12,13). This theory suggests that CF activity may encode more gradual transitions in the use of groups of muscles for postural control, and that this general control is superimposed on more specific sensory information conveyed by the mossy fiber system. This "corticonuclear strip" proposal is explicit in its predictions about the influence of climbing fiber function on cerebellar activity. However, it is less specific about the consequences of this activity for the control of movement.

Heterosynaptic Modulation of Purkinje Cells by Climbing Fibers

It has been suggested that climbing fiber discharge alters the excitability of Purkinje cells to temporally correlated mossy fiber-granule cell inputs, and that this altered state of excitability persists in the absence of further CF input (1,69). In essence, these theories hypothesize a long-term heterosynaptic modification induced in the parallel fiber synapse as a consequence of coincident CF activation. This idea is at once both appealing and totally unsubstantiated by any direct experimental observations. However, there is indirect evidence which supports the idea that cerebellar climbing fibers may participate in some form of long-term modulation of motor control. An increase in climbing fiber discharge is evoked in monkeys performing a repetitive postural task when the monkey is confronted by a sudden change in one of the parameters of the task. This increase in climbing fiber discharge is accompanied by a decrease in "simple spike" discharge. As the new task becomes "routine," then the climbing fiber discharge returns to a "spontaneous" level (34). Both CFRs and SSs of floccular Purkinje cells are altered by vestibuloocular reflex stimulation in the light, and these alterations persist even when vestibuloocular stimulation is continued in the dark (33). These data indicate that climbing fibers *may* be involved in long-term cerebellar reflex modulation. However, the mechanism of *heterosynaptic modulation* is only one of several mechanisms by which climbing fiber activity could induce a change in the simple spike activity of Purkinje cells. The origin of this change in climbing fiber discharge, and its relationship to motor performance remain to be determined.

Trophic Regulation of Purkinje Cell Dendrite Formation by Climbing Fibers

It has also been suggested that "learning" might occur by means of a trophic effect of climbing fibers in inducing or maintaining dendritic spines on Purkinje cell dendrites. The evidence for such a trophic effect comes from developmental ultrastructural studies, in which it has been observed that climbing-fiber synapse formation was antecedent to dendritic spine formation in tertiary Purkinje cell dendrites (41). This implies that dendritic spine formation might be induced by the climbing fiber synapse. Furthermore, destruction of climbing fibers in neonatal kittens by electrolytic lesions of the inferior olive, or by transection of the inferior cerebellar peduncle, causes a loss of Purkinje cell dendritic spines. These findings imply that the CF synapse is necessary for both the *induction* and *maintenance* of Purkinje cell dendritic spines (41,56,58). However, surgical techniques for producing climbing fiber destruction invariably destroy mossy fiber projections to the cerebellum as well. Therefore, the observations of reduced numbers of dendritic spines following climbing fiber depletion by surgical techniques might be attributed to a more general deafferentation. Chemical destruction of the inferior olive by 3-acetylpyridine in the rat causes not a decrease but an actual increase in the number of dendritic spines (14,85). Thus, at present, the idea that the climbing fiber exerts a positive trophic influence of Purkinje cell spine formation is not well substantiated. Although there undoubtedly are trophic influences of both climbing fibers and parallel fibers on the development of the Purkinje-cell dendritic tree (15,85,86), it seems unreasonable to suppose that such trophic influences would be under the exclusive control of climbing fibers. Even assuming that degeneration of climbing fibers causes a change in the morphology of the Purkinje cell dendritic tree, it would seem arbitrary to conclude that the normal synaptic modulation of climbing fiber activity would exert a similar trophic influence.

Phasic Motor Generator

It has been suggested that CFs might exert a "phasic effect" on the cerebellar cortex, rather than a "tonic effect" (10,62). This notion derives from the demonstrated role of the inferior olive in the production of the tremor induced by the administration of harmaline (21,59,63,83). It has been argued that this 10/sec tremor is an experimental exaggeration of normal climbing fiber function, which is the phasic activation of skeletal muscle groups for the "sudden initiation or sudden cessation of a motor act" (63). This notion has received support from experiments in which forelimb proprioceptors have been activated by passive limb displacements (74,79). In these experiments, a single CFR was evoked for each cycle of muscle stretch or forepaw movement. The apparent lack of correlation of CF discharges with other parameters of forepaw movement led to the postulate that climbing fibers acted as "threshold devices" (74), or "event detectors" (79). However, the absence of tonic CFRs in these experiments might be attributed to the lack of appropriate stimulation. Even if one were to assume that climbing fibers encode transient or high frequency sensory events, it would not necessarily follow that this information, once encoded, would be used to evoke motor behavior in which these high frequencies of sensory origin would be reproduced. Rather than evoking transient postural adjustment, the low discharge rates of climbing fibers might be responsible for encoding low frequency sensory events and/or regulating low frequency postural adjustments. Regardless of the domains of frequencies of sensory stimulation which are effective in triggering CFRs and frequencies of postural modifications influenced by these CFRs, the "phasic motor generator" hypothesis suffers from a lack of precision in its formulation and consequently cannot be tested rigorously in a behavioral context.

Description of a Visual Olivocerebellar System

In order to study the contributions of the olivocerebellar pathway to the control of movement, it is necessary to relate the activity in this pathway to a movement which can be accurately measured, easily interpreted, and evoked without recourse to tedious behavioral training paradigms. The visual climbing fiber input to the cerebellar flocculus is particularly appealing to study because of its well-defined afferent connections to the visual system and its efferent connections via the vestibular nuclei to the oculomotor system. The major anatomical connections which are likely to mediate visual control of eye movement via the olivocerebellar pathway are illustrated in Fig. 1. Ganglion cells, encoding information which is destined for the ipsilateral cerebellar flocculus, send a predominantly contralateral projection to two pretectal nuclei: the nucleus of the optic tract and the dorsal terminal nucleus (9,35,55). In addition, two nuclei of the posterior accessory optic system, the medial terminal nucleus and the lateral terminal nucleus, also receive afferents predominantly from the contralateral retina (35,43,92). The nucleus of the optic tract, the dorsal terminal nucleus, and the lateral terminal nucleus, project ipsilaterally to the dorsal cap of Kooy, a subdivision of the medial accessory division of the inferior olive (48,57,71,72,88). The neurons of the nucleus of the optic tract are both *directionally selective* and *velocity-sensitive,* discharging preferentially in response to large, contour-rich, optokinetic stimuli moved in the posteroanterior direction with respect to the viewing eye at velocities of 1 to 10 deg/sec (18). Climbing fibers which originate from the dorsal cap re-cross the brainstem to synapse in the contralateral flocculus (2,47,65).

Physiological data indicate that the flocculus receives a visual mossy fiber projection (66,67). The flocculus also receives mossy fiber vestibular inputs from the ipsilateral and contralateral semicircular canals (17,78,82,87) and vestibular nuclei (82,94). Although at present it is not certain how the visual mossy fiber input reaches the flocculus, preliminary evidence suggests that the medial terminal nucleus might send a visual mossy fiber projection directly to the flocculus (43). However, it is also possible that the visual mossy fiber input may indirectly reflect visual climbing fiber–induced modulation of Purkinje cells Purkinje cells could modulate the activity of secondary vestibular cells whose axons would return to the cerebellum as "visually driven" mossy fibers. Neck proprioceptive inputs are conveyed to the flocculus by both climbing fibers and mossy fibers (93). The inhibitory Purkinje cell output of the flocculus projects onto a number of ipsilateral brainstem nuclei, but most profusely onto the ipsilateral medial

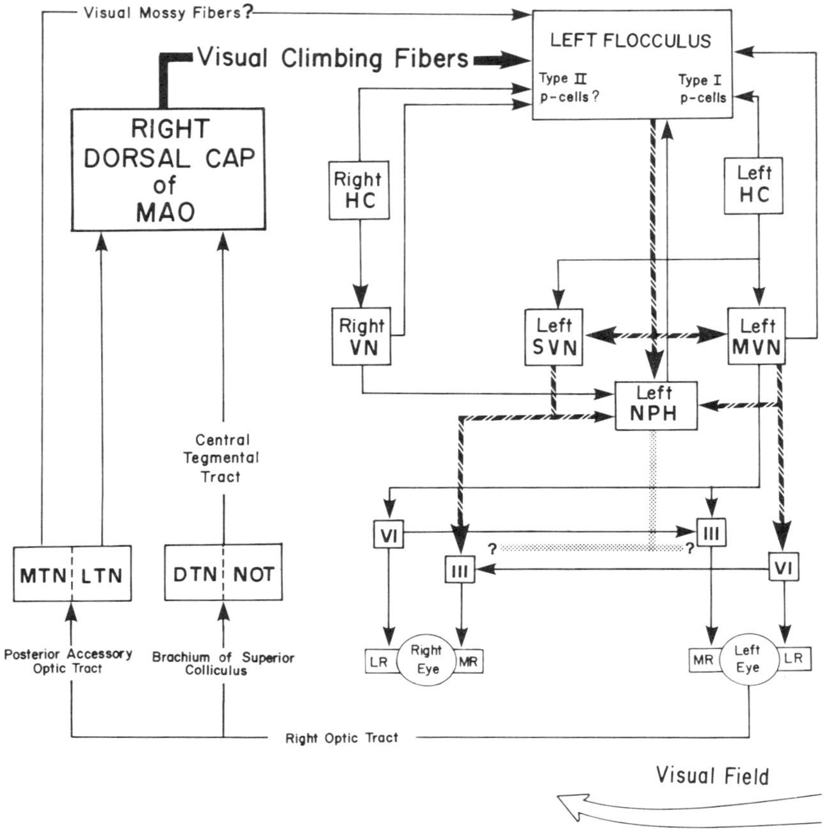

FIG. 1. Origins and central projections of the visual olivocerebellar pathway relevant for the control of horizontal eye movements. Abbreviations: MTN, medial terminal nucleus; LTN, lateral terminal nucleus; DTN, dorsal terminal nucleus; NOT, nucleus of the optic tract; MAO, medial accessory division of the inferior olive; HC, horizontal semicircular canal; SVN, superior vestibular nucleus; MVN, medial vestibular nucleus; NPH, nucleus praepositus hypoglossi; III, oculomotor nucleus; VI, abducens nucleus; LR, lateral rectus muscle; MR, medial rectus muscle.

vestibular nucleus and superior vestibular nucleus (3,16,40,53,95). These nuclei receive primary afferents from the ipsilateral horizontal semicircular canal (87). The rostral pole of the medial vestibular nucleus projects to the ipsilateral division of the oculomotor nucleus (III), which innervates ipsilateral medial rectus motoneurons (31,45,70). The medial vestibular nucleus also projects to both the ipsilateral and contralateral abducens nuclei (VI), which innervate the lateral rectus muscles (31,44,45,70). The superior vestibular nucleus projects to the ipsilateral and contralateral subnuclei of the oculomotor complex (31,45,70,90).

There is a crossed ascending excitatory projection from the abducens nucleus to the medial rectus subdivision of the contralateral oculomotor nucleus (6,

11,32,37,38). The functional polarity of these projections has been determined primarily by experiments which have combined the technique of electrical stimulation with intracellular recording and/or EMG recording from extraocular muscles (6,44,45,46,50,51,52,53). In addition, there are vestibular projections onto the nucleus praepositus hypoglossi (5,6), as well as reciprocal connections of the nucleus praepositus hypoglossi with the flocculus (2,5,95). The nucleus praepositus hypoglossi also projects onto oculomotor and abducens motoneurons (6,32,37,38,64), but the functional polarity and extent of these connections are not yet known.

Although the various anatomical components of the schematic visuo-olivo-cerebello-vestibulo-oculomotor circuit illustrated in Fig. 1 have been revealed by more than one experimental technique, it should be remembered that this schematic is in all probability incomplete. It is *not* inclusive of all cerebellar areas related to vision. It excludes both the nodulus and posterior vermis, and it ignores possible reciprocal connections of the vestibular nuclei with the cerebellar nuclei. In addition, there are extracerebellar pathways by which visual information can influence the discharge of secondary vestibular neurons (Barmack and Pettorossi, *unpublished observations*). Nevertheless, it would be useful to hold in abeyance our despair at the possible overwhelming complexity of the system, to assume that the schematic of Fig. 1 is a relatively complete description of how visual information is transduced by the olivocerebellar system, and to examine how the system might use this information to control eye movements. Posteroanterior movement of the visual field, as viewed by the left eye, would cause an increased discharge of climbing fibers which originate from the right dorsal cap of the inferior olive. If these climbing fibers, which synapse on Purkinje cells of the left flocculus, *increased* the net Purkinje cell discharge (CFRs + SSs), then the consequent increased inhibition of the left vestibular nuclei would cause a conjugate eye movement to the left. If activity of visual climbing fibers *decreased* the inhibitory output of floccular Purkinje cells, then the consequent disinhibition of the subjacent vestibular nuclei would cause the eyes to move conjugately to the right and would be functionally equivalent to directly exciting the left medial vestibular nucleus via the left horizontal semicircular canal by rotation of the head to the left.

EXPERIMENTAL APPROACH

In order to understand the contribution of the olivocerebellar system to the control of eye movement, the responses of climbing fibers have been examined at the level of the inferior olive in anesthetized and unanesthetized rabbits during vestibular and optokinetic stimulation. Rabbits under anesthesia were prepared for microelectrode recording by implantation of a head plug, which was fixed to a rigid support at the center of a biaxial rate table. The dorsal surface of the brainstem was surgically exposed at the level of the obex while the rabbit was anesthetized. In experiments on unanesthetized rabbits, surgical preparations were completed while the rabbit was temporarily anesthetized with halothane. Vestibular stimulation was delivered by oscillation of the rate table about the vertical axis, evoking the horizontal vestibuloocular reflex (HVOR) or about the longitudinal axis, evoking the vertical vestibuloocular reflex (VVOR). Optokinetic stimulation was provided by a projector-mirror galvanometer system which beamed a random-dot, contour-rich pattern onto a rear projection tangent screen, subtending a visual angle of 72 × 72 deg. The rabbit was aligned so that the tangent screen occupied the central visual field of one eye. In experiments on unanesthetized rabbits, the eye position of the eye which was not being optokinetically stimulated was monitored by an infrared-light-projection technique.

RESULTS

Trigger Features of Climbing Fibers Originating from the Dorsal Cap of the Inferior Olive. Flash-evoked Dorsal-Cap Field Potential

A large flash-evoked field potential was used as a basis for localizing a recording microelectrode in the dorsal cap of the inferior olive. The left eye was typically stimulated at a rate of .3 Hz as a microelectrode was advanced towards the right dorsal cap. At a depth of about 3,300 μ from the dorsal surface of the brainstem, a large, negative field potential with a 0.5 to 2.0 mV amplitude and a latency of 25 to 28 msec was recorded (Fig. 2). This field potential approached its maximal amplitude in the caudal portion of the dorsal cap, about 200 μ anterior to the caudal pole. Thus, the flash-evoked field potential serves as a convenient way of localizing a recording microelectrode. However, this field potential, which can be easily evoked in rabbits anesthetized with either sodium pentobarbital or chloralose-urethane, is severely attenuated in unanesthetized rabbits and in rabbits anesthetized with halothane.

Multiple Olivary-Unit Activity

Combined visual and vestibular stimulation were used to explore the sensory specificity of multiple-unit activity of the dorsal cap. While it is possible to isolate single olivary neurons, in practice this is not easily achieved because of the variable amplitude of climbing fiber action potentials and the relatively homogeneous trigger features of the population of dorsal cap neurons. Therefore, multiple-unit activity was routinely recorded and is illustrated in subsequent figures. This activity represents the conjoint activity of 2 to 8 dorsal cap neurons. The technique of multiple-unit recording has the potential liability of yielding a negative result when one records from a neuronal population having subsets of neurons which are oppositely modulated by a particular mode of stimulation. However, this logical possibility is in fact rather remote in practice. The benefit of the multiple-unit technique is its *lack* of sensitivity to the nuances in discharge characteristics of individual units. This lack of sensitivity makes a positive result

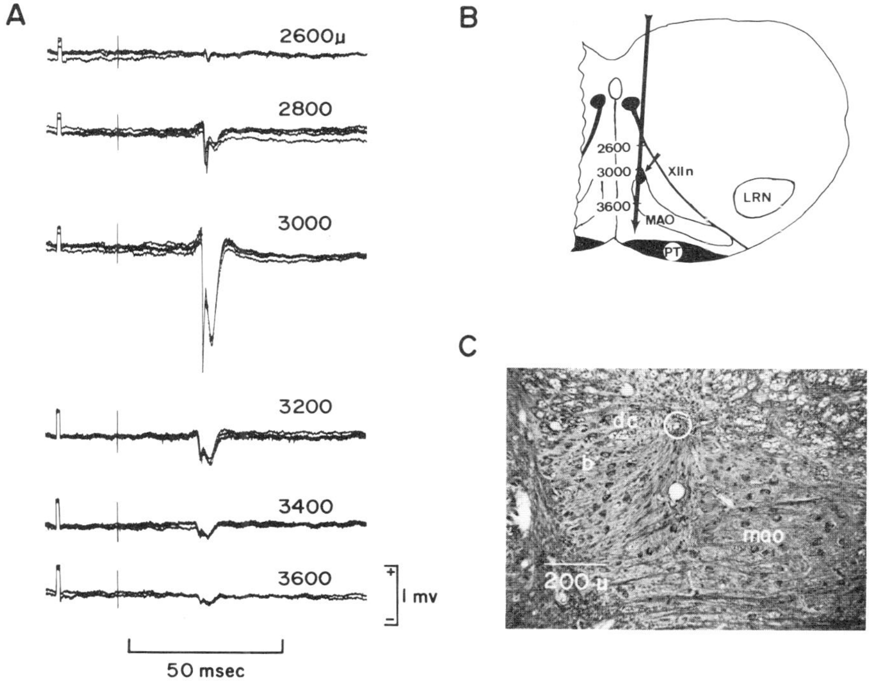

FIG. 2. Field potential profile evoked in the right dorsal cap by photic stimulation. The rabbit was anesthetized with chloralose-urethane, and a microelectrode was advanced through the region of the right dorsal cap **(B),** while the left eye was stimulated with a strobe light. Multiple traces were recorded at each depth **(A).** An electrode impedence measuring pulse and a stimulus artifact precede the evoked potential in each trace. A small lesion (20 μa, 30 sec) was made at the level of the maximum negative potential (3,000 μ) **(C).** In this experiment the animal was allowed to recover from the operation and was sacrificed 2 weeks later. The center of the lesion is indicated by the *open circle.* Glial processes radiate from the lesion. A capillary lies about 200 μ below the dorsal cap. Abbreviations are: dc, dorsal cap; b, β nucleus; mao, medial accessory olive; LRN, lateral reticular nucleus; XIIn, hypoglossal nerve; PT, pyramidal tract.

a more convincing description of the population than a sequential, idiosyncratic accumulation of single-unit recordings from the same population, in which attempts to infer the properties of the population would seem to be much more susceptible to technical problems and to sampling bias.

When an anesthetized rabbit was oscillated in the horizontal plane, with unobscured vision of the left eye in the dimly lit laboratory environment, the activity of neurons of the right dorsal cap was modulated. This activity was characterized by negative-polarity action potentials of low discharge frequency and variable amplitude. When the rate table was rotated towards the left, sweeping the visual field across the immobile left eye in the posteroanterior direction, the discharge frequency of dorsal cap neurons increased above the spontaneous level. When the rate table was rotated towards the right, moving the visual field in the anteroposterior direction with respect to the left eye, the discharge frequency of dorsal cap neurons fell below the spontaneous level (Fig. 3A). Movement of the table about the longitudinal axis (in the vertical plane) failed to modulate the activity of dorsal cap neurons, and movement of the table in the horizontal plane with vision of the left eye obscured also failed to modulate the discharge (Fig. 3B,C). These combinations of vestibular and visual stimulation revealed that dorsal cap activity was modulated primarily by horizontal visual stimulation at relatively low velocities.

The parameters of visual stimulation most effective in modulating dorsal cap activity were studied in isolation by presenting optokinetic targets at different velocities to anesthetized rabbits. Monocular, triangular-waveform, constant-velocity optokinetic stimulation confirmed the effectiveness of posteroanterior stimulation in increasing the discharge of dorsal cap neurons. Anteroposterior stimulation was less effective in *reducing* the discharge, presumably because the already relatively low spontaneous level of activity limited the depth of modulation that could be evoked by anteroposterior stimulus movement (Fig. 4). The *directional selectivity* of these dorsal cap neurons was maximal at stimulus

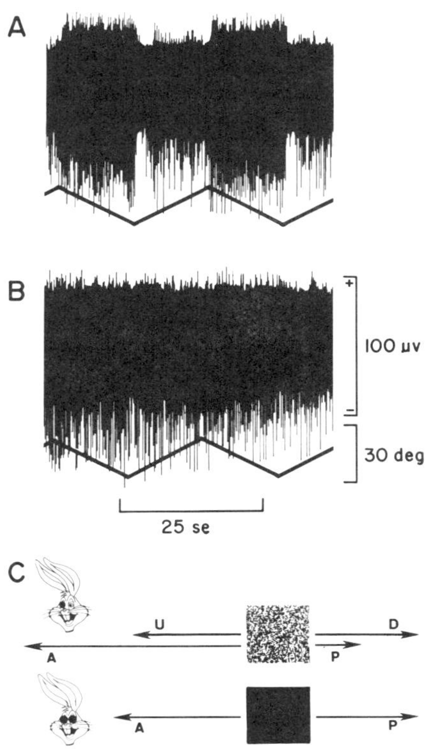

FIG. 3. The effect of vision on the multiple-unit activity of the right dorsal cap during vertical and horizontal VOR stimulation. **A:** The rabbit was oscillated horizontally and had unobscured vision of the left eye. **B:** The left eye was occluded, and the rabbit was tested with the same VOR stimulation. An upward displacement of the stimulus trace indicates an anteroposterior movement of the visual field referenced to the left eye. The multiple-unit activity was averaged for 5 half-cycles of VOR stimulation in the horizontal and vertical planes with the left eye unoccluded and in the horizontal plane with the left eye occluded. **C:** The *arrows* indicate the direction of *visual field movement* referenced to the left eye. The length of the arrows is proportional to the average discharge frequencies evoked by the indicated movements: A, anterior; P, posterior; D, down; U, up. The rabbit was anesthetized with chloralose and the VOR stimulation evoked no eye movements.

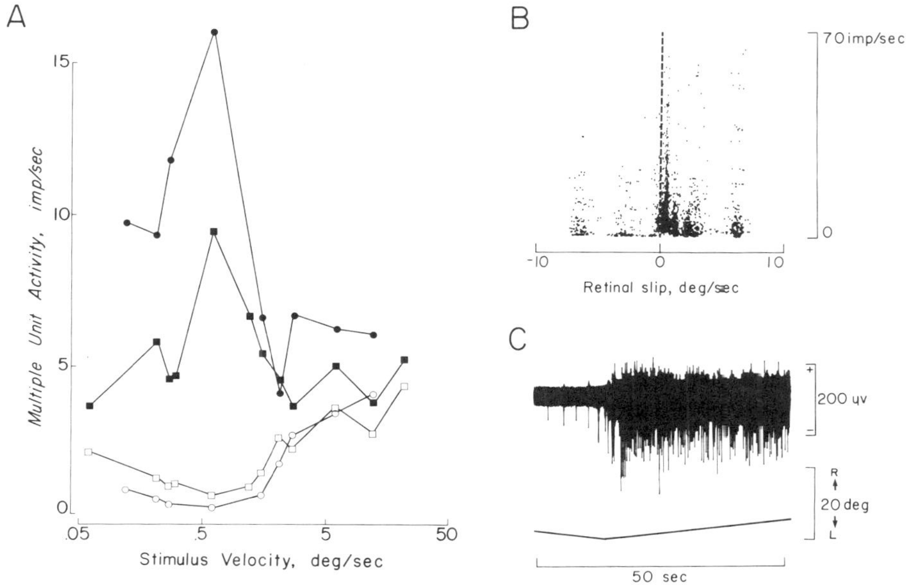

FIG. 4. Comparison of directional selectivity and velocity-sensitivity of unit activity in the right dorsal cap evoked by optokinetic stimulation of the left eye. **A:** Average multiple-unit activity is plotted as a function of optokinetic stimulus velocity for 2 separate experiments. The *circles* represent data obtained from a chloralose-urethane anesthetized rabbit, and the *squares* represent data from a sodium-pentobarbital anesthetized rabbit. The *filled symbols* represent stimulus movement in the posteroanterior direction, and the *open symbols* represent stimulus movement in the anteroposterior direction. **B:** Instantaneous rate of discharge for right dorsal cap unit activity evoked by optokinetic stimulation of the left eye. These data were obtained from the chloralose-urethane anesthetized rabbit whose average discharge rates are illustrated in **A.** Constant-velocity optokinetic stimuli were presented for equal durations at .28, .8, 2.0, 2.8, 3.7, and 7.4 deg/sec. **C:** Multiple-unit activity recorded from the right dorsal cap of the chloralose-urethane anesthetized rabbit at an optokinetic velocity of .22 deg/sec.

velocities of about 1 deg/sec. The multiple-unit activity evoked by optokinetic stimulation was dependent on the *acceleration* of the stimulus as well as its velocity. This was particularly evident when the optokinetic stimulus made the transition from the posteroanterior direction to the anteroposterior direction. Such a transition would temporarily silence dorsal cap activity, which would then slowly increase to a level of activity which was proportional to the anteroposterior stimulus velocity (Fig. 5A). The time course for the initiation of olivary discharge for posteroanterior optokinetic stimulation and for the cessation of discharge for anteroposterior movements was examined by superimposing several half-cycles of stimulation on a CRO (Fig. 5). This display makes it easier to distinguish individual-unit activity which is time-locked to the optokinetic stimulation. Dorsal cap neurons were recruited roughly in accordance with a "size principle." The units with the smallest action potentials were recruited at the shortest latencies (Fig. 5B). The earliest recruitment latencies were 40 to 45

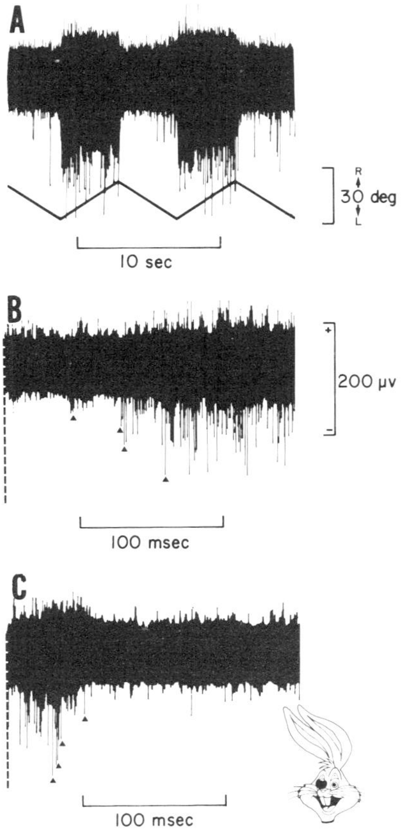

FIG. 5. Recruitment and latency of multiple-unit activity evoked by optokinetic stimulation. **A:** Multiple-unit activity was recorded in the right dorsal cap during constant-velocity, triangular-waveform, optokinetic stimulation of the left eye at .05 Hz ± 10 deg. Note that unit activity is suppressed immediately following the transition of the stimulus to the anteroposterior direction, but that some of the units resume discharging during continued anteroposterior movement. **B:** Five superimposed traces of multiple-unit activity were synchronized with the transition of a .05-Hz optokinetic stimulus to the posteroanterior direction. The *dashed line* indicates the time of transition. The *filled triangles* indicate the recruitment of individual units. The clustering of action potentials of the same amplitude indicates the variability in latency with which the same unit was recruited. **C:** Ten superimposed traces of multiple-unit activity were synchronized with the transition of a .3-Hz optokinetic stimulus to the anteroposterior direction. The *filled triangles* indicate the drop-out of individual units.

msec. Units were recruited over a wide range of latencies (40 to 180 msec). The initiation of optokinetic stimulus movement in the anteroposterior direction caused units with the largest action potentials to drop out first. The earliest drop-out latency was also 40 msec (Fig. 5C). The range of latencies over which dorsal cap neurons were silenced was narrower than the range over which they were recruited (40 to 120 msec).

The depth of optokinetic modulation of dorsal cap unit activity was greatest in the caudal region, where the field potential evoked by photic stimulation of the contralateral eye was also maximal. Because optokinetic stimulation was effective in modifying the activity of a variable number of neurons in the recorded population, depending on the location of the electrode within the dorsal cap, the directional selectivity of multiple-unit activity recorded in 8 separate experiments was normalized with respect to the optokinetic velocity at which the difference in average discharge frequency in the posteroanterior and anteroposterior directions was maximal (Fig. 6). These curves indicate that the peak direc-

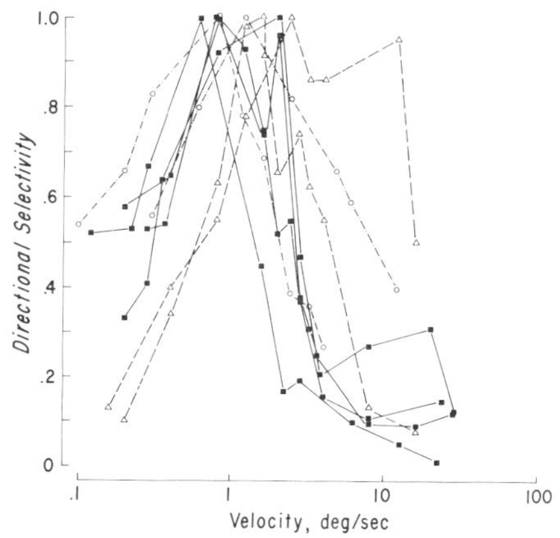

FIG. 6. Directional selectivity of multiple-unit activity recorded from the right dorsal cap of 8 rabbits. Directional selectivity was derived from the difference in average discharge frequency in the A-P and P-A directions, and normalized with respect to the velocity at which this difference was maximal.

tional selectivity coincided with optokinetic stimulus velocities of .5 to 1.5 deg/sec (7,84).

Optokinetically Evoked Dorsal Cap Multiple-Unit Activity in Unanesthetized Rabbits

In anesthetized rabbits, it was evident that a major parameter encoded by dorsal cap neurons was optokinetic stimulus velocity. However, of greater interest is how the dorsal cap encodes information in unanesthetized rabbits. In anesthetized rabbits, the optokinetic stimulus velocity was equivalent to the retinal slip velocity, because the anesthesia eliminated eye movements normally evoked by vestibular or visual stimulation. However, because of these reflexive eye movements in unanesthetized rabbits, optokinetic stimulus velocity would not necessarily be equivalent to retinal slip velocity. One way to eliminate the confounding influence of eye movements in unanesthetized animals is to immobilize the optokinetically stimulated eye. This immobilization was achieved surgically in two rabbits in which previously, under temporary anesthesia, the muscles of the optokinetically stimulated eye were cut. In one of the rabbits this operation was incomplete so that upon recovery from anesthesia there were some small, optokinetically evoked, residual eye movements. These small, residual eye movements displaced the velocity sensitivity of the multiple-unit activity recorded from the dorsal cap toward higher values than would have been obtained if the conditions of optokinetic stimulation had been "open loop." Nevertheless, the normalized velocity-sensitivity curves for these two rabbits indicated that, in unanesthetized rabbits, dorsal cap neurons were encoding retinal slip velocity (open triangles, Fig. 6). Dorsal cap multiple-unit activity was also recorded in unanesthetized rabbits with intact peripheral oculomotor systems. In these experiments it was possible to derive a measure of retinal slip velocity by electronically differentiating the difference between the optokinetic stimulus position and the position of the nonstimulated eye. During monocular optokinetic stimulation of the unanesthetized rabbit, the eyes followed the target better when it moved in the posteroanterior direction, rather than the anteroposterior direction. Hence the range of retinal slip velocities sampled during optokinetic stimulation of the unanesthetized rabbits was different for the anteroposterior and posteroanterior directions. Nevertheless, there was a clear relationship between the computed retinal slip velocity and the discharge of dorsal cap neurons. (Fig. 7A). However, since in this instance retinal slip velocity was correlated with a posteroanterior eye movement, it is possible that the dorsal cap discharge in the unanesthetized rabbit was in fact correlated with the eye movement or the "intention" to move the eyes in the posteroanterior direction. A clear experimental distinction between these alternatives was possible by comparing the discharge of dorsal cap neurons during HVOR stimulation in the dark (with no retinal slip) and in the light, with vision unobscured. During HVOR stimulation in

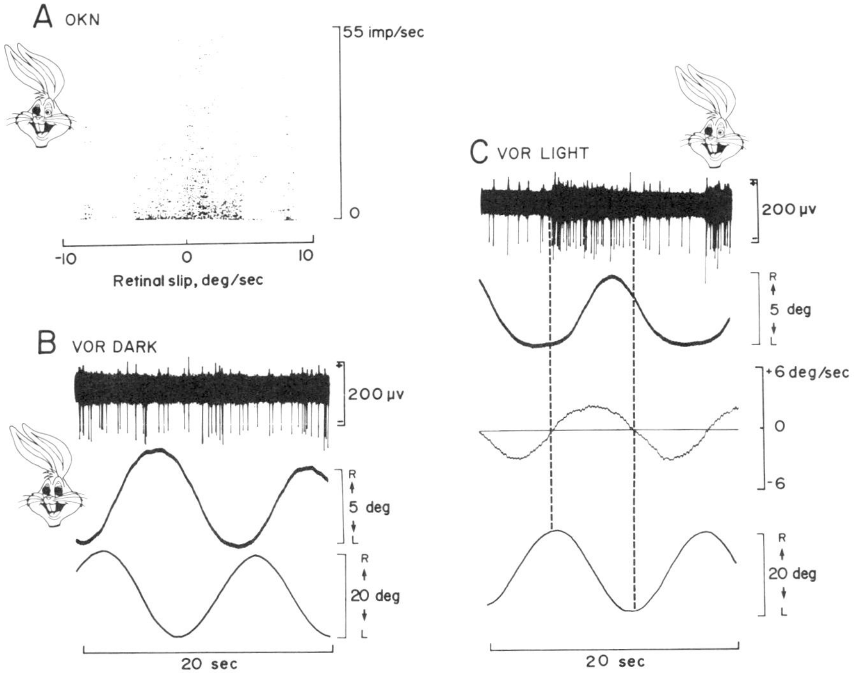

FIG. 7. Multiple-unit activity recorded in the right dorsal cap of an unanesthetized rabbit. **A:** The instantaneous discharge rate was plotted as a function of retinal slip velocity of the optokinetic stimulus relative to the left eye. **B:** The discharge rate was unmodified during horizontal VOR stimulation during which vision was occluded. The positon of the right eye and the table position are illustrated in the second and third traces. **C:** The discharge rate was recorded during horizontal VOR stimulation, as in **B,** but with left eye unoccluded. Eye position, retinal slip velocity, and table position are illustrated in the second, third, and fourth traces. Note that the olivary activity is modulated in phase with posteroanterior retinal slip velocity.

the dark, dorsal cap multiple-unit activity in the unanesthetized rabbit was related to neither the stimulus nor the eye movement (Fig. 7B). However, multiple-unit activity of the right dorsal cap was related to retinal slip velocity during optokinetic stimulation of the left eye (Fig. 7A) and to retinal slip velocity created by HVOR stimulation with the left eye unobscured (Fig. 7C).

Microstimulation of the Dorsal Cap

Although the above observations implicate the dorsal cap in the encoding of information concerning retinal slip velocity, they do not address the question of how the cerebellum uses this information. Using optokinetic stimulation, it is not possible to activate these visually sensitive dorsal cap cells without also activating other circuitry which might contribute to optokinetic reflexes. Nor can it be assumed that because both dorsal cap unit activity and eye movements are evoked by optokinetic stimulation, dorsal cap neurons contribute functionally to the production of these optokinetically evoked eye movements. In order to study how the activation of dorsal cap neurons could influence subsequent processing of visual information by the cerebellum, to which dorsal cap neurons project as CFs, and how this information might influence the control of eye movement, it would be necessary to selectively stimulate dorsal cap neurons. The technique of microstimulation at the level of the dorsal cap would permit this visual CF pathway to be activated selectively. Therefore, in subsequent experiments on unanesthetized rabbits, microstimulation was delivered at the source of the visual CFs in the dorsal cap, and the effects of this microstimulation on eye movements were examined.

Surgical exposure of the fourth ventricle from the obex to the posterior aspect of the cerebellar vermis by a midline incision and retraction of the overlying neck muscles was performed in rabbits which were anesthetized with halothane. The dura was incised, the cisterna magna was drained, and the exposed brainstem was bathed with a silicone fluid (38°C). The wound margin was infiltrated locally with xylocaine-epinephrine, and a topical cotton compress of xylocaine ointment ringed the incision. The rabbit was then permitted to recover from the halothane anesthesia for at least 1 hr before the experiment began. A tungsten microelectrode was advanced toward the right dorsal cap while the left eye of the rabbit was exposed to a triangular-waveform, constant-velocity optokinetic target. When the recorded electrical activity indicated that the microelectrode entered the dorsal cap, microstimulation was delivered through the recording microelectrode (20 to 50 μa, .1 to .2-msec pulses, 10 to 60 pulses/sec, 2 to 8-sec trains). If vision with one or both eyes was allowed, microstimulation of the dorsal cap evoked only small eye movements, with an initial phase toward the right (Fig. 8A,C,D). However, if the rabbit was placed in the dark, then microstimulation of the right dorsal cap consistently evoked low-velocity, conjugate eye movements toward the right (Fig. 8B). These eye movements had two distinctive features: (a) Their velocity increased during the period of stimula-

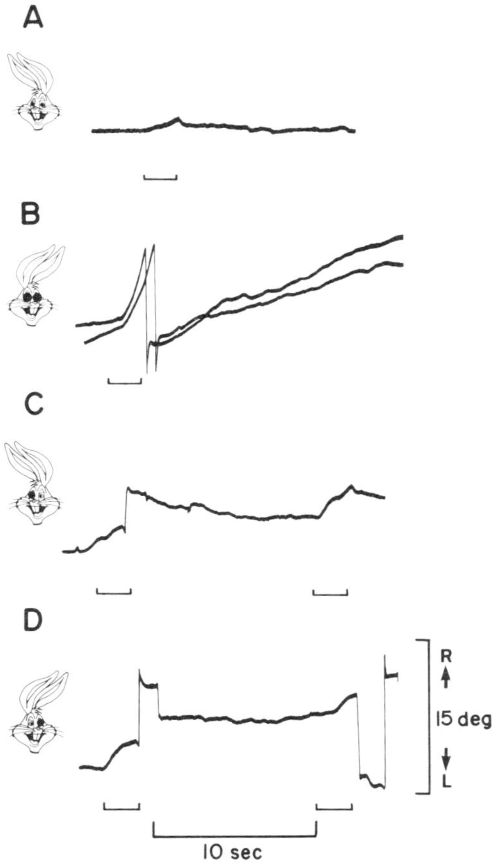

FIG. 8. The effect of visual feedback on eye movements evoked by microstimulation. Microstimulation (2-sec train, 30 pulses/sec, .1-msec pulses, 50 μa, electrode negative) of the right dorsal cap evoked low-velocity horizontal eye movements to the right only when visual feedback was prevented by occluding both eyes. **B:** Note that the evoked eye movement outlasted the stimulus train (indicated by *brackets*). In **A,C,** and **D** visual feedback from one or both eyes suppressed the effect of microstimulation.

tion until a peak velocity was attained (Figs. 8B,9A); (b) the eye movements often outlasted the stimulus and continued toward the right at decreasing velocity for 20 to 60 sec after the cessation of microstimulation (Fig. 8B).

Depth profiles were constructed, demonstrating the vertical extent of the region through the dorsal cap from which eye movements could be evoked (Figs. 9 and 10). As the electrode neared the dorsal cap, eye movements of greater peak velocities, of shorter latencies, and of lower current thresholds could be evoked. The area from which eye movements of highest velocity could be evoked corresponded to the middle of the dorsal cap in the vertical plane (Fig. 10C). The anteroposterior extent of the region from which eye movements could be evoked was not mapped so systematically, but the region about 300 to 400 μ anterior to the caudal pole of the dorsal cap appeared to yield eye movements of the highest peak velocities (4 to 12 deg/sec) and lowest stimulating

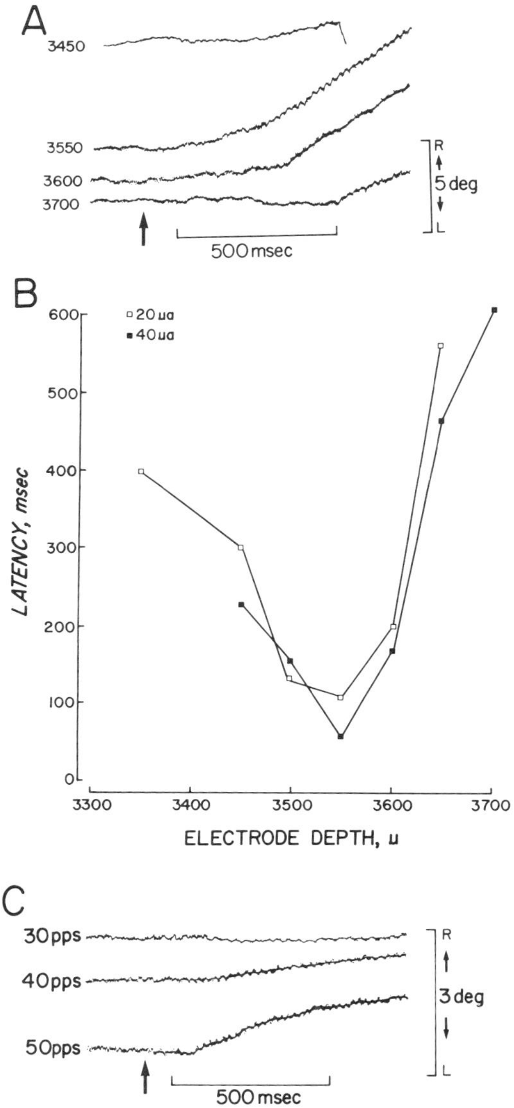

FIG. 9. Latency of eye movements evoked by microstimulation of the dorsal cap. **A:** Eye movements were evoked by microstimulation at different depths in an electrode penetration through the dorsal cap (2-sec train, 50 pulses/sec, .2-msec pulses, 40 μa, electrode negative). **B:** The latencies of the evoked eye movements were plotted as a function of electrode depth, at two different intensities of stimulating current (2-sec train, 50 pulses/sec, .2-msec pulses, 20 or 40 μa, electrode negative). A marking lesion was placed at the electrode location from which eye movements of the highest velocity and shortest latency could be evoked (3,550 μ). This lesion is illustrated in Fig. 10C. **C:** The effect of varying the frequency of the stimulating pulses was examined at an electrode depth of 3,400 μ (2-sec train, 30,40,50 pulses/sec, .2-msec pulses, 50 μa, electrode negative). No eye movement was evoked by a stimulus of 30 pulses/sec at this electrode depth.

current thresholds (15 to 20 μa). Because the dorsal cap is roughly cylindrical, with a diameter of 200 to 300 μ and a length of about 1.8 mm, it is not possible to stimulate electrically all of the cells of the dorsal cap without also stimulating other regions of the ipsilateral inferior olive, and, at higher stimulus currents, decussating CFs which originate from the contralateral inferior olive. Thus, the efficacy of microstimulation of the right dorsal cap in evoking conjugate

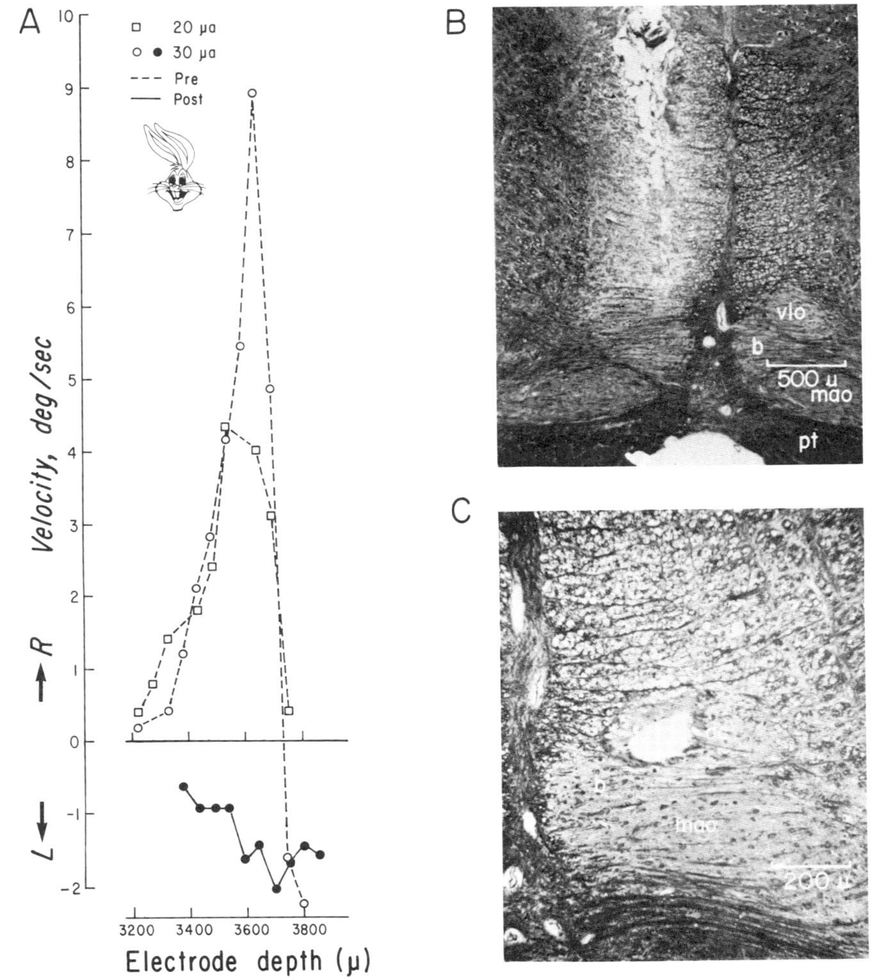

FIG. 10. The effect of severing the olivocerebellar pathway on eye movements evoked by microstimulation of the right dorsal cap. **A:** A depth profile of the velocity of eye movements evoked by microstimulation (2-sec train, 50 pulses/sec, .2-msec pulses, 20 or 30 μa, electrode negative) is indicated for two different stimulus intensities (*open square* = 20 μa, *open circles* = 30 μa) prior to making a sagittal cut in the brainstem, severing the crossed olivocerebellar pathway. The depth profile measurements were repeated following the cut (*filled circles* = 30 μa). **B:** The sagittal cut extended from .5 mm rostral to the ventrolateral outgrowth, posteriorly for 1 mm, destroying, in part, the contralateral dorsal cap. **C:** A marking lesion was made at the depth at which the maximal preoperative stimulating effect was obtained (3,550 μ). Abbreviations: dc, dorsal cap; b, β nucleus; mao, medial accessory olive; vlo, ventrolateral outgrowth; pt, pyramidal tract.

rightward eye movement depends critically on an optimal placement of the stimulating electrode.

In order to determine whether the eye movements evoked by dorsal cap microstimulation were in fact caused by selective activation of the visual CF pathway, the effects of the surgical destruction of this pathway on eye movements evoked by microstimulation were examined. Microstimulation depth profiles were obtained for the right dorsal cap. Then the rabbit was temporarily anesthetized and a sagittal cut was made in the left brainstem passing through the medial aspect of the left inferior olive (Figure 10B). This cut severed the crossed olivocerebellar output of the right dorsal cap, but left intact the visual projection to the right dorsal cap. The microstimulation depth profiles were again measured after the rabbits recovered from the anesthesia. The effect of the transection of the olivocerebellar pathway was to eliminate the rightward eye movements previously evoked by microstimulation. In some instances, low-velocity eye movements to the left were evoked following the transection (Fig. 10A). The stimulus depth profile of these electrically evoked movements suggested that they were caused by stimulation of the CFs originating from the left dorsal cap and projecting to the right flocculus. These experiments ruled out the possibility that eye movements evoked by microstimulation were the result of antidromic activation of dorsal cap afferents by demonstrating that an intact olivocerebellar pathway was essential to evoke these eye movements.

The importance of the flocculus as a target site for the CFs originating from the dorsal cap was emphasized by similar control experiments in which the effects of the left flocculus lesion on eye movements evoked by microstimulation of the right dorsal cap were measured. Unilateral lesions of the flocculus, in conjunction with partial damage to the overlying paramedian cerebellar cortex, produced a nystagmus with the slow phase to the right within 5 min after anesthesia was discontinued, if the animal was placed in darkness (Fig. 11B,D). The rightward-directed slow phase attained velocities of 20 to 40 deg./sec. Microstimulation of the right dorsal cap following left floccular lesions failed to increase the velocity of these rightward-directed eye movements. However, in one experiment, the electrode used for microstimulation was localized to a region in the rostral right dorsal cap near the β nucleus, just below a bundle of decussating fibers originating from the left dorsal cap (Fig. 11E). This stimulating electrode location was 330 μ lateral to an electrode track in which visually driven multiple-unit activity was previously recorded (*arrow,* Fig. 11E). Before the left flocculus lesion was made, microstimulation in this location caused a rightward, low-velocity eye movement (Fig. 11A). After the left floccular bundle was severed (leaving intact most of the floccular cortex) and the rabbit was allowed to recover from the anesthesia, the familiar nystagmus with a rightward-directed slow phase developed (Fig. 11B,D). Postoperative microstimulation of the same olivary location now evoked leftward-directed eye movement (Fig. 11B), probably through activation of decussating CFs projecting to the right flocculus. The electrically evoked activity of these decussating CFs which originated from the

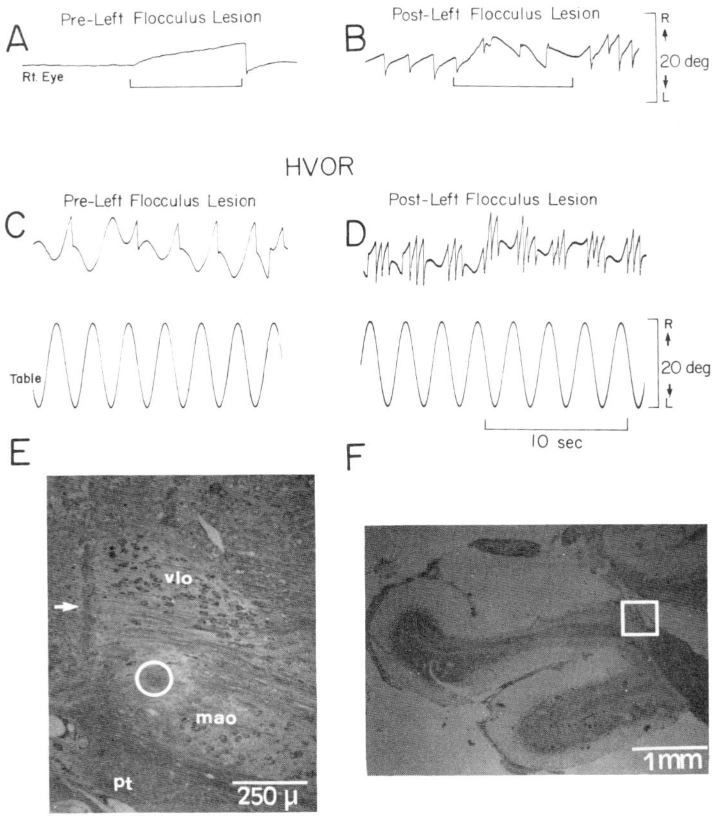

FIG. 11. Microstimulation of the right dorsal cap before and after lesioning the contralateral flocculus. **A:** Microstimulation of the right anterior dorsal cap evoked a slow deviation of the eyes toward the right. The stimulus was a train of 50 pulses/sec, .2-msec, 30-μa pulses lasting 8 sec and is indicated by *bracket*. The rabbit was temporarily anesthetized with halothane (5 min) and a lesion severing the floccular bundle as it leaves the brachium pontis was made **(E)**. Fifteen minutes later a nystagmus caused by the left floccular lesion was evident. The rightward-directed slow phase of this spontaneous nystagmus was interrupted and *reversed* by microstimulation of the right dorsal cap **(B)**. **C** and **D** illustrate the effect of the left floccular lesion on the horizontal vestibuloocular reflex. **F** illustrates the position of the stimulating microelectrode, at the rostral portion of the dorsal cap at the level of the ventrolateral outgrowth (vlo). A microlesion (indicated by *circle*) was made at this location (5 μa, 10 sec). This electrode location lies just lateral to an electrode track in which posteroanterior directionally selective cells were recorded *(arrow)*. The level caudal-rostral location of the stimulating microelectrode corresponds to sections at 1,850 μ and 2,050 μ in Fig. 12.

left inferior olive had probably been masked preoperatively by the relatively greater preoperative activation of CFs projecting to the left flocculus.

These data support the hypothesis that the rightward-directed eye movements evoked by microstimulation of the dorsal cap were the result of activity conveyed to the left flocculus by the crossed CF projection. If the effect of this CF activity were to *increase* the net activity of floccular Purkinje cells, then stimulation of the right dorsal cap should have evoked eye movements to the left, not the right. The rightward eye movements which were observed suggest that the activation of visual CFs causes a net suppression of Purkinje cell discharge in the contralateral flocculus. This resultant suppression of Purkinje cell discharge would cause a disinhibition of cells in the contralateral vestibular nuclei and a consequent reciprocal activation of the oculomotor nuclei, evoking a rightward eye movement (Fig. 1). These microstimulation experiments further implicate the olivocerebellar system in the control of eye movements based on visual information. However, it would be useful to have a direct measure of the functional contribution of the dorsal cap to the eye movements based on visual information. This contribution was examined directly by unilaterally destroying the dorsal cap and measuring the effects of the destruction on eye movements evoked by visual and vestibular stimulation.

Microlesions of the Dorsal Cap

As previously indicated, field potentials can be evoked in the dorsal cap by photic stimulation of the contralateral eye only in the caudal half of the dorsal cap. Nevertheless, more than 50% of the entire dorsal cap (including the dorsomedial aspect of the β nucleus) degenerates following lesions of the contralateral flocculus (Fig. 12). Furthermore, in the rostral half of the dorsal cap, there are decussating fibers which originate in the contralateral dorsal cap. Therefore, in this experiment, unilateral lesions were restricted to the caudal half of the right dorsal cap. The effects of these unilateral lesions on eye movements evoked by optokinetic and vestibuloocular reflexes were examined. The sites of microlesions were determined by using the large field potential evoked by photic stimulation of the left eye to guide the placement of the microelectrode in the right dorsal cap of sodium-pentobarbital anesthetized rabbits. The microelectrode was then connected to a constant current source and a microlesion was made by passing 10 to 100 μa for 10 to 40 sec with the electrode negative. The immediate effects of this initial lesion on the photically evoked field potential were examined. If there was a residual photically evoked field potential, then one or two additional microlesions were made (10 to 30 μa) after the microelectrode was moved rostrally or caudally. Following this operation, the rabbits manifested no obvious motor abnormalities. Within 2 hr they could locomote down a carpeted hallway at their accustomed rate with no asymmetry of gait or posture.

The Effect of Dorsal Cap Lesions on the HVOR

If the rabbits were placed in total darkness and the eye movements were recorded, the consequences of the microlesion of the right dorsal cap became immediately evident. A sustained conjugate drift of the eyes to the left was observed (Fig. 13). This drift did not affect the gain of the HVOR. (Gain =

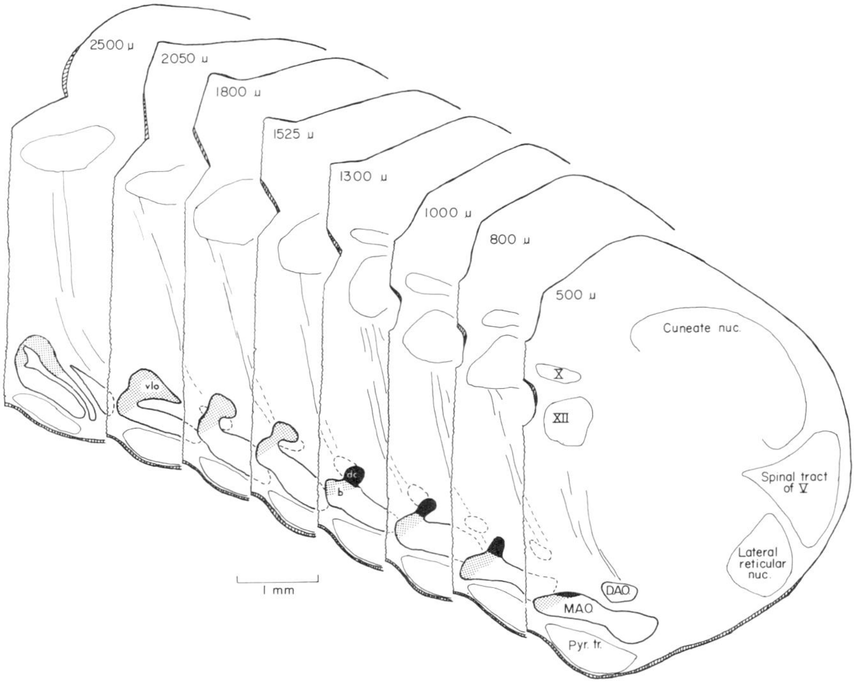

FIG. 12. Illustration of the right brainstem and caudal inferior olive of the rabbit. The area of the inferior olive marked by stippling and solid black, indicates that part of the right inferior olive in which more than 50% of the olivary cells degenerated following destruction of the contralateral flocculus. The area of the inferior olive marked in solid black indicates that part of the inferior olive from which field potentials can be evoked by stimulation of the contralateral eye. The numbers in the upper left of each section indicate, in microns, the distance of the section from the caudal pole of the inferior olive. Abbreviations: b, β nucleus; DAO, dorsal accessory olive; dc, dorsal cap; MAO, medial accessory olive; Pyr. tr., pyramidal tract; X, vagus nucleus; vlo, ventrolateral outgrowth; XII, hypoglossal nucleus.

$(V_R + V_L)/2V_T$, where V_R = peak compensatory eye velocity to the right; V_L = peak compensatory eye velocity to the left; V_T = peak table velocity.) However, the postoperative HVOR was markedly asymmetric due to the leftward drift of the eyes. A derived measure of eye-movement bias was compared before and after dorsal cap lesions. [Bias $= (V_R - V_L)/(V_R + V_L)$, where V_R = peak compensatory eye velocity to the right and V_L = peak compensatory eye velocity to the left.] Preoperatively there was no marked velocity bias of the HVOR. Postoperatively, the velocity bias was greatest at the lowest stimulus frequencies tested (Fig. 14). In fact, in some animals tested within 24 hr postoperatively, the leftward bias was so great that the eyes continued to move toward the left as the table was rotated toward the left. Although the velocity bias was attenuated

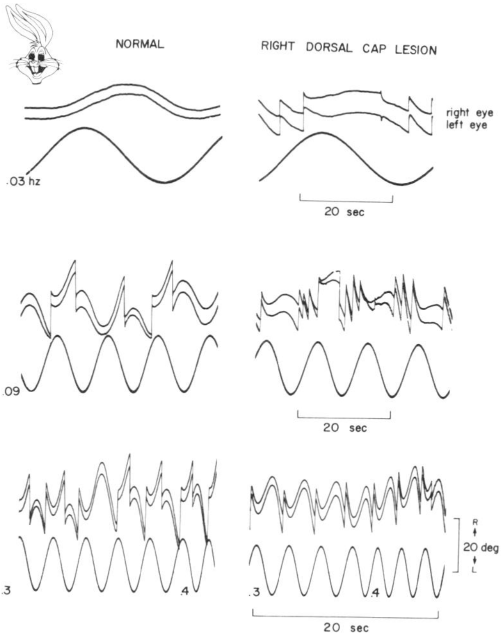

FIG. 13. Conjugate horizontal eye movements evoked by VOR stimulation before and after a partial electrolytic lesion of the right dorsal cap of I.O. #22. The *upper two traces,* indicating the positions of the right and left eye, have been displaced vertically by an arbitrary amount for graphic presentation. The *third trace* indicates the table (head) position. Note the imbalance in the postoperative eye movements; the eyes move at a higher velocity to the left.

at high stimulus frequencies, the absolute value of the disparity in the velocities of the compensatory eye movements in the left and right directions $(V_R - V_L)$ actually increased. Hence the postoperative velocity bias could not be attributed to a simple constant leftward drift of the eyes superimposed on the HVOR. In most rabbits, the spontaneous postoperative leftward drift of the eyes disappeared over a period of 1 to 4 days. However, even when this drift of the eyes had disappeared, the HVOR velocity bias still persisted. In all rabbits, the velocity bias was greatly diminished after 1 week postoperatively. However, in rabbits which sustained the most extensive damage to the right dorsal cap, the leftward velocity bias persisted for several weeks.

The Effect of Dorsal Cap Lesions on the Optokinetic Reflex

In normal rabbits, monocular, sinusoidal optokinetic stimulation evokes asymmetric, conjugate eye movements. This asymmetry in the optokinetic reflex is dependent on the velocity of the optokinetic stimulus. At optokinetic stimulus velocities of less than .8 deg/sec, the eyes follow optokinetic stimuli moving in either direction almost equally well, but at higher stimulus velocities eye movements in the posteroanterior direction, with reference to the stimulated eye, are evoked at a higher optokinetic gain. (Gain = eye velocity/stimulus velocity) (Figs. 15 and 16).

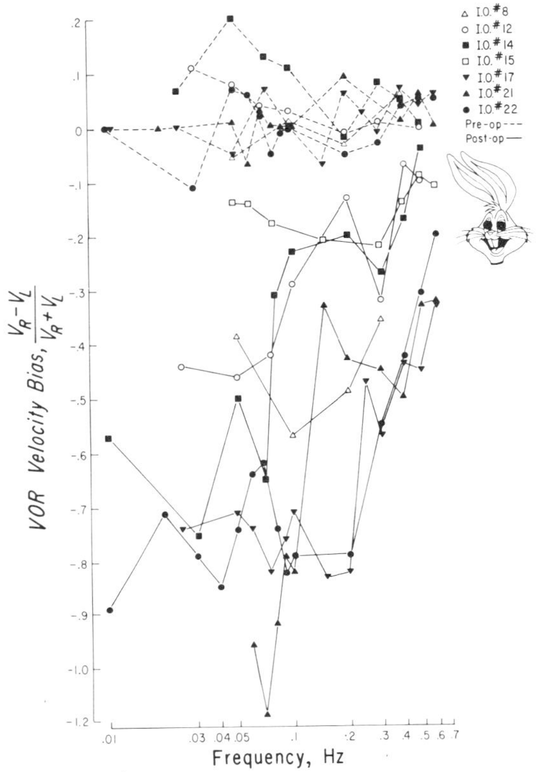

FIG. 14. VOR velocity bias before and after partial electrolytic destruction of the right dorsal cap. The horizontal movements of the right and/or left eyes were recorded during sinusoidal VOR stimulation, ±10 deg, at different frequencies. The peak eye velocities in each direction at each frequency were measured. The derived measure of velocity bias was computed according to the relationship $\frac{V_R - V_L}{V_R + V_L}$.

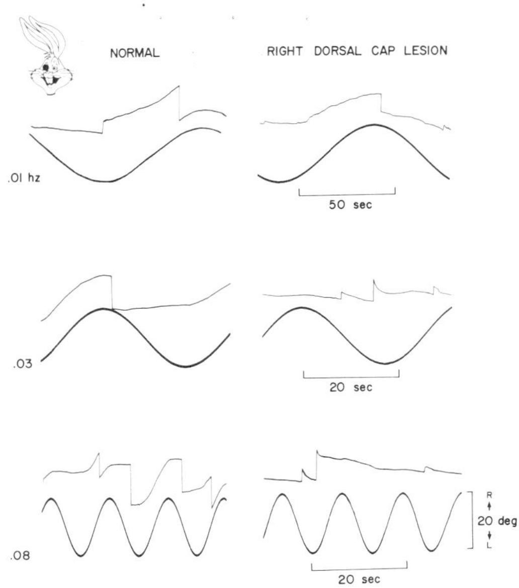

FIG. 15. Eye movements evoked by optokinetic stimulation of the left eye of I.O. #22 before and after partial electrolytic destruction of the right dorsal cap. The eye position of the occluded right eye was recorded while sinusoidal stimulation, ±10 deg, was delivered to the left eye at the indicated frequencies.

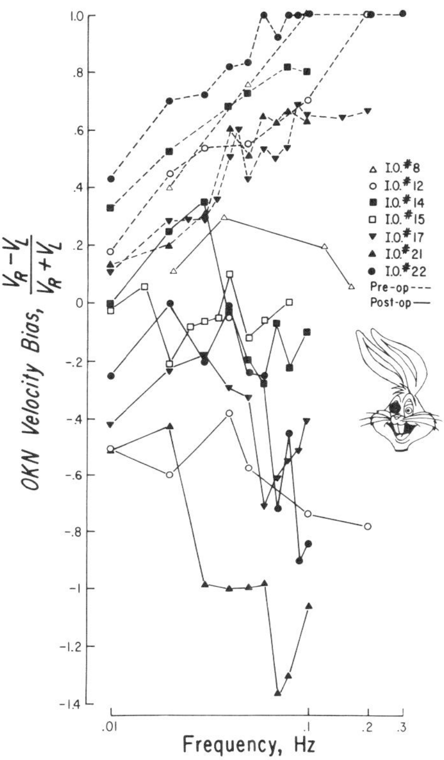

FIG. 16. Optokinetic velocity bias as a function of stimulus frequency. The left eye was exposed to sinusoidal ±10-deg optokinetic stimulation at different frequencies, and the horizontal movements of the occluded right eye were recorded. The peak eye velocity in each direction of stimulation at each frequency gave a derived measure of optokinetic bias according to the relationship $\frac{V_R - V_L}{V_R + V_L}$.

Following partial damage to the right dorsal cap, the normal rightward bias for monocular, sinusoidal, optokinetic stimulation of the left eye was reversed (Figs. 15 and 16). The measurement of optokinetic gain is complicated by the use of sinusoidal stimuli to describe a system which is inherently nonlinear. Furthermore, since the dynamics of the optokinetic reflex include a relatively long time-constant integrative process, it takes several seconds of constant-velocity optokinetic stimulation for the eyes to achieve a steady-state velocity. Consequently, constant-velocity, unidirectional, optokinetic stimulation was used in the present experiment to measure the steady-state optokinetic gain of the optokinetic reflex in the posteroanterior direction, before and after microlesions were placed in the dorsal cap. A series of optokinetic stimuli at slowly incremented velocities were used to make this measurement. The eye movements evoked by constant-velocity optokinetic stimulation of the left eye of normal rabbits had a gain which was near unity at stimulus velocities of less than 5 deg/sec and which decreased at higher stimulus velocities (Figs. 17A and 18).

Following a microlesion of the right dorsal cap, the optokinetic gain was impaired when stimulation was delivered to the left eye, the eye which projected to the damaged right dorsal cap (Figs. 17B and 18). However, eye movements

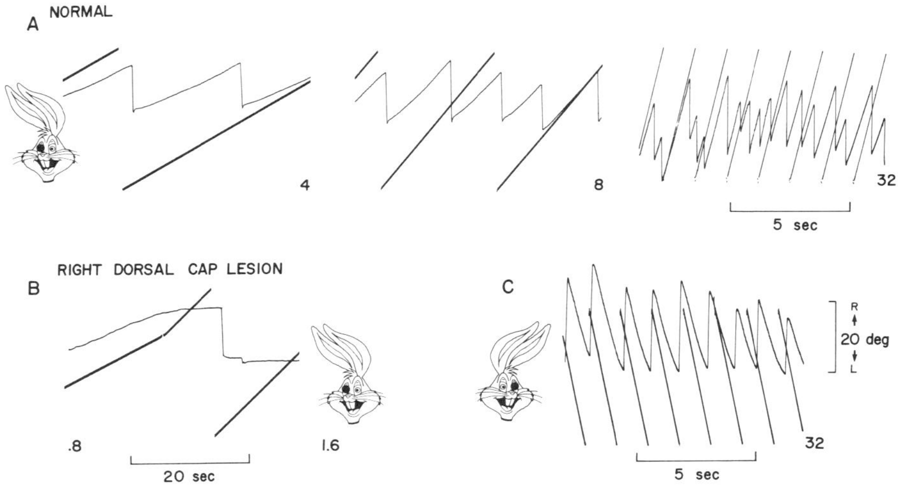

FIG. 17. Eye movements evoked by unidirectional, constant-velocity, optokinetic stimulation before and after partial electrolytic destruction of the right dorsal cap of I.O. #22. **A:** Eye movements evoked preoperatively by exposure of the left eye to the optokinetic target, moving in the posteroanterior direction at 4, 8, and 32 deg/sec. **B:** Eye movements evoked postoperatively by exposure of the left eye to the optokinetic target moving in the posteroanterior direction at .8 and 1.6 deg/sec. Note that the eye movement does not follow the 1.6-deg/sec target. **C:** Eye movements evoked postoperatively by exposure of the *right* eye (recording eye movement of the left eye) to an optokinetic target moving at 32 deg/sec in the posteroanterior direction (as seen by the right eye).

evoked by postoperative stimulation of the right eye were normal (Fig. 17C). No rabbit with a right dorsal cap lesion could follow optokinetic targets presented to the left eye, moving in the posteroanterior direction at velocities greater than 5 deg/sec, with a gain greater than .3. Comparably low gains were evinced by normal rabbits only at optokinetic stimulus velocities of 20 to 50 deg/sec (Fig. 18). This deficit in optokinetic gain persisted even after the HVOR velocity bias had been compensated.

Histological reconstruction of the microlesions and subsequent cell counts revealed that the caudal 800 μ of the damaged dorsal caps contained 496 $\pm$ 303 intact neurons. The unlesioned dorsal caps contained 936 $\pm$ 158 intact neurons, and the dorsal caps of 4 unoperated control rabbits contained 1,195 $\pm$ 127 neurons. The unilateral microlesions were confined substantially to the dorsal cap and destroyed 10 to 80% of the olivary cells normally contained within the region from which field potentials could be evoked by photic stimulation of the contralateral eye. However, the possibility remained that the decrease in optokinetic gain and leftward bias of HVOR might be attributable to partial destruction of the region surrounding the dorsal cap. This possibility was examined by making a large lesion in the region dorsal and slightly lateral to the dorsal cap. This lesion did not produce a HVOR bias, nor did it reduce the

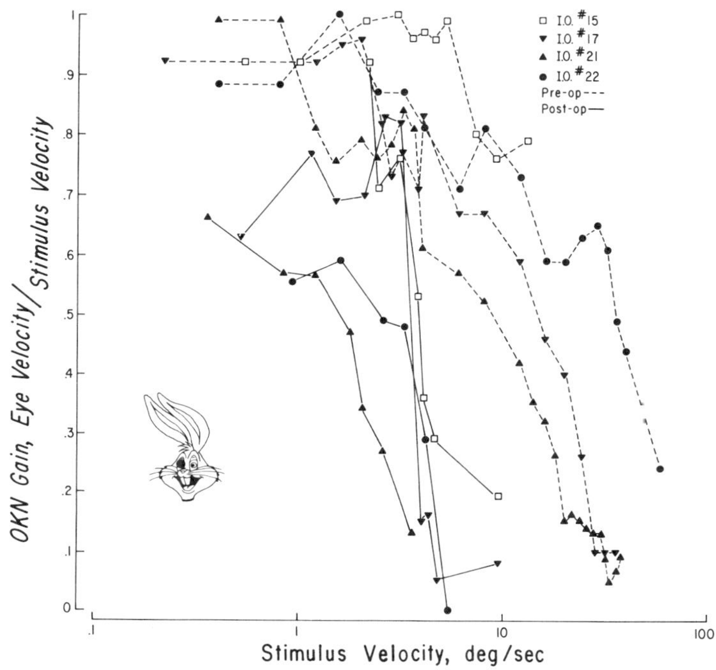

FIG. 18. Optokinetic gain before and after partial electrolytic destruction of the right dorsal cap. Constant-velocity, unidirectional, optokinetic stimuli were presented to the left eye (posteroanterior direction) and the movement of the right eye was measured. The peak eye velocity evoked by each stimulus velocity was measured, and the derived measure of optokinetic gain was computed according to the relationship, $G=$ eye velocity/optokinetic stimulus velocity. Pre- and postoperative gains are plotted for each rabbit except I.O. #15. The "pre-op" values for I.O. #15 were actually postoperative gain values measured for optokinetic stimulation of the opposite eye.

optokinetic gain, even though it destroyed a volume of tissue which was at least twice the volume of tissue destroyed by even the largest microlesions of the dorsal cap. (Compare Fig. 19A and B.) Because the microlesions were restricted to the caudal aspect of the dorsal cap, they caused only a modest retrograde cell loss in the contralateral dorsal cap. This finding suggests that the crossed olivocerebellar projection probably projects rostrally on the ipsilateral side for 500 to 1,000 μ before it crosses the midline, passing above, below, and through the contralateral dorsal cap before it reaches the cerebellum. These results suggest that the dorsal cap participates in the visual guidance of low-velocity eye movements. The low frequency of discharge of CFs would seem entirely appropriate for the continuous modulation of such slow eye movements. If the general sense of the schematic of Fig. 1 is correct, it would appear that increased activity in the right dorsal cap increases the net excitation of the

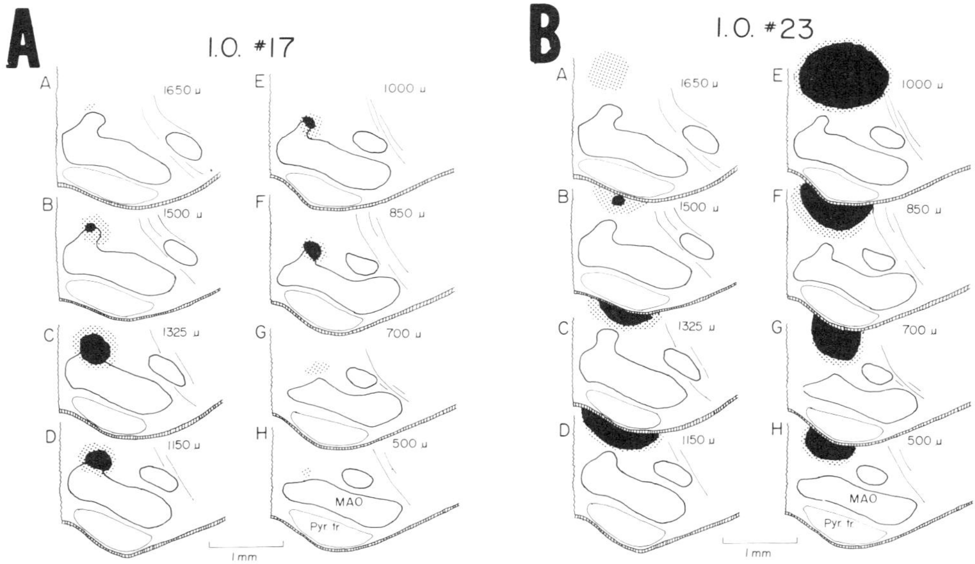

FIG. 19. Illustrations of two microlesions. The black areas indicate the regions of the microlesions in which there was an absence of olivary cells and/or nerve fibers. The stippled areas indicate the presence of some olivary cells and/or nerve fibers mixed with microglia and glial processes. **A:** Microlesion of I.O. #17, made with three penetrations spaced 300 μ apart (rostrocaudally); caudal and rostral lesions, 10 μa—15 sec; middle lesion, 20 μa—20 sec. **B:** Microlesion of I.O. #23. Optokinetically evoked multiple-unit activity was used to guide the placement of the microelectrode. The microelectrode was then raised 350 μ and a lesion was made; 40 μa—30 sec. Immediately after the lesion was made, the electrode was driven back into the dorsal cap where optokinetically evoked multiple-unit activity was recorded again.

left vestibular nuclei, causing rightward eye movements, by *decreasing* the net Purkinje cell inhibitory discharge of the left flocculus.

Visually Modulated Purkinje Cells of the Flocculus

The paradox that the powerful excitatory action of CFs on Purkinje cells could be functionally less important than the pause in SS activity evoked by CFs is sufficiently grotesque to warrant a direct examination. This has been achieved by recording from Purkinje cells of the flocculus in anesthetized and unanesthetized rabbits and using the same optokinetic stimulation previously used to study unit activity in the dorsal cap. Optokinetically driven Purkinje cells of the flocculus constitute a minority of the total population. However, one can increase the likelihood of recording from an optokinetically modulated Purkinje cell by restricting the search for such cells to regions of the flocculus from which a field potential can be evoked by photic stimulation of the eye *ipsilateral* to the flocculus. Climbing fiber field potentials were clearly present in the ventral aspect of the caudal flocculus. Rostrally, in the dorsal aspect of the flocculus, a complicated field potential was evoked which may be due to the activation of both CF and MF inputs (66,67). Less than 50% of the Purkinje cells in or near these regions had CFs which could be optokinetically modulated. However, when a Purkinje cell which was optokinetically sensitive in the horizontal plane was encountered, the CFRs were always evoked by stimulation of the ipsilateral eye in the posteroanterior direction (Fig. 20). The SSs recorded from the same Purkinje cell were always modulated in an antiphasic manner, increasing their rate of discharge during anteroposterior movement of the optokinetic stimulus (Fig. 20). Since the spontaneous rate of SSs was 20 to 60 imp/sec and their absolute depth of modulation by optokinetic stimulation was greater than that of the CFRs, the net effect of optokinetic stimulation was to *increase* the total Purkinje cell discharge (CFRs + SSs) during *anteroposterior* optokinetic stimulation, and to *decrease* Purkinje cell discharge during *posteroanterior* optokinetic stimulation. The depth of modulation of SSs did not appear to be dependent on the depth of modulation of the CFRs of the same Purkinje cell. In fact, in regions of the flocculus where CF field potentials could be evoked, it was possible to record CFRs which were strongly modulated with only nominal

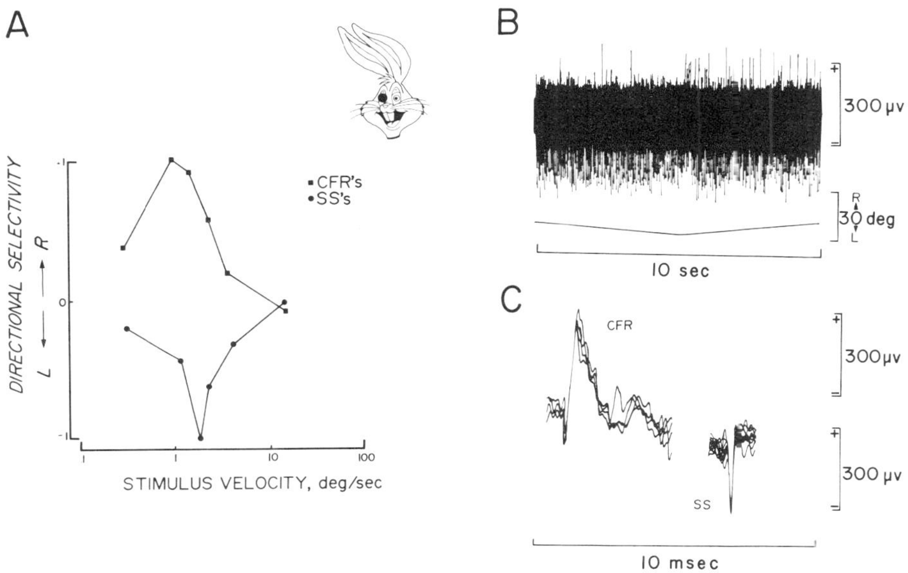

FIG. 20. The effect of OKN stimuli on the modulation of CFRs and SSs in a left floccular Purkinje cell. The OKN stimuli were constant-velocity targets moving horizontally in the posteroanterior and anteroposterior directions, viewed by the left eye. **A:** The directional selectivities of the CFRs and SSs were antiphasic. The discharge frequency of CFRs increased and SSs decreased for posteroanterior OKN stimulus velocities. Both CFRs and SSs had peak velocity-sensitivities at 1 to 2 deg/sec. **B:** Example of positive CFRs and negative SSs to an OKN target *(second trace)* moving at ± 2 deg/sec. **C:** Expanded record of superimposed CFRs and SSs.

modulation of SSs. Conversely, in these same regions it was possible to record strongly modulated SSs (with anteroposterior directional selectivity) but with no CFR modulation. An example of the type of modulation is illustrated in Fig. 21, a recording from a floccular Purkinje cell whose SSs were modulated by optokinetic stimulation, but whose CFRs were not. It was also possible to record CFRs from Purkinje cells in unanesthetized rabbits which were modulated by posteroanterior retinal slip velocity of the optokinetic stimulus, but in which there was no apparent modulation of SSs (Fig. 22). There was a profound difference in the manner by which CFRs and SSs were modulated by optokinetic stimuli. CFRs were sensitive to retinal slip acceleration as well as retinal slip velocity. Consequently, when CFRs were evoked by triangular optokinetic velocity waveforms, they burst or were transiently silenced when the stimulus reversed direction, or when the velocity of the retinal image was changed by a saccadic

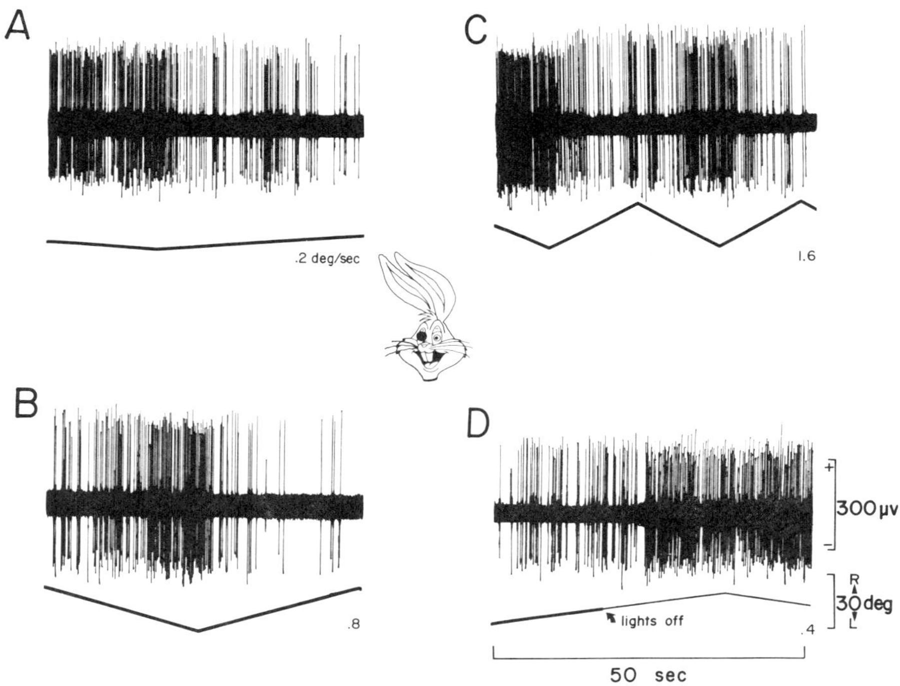

FIG. 21. Recording from a directionally selective Purkinje cell in the left flocculus. OKN stimulation of the left eye evoked directionally selective SSs, but did not change CFRs. The SSs decreased in frequency during posteroanterior stimulation and had a peak sensitivity at .8 deg/sec. **A:** OKN stimulation at .2 deg/sec. **B:** OKN stimulation at .8 deg/sec. Note the lag in the modification of SS activity when the OKN stimulus reversed direction. **C:** At OKN velocities of 1.6 deg/sec the directional selectivity was impaired by the sluggish response of the SS to changes in velocity. **D:** The OKN stimulus was turned off during movement in the posteroanterior direction. Six seconds elapsed before this removal of inhibition was detected in the SS discharge.

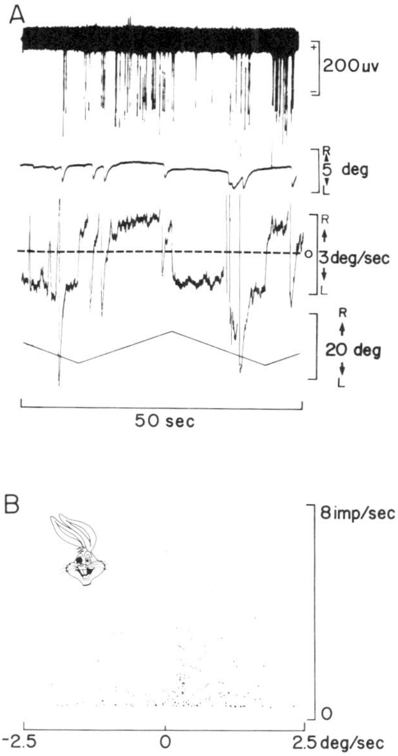

FIG. 22. A recording of a directionally selective CFR from a left floccular Purkinje cell in an unanesthetized rabbit. **A:** Records of CFR *(top trace)*, eye position of the right eye *(second trace)*, retinal slip velocity *(third trace)*, and OKN stimulus position *(fourth trace)*. Note that the CFR was related to retinal slip velocity. **B:** Instantaneous discharge frequency of CFR plotted as a function of retinal slip velocity.

eye movement (Figs. 20,22). By contrast, SS discharge was not sensitive to retinal slip acceleration. The modulation of SS activity lagged by several seconds any change in retinal slip velocity. This delay caused a significant phase lag of SSs at higher frequencies of stimulation which could even make the SS directional selectivity appear to reverse.

The antiphasic behavior of CFRs and SSs and their common velocity sensitivities suggest that directionally selective CF activity may be causally related to directionally selective SS activity. However, CFRs in a particular Purkinje cell are neither necessary nor sufficient for strong, antiphasic modulation of SSs in the same Purkinje cell. These data rule out the possibility that the SS pause which follows CFRs can be attributed to the "inactivation" of the Purkinje cell membrane by the large amplitude, but short duration depolarization evoked by the CF synapse. CF activity could modulate SS activity by means of: (a) a cerebellar inhibitory interneuronal network, or (b) the inhibitory action of Purkinje cells on extracerebellar sources of mossy fibers. The first possibility, that the antiphasic SS activity is caused by an inhibitory interneuronal network, is illustrated in Fig. 23. The only inhibitory interneuron for which there is unequivo-

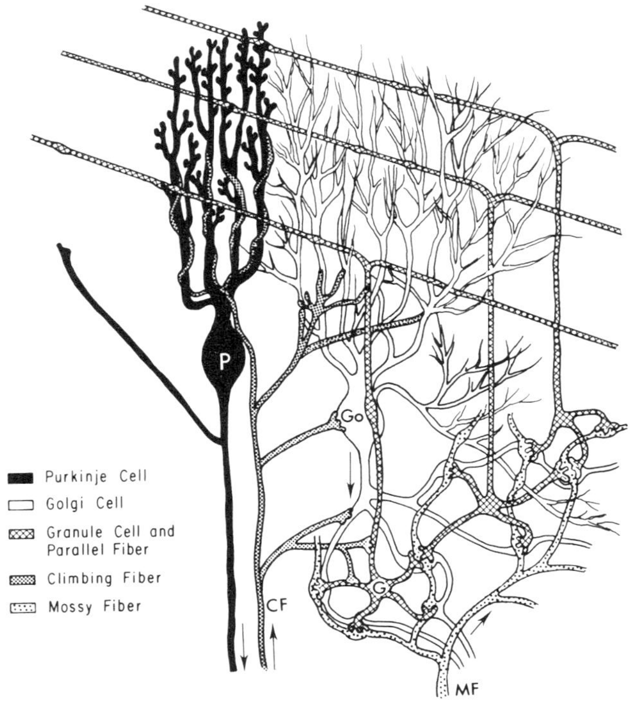

FIG. 23. Illustration of cerebellar neurons innervated by climbing fibers. The climbing fiber synapse on Golgi cells could account for the antiphasic CFR and SS modulation evoked by optokinetic stimulation, since Golgi cell inhibition of granule cells would reduce the excitation of Purkinje cells by mossy fiber inputs.

cal anatomical evidence of CF synapses is the Golgi cell (23). Golgi cell inhibitory modulation of granule cells has been documented electrophysiologically (25). Therefore, CF-evoked Golgi cell discharge could account for the antiphasic behavior of the CFRs and SSs. The second possibility, that CFRs cause a decrease in SSs by the inhibitory action of Purkinje cells on extracerebellar sources of mossy fibers, is suggested by the presence of reciprocal connections between the flocculus and the vestibular nuclei (Fig. 1). It is possible that there is a visual MF input to the flocculus which is independent of CF-evoked Purkinje cell activity. Such an input would have to be antiphasic to the CF input and have an identical velocity sensitivity. The presence of an independent visual MF input has been suggested by recent anatomical and physiological experiments (43,66,67). However, the kind of information conveyed by this source of visual MFs remains to be determined. At present, it seems likely that the modulation of SS activity in the flocculus reflects: (a) the indirect influence of visually excited CFs on floccular interneurons; (b) the indirect influence of CF-evoked Purkinje cell activity on nonvisual inputs to the flocculus; (c) the direct influence of visually evoked MF inputs to the flocculus.

Unilateral Lesions of the Flocculus

We have hypothesized that posteroanterior optokinetic stimulation of the left eye *increases* the CF activity originating from the right dorsal cap and *decreases* the inhibition by left floccular Purkinje cells of the subjacent medial and superior vestibular nuclei, thereby evoking rightward eye movements. Part of this hypothesis can be tested behaviorally by examining the effects of unilateral removal of the flocculus on vestibuloocular and optokinetic reflexes. If rightward eye movements are produced by Purkinje cell disinhibition of the left vestibular nuclei, then a lesion of the left flocculus should likewise evoke rightward eye movements, since such a lesion should be equivalent to the maximum possible Purkinje cell disinhibition of the vestibular nuclei. In other words, a lesion of the left flocculus should duplicate the effects of maximal right dorsal cap stimulation, since both procedures would cause a decrease in left floccular Purkinje cell activity.

Floccular lesions were made by aspiration in a series of rabbits using three different surgical approaches: (a) a dorsal approach, destroying, in addition to the flocculus, parts of the overlying ansiform lobes and lateral anterior lobes; (b) a caudal approach through the foramen of the anterior semicircular canal, destroying portions of the dorsal paraflocculus as well as part of the lateral cerebellar nucleus; (c) a lateral approach, through the middle ear, just rostral to the anterior semicircular canal. This latter operation could be done without damaging the vestibular apparatus or other parts of the cerebellum

The consequences of unilateral floccular lesions made with each of these surgical approaches were qualitatively similar. Upon recovery from the anesthesia, the rabbits which had sustained damage to the ansiform and anterior lobes extended their forepaws rigidly to the left, and, if provoked by some external stimulus, would tumble onto their right sides, rolling over several times. This tumbling was fully compensated within 48 hr. If the lesion was restricted to the flocculus, the rabbits evinced near-normal posture and locomotion immediately after recovery from anesthesia. However, the postoperative eye movements of all rabbits evinced the same general pattern when the rabbits were placed in the dark; a conjugate rightward drift of the eyes. In most rabbits, this drift had a nystagmic character with a slow phase of 1 to 40 deg/sec. This rightward drift was evident in rabbits with lesions restricted to the flocculus, but its velocity and duration were greater if there was also damage to the ansiform lobes or to the lateral part of the anterior lobe. In some rabbits, this rightward drift was not spontaneously nystagmic. Rather, the eyes would slowly (.5 to 2 deg/sec) deviate to the right and maintain an eccentric position. The eyes would be reset to near primary position and renew their rightward drift only if the rabbit was provoked by auditory or cutaneous stimuli. This rightward drift, even in animals in which it attained velocities of 20 to 40 deg/sec, could be suppressed by exposure of *either* eye to patterned visual stimuli. The drift abated over a period of 1 to 10 days after the operation.

The Effects of Left Flocculus Lesions on the Horizontal Vestibuloocular Reflex

As might be expected, this rightward drift caused a pronounced rightward bias of the HVOR. This bias was particularly evident at low frequencies of vestibular stimulation (Figs. 24 and 25). These data suggest that the postoperative nystagmus was the consequence of an interaction between primary afferent inputs and inputs from the flocculus to the vestibular nuclei. This point is further supported by our observations of unilaterally flocculectomized rabbits which had previously received bilateral labyrinthectomies. In these rabbits, after the HVOR had been abolished by the labyrinthectomy, removal of the left flocculus did not evoke a rightward drift of the eyes.

The unilateral flocculectomy had a complex influence on the gain of the HVOR when the operated rabbits were tested within 1 week postoperatively. At low frequencies of stimulation (.02 to .05 Hz), the gain of the HVOR was increased, primarily because of the spontaneous drift of the eyes to the right (Fig. 1). At intermediate frequencies (.05 to .10 Hz), the gain was reduced. This reduction could also be attributed to the rightward bias of the eyes. Because

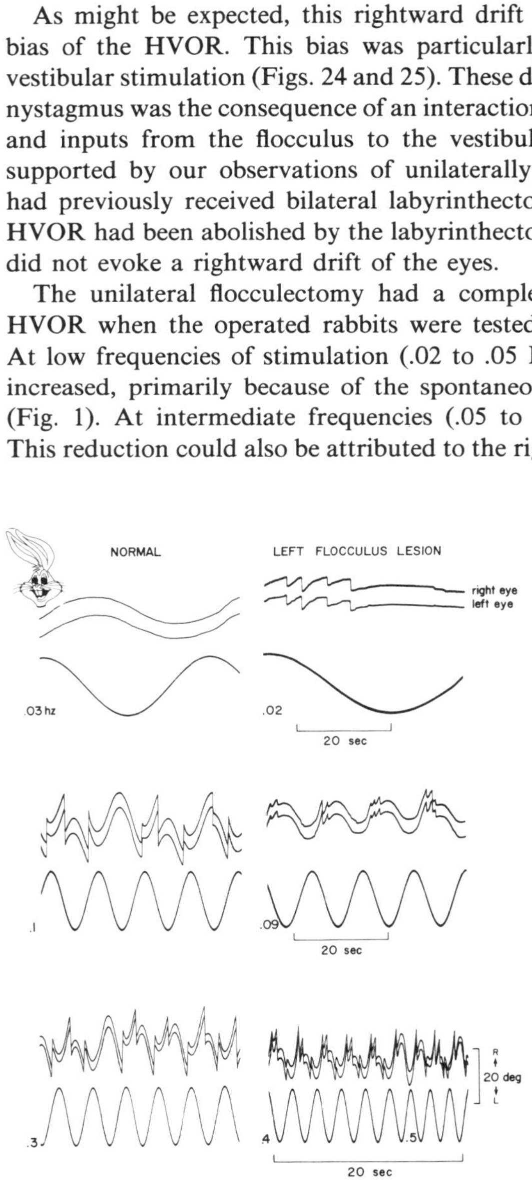

FIG. 24. VOR. The effect of lesions of the left flocculus on eye movements. The VOR was evoked in the dark by sinusoidal horizontal rotation of the rate table (± 10 deg). Representative conjugate eye movement records are illustrated. The records were vertically displaced for illustrative purposes. Each series of traces, from top-to-bottom, consists of right eye position, left eye position, table position. Note that: (a) The lesion does not disrupt the conjugacy of the VOR. (b) There is a pronounced rightward velocity bias caused by the left floccular lesion. (c) The velocity bias is more pronounced at low frequencies of stimulation.

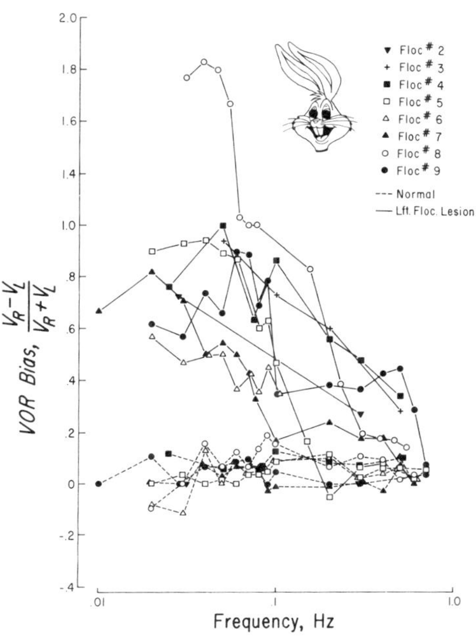

FIG. 25. The effect of lesions of the left flocculus on VOR velocity bias. The VOR was evoked in the dark by a sinusoidal ± 10 deg stimulus. It was measured in 6 rabbits preoperatively and 8 rabbits postoperatively. Note that the effect of the left flocculus lesion was to cause a profound velocity bias to the right. Note also that the velocity bias was attenuated at higher stimulus frequencies.

of this bias, many more resetting saccades than usual were made when eye movements were evoked by leftward rotation of the table (Fig. 24). We have observed that a higher incidence of saccades produces a lower gain of the HVOR even in normal rabbits (Fig. 26). Therefore, it is likely that the reduction in

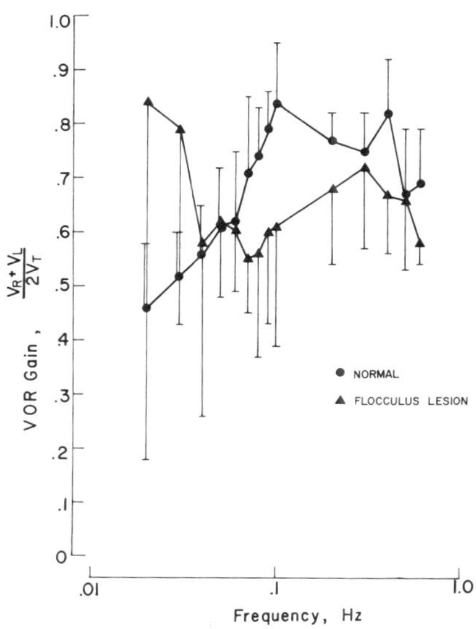

FIG. 26. The effect of lesions of the left flocculus on VOR gain. The VOR gain was measured for 8 rabbits before *(closed circles)* and after *(triangles)* lesions were made in the left flocculus. The effect of the lesion was to increase the gain of the VOR at low frequencies and to reduce the gain at intermediate frequencies. The lesions had virtually no effect on the VOR gain at higher frequencies.

HVOR was secondary to the rightward bias. At higher frequencies of stimulation (.2 to .8 Hz), both the bias and the gain of the HVOR were near normal.

The Effect of Lesions of the Left Flocculus on the Optokinetic Reflex

The influence of left flocculus lesions on the optokinetic reflex was paradoxical. Removal of the left flocculus was the surgical equivalent of exposing the left eye to an optokinetic stimulus moving in the posteroanterior direction. Both procedures could cause a rightward drift of the eyes by disinhibition of the subjacent vestibular nuclei. Therefore, it was not surprising to observe postoperatively that optokinetic stimulation of the left eye (the eye projecting primarily to the damaged flocculus) evoked eye movements which were only slightly reduced in gain (Figs. 27 and 28). Conversely, posteroanterior optokinetic stimulation of the right eye (the eye projecting primarily to the intact right flocculus) evoked eye movements of very low gain, if the animals were tested during the 1st postoperative week (Figs. 27C,28). In essence, the efficacy of the optokinetic stimulation could not be dissociated from the rightward bias caused by the left flocculectomy. The rightward drift masked a reduced gain for eye movements evoked by posteroanterior optokinetic stimulation of the left eye and caused an apparent reduction in gain for eye movements evoked by posteroanterior stimulation of the right eye. Optokinetic reflex testing performed subsequent to compensation of the rightward bias of the eyes demonstrated a permanent

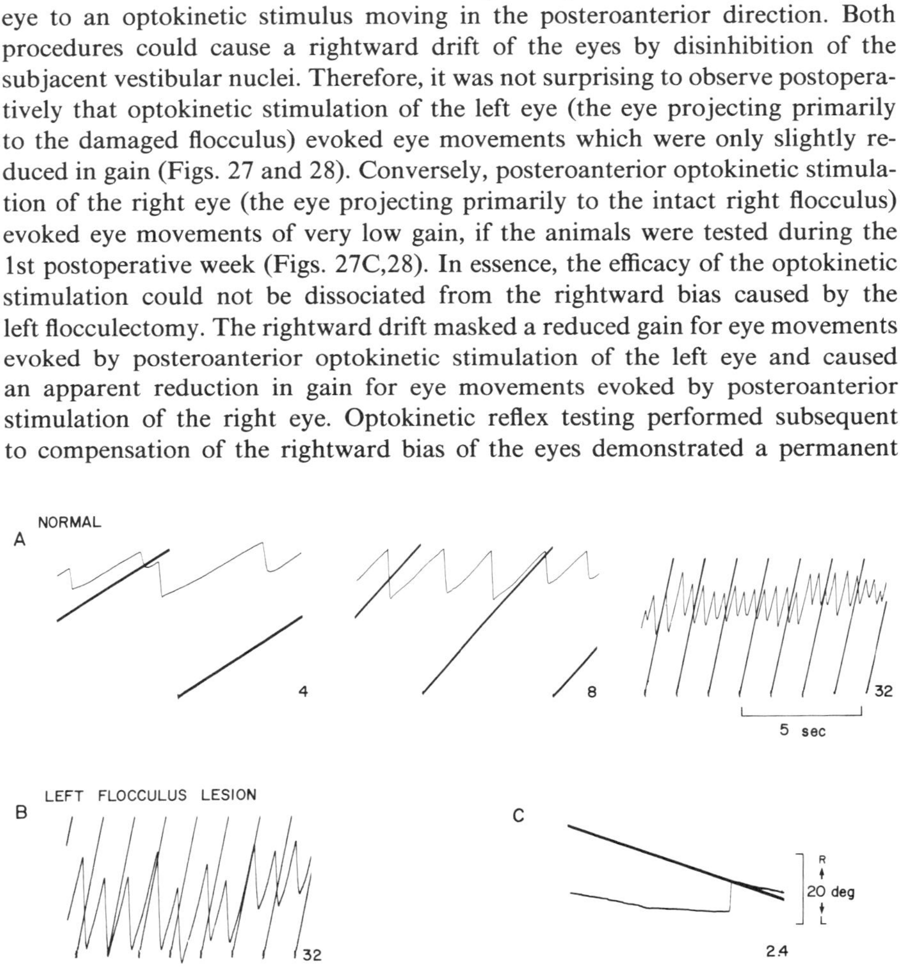

FIG. 27. The effect of lesions of the left flocculus on eye movements evoked by unidirectional, constant-velocity, monocularly viewed OKN stimuli. **A:** Preoperative movements were recorded from the right eye while the left eye viewed a target moving in the posteroanterior direction at velocities of 4, 8, and 32 deg/sec. **B:** Postoperatively the rabbit could still follow high-velocity OKN targets viewing with the left eye. **C:** Postoperatively the rabbit could not follow targets moving in the posteroanterior direction when those targets were viewed by the right eye, the eye which projected to the intact flocculus.

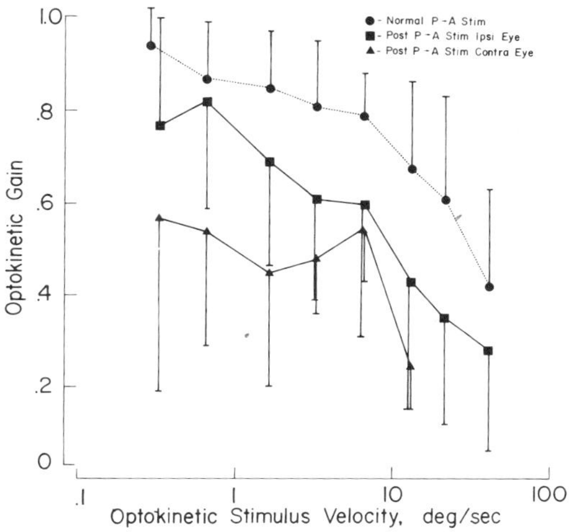

FIG. 28. Optokinetic gain before and after unilateral destruction of the flocculus. A monocular, constant-velocity, unidirectional, posteroanterior optokinetic stimulus was presented to one eye, and the movement of the opposite eye was measured. The peak eye velocity evoked at each stimulus velocity was measured, and the derived measure of optokinetic gain was computed according to the relationship: G = eye velocity/optokinetic stimulus velocity. The data were derived from 8 rabbits before *(filled circles)* unilateral floccular destruction, and after this destruction (*filled squares,* stimulation of eye ipsilateral to lesion; *filled triangles,* stimulation of eye contralateral to lesion). The error bars indicate 1 SD.

deficit in the posteroanterior optokinetic gain of the *left* eye and a recovery to normal values of the gain for eye movements evoked by posteroanterior stimulation of the *right* eye. In sum, the major effect of a unilateral flocculus lesion is to produce an immediate rightward bias of eye movement. This surgically induced bias causes transient abnormalities in both the HVOR and optokinetic reflex. After the surgically induced bias is compensated, the HVOR returns to normal, but the optokinetic reflex evoked by stimulation of the eye which projects to the damaged flocculus remains impaired.

Sustained Influences of Optokinetic Stimulation of the HVOR

In the experiments described above, optokinetic and vestibuloocular reflexes were tested routinely in the same rabbit during the same testing session. In normal rabbits, the response to posteroanterior optokinetic stimulation of the left eye was measured prior to the measurement of the HVOR. Immediately upon cessation of optokinetic testing, there was frequently an optokinetic afternystagmus lasting from 5 to 30 sec. However, even after this optokinetic afternystagmus had subsided, there was an apparent residual influence on the bias of the HVOR which was tested minutes later. This bias was small but was in the same direction as the previous optokinetic testing (Fig. 14; Fig. 25). These data suggested that an HVOR bias was being optokinetically induced and that it might be increased by prolonged unidirectional optokinetic stimulation. The possibility that this HVOR bias might also reflect an interesting long-term influence of the olivocerebellar system on the HVOR was implied by experiments in which microstimulation of the dorsal cap frequently evoked eye movements which outlasted a single train of microstimulation by 10 to 30 sec (Fig. 8C).

Therefore, the effects of prolonged, posteroanterior, monocular, optokinetic stimulation on spontaneous eye movements, and eye movements evoked by the HVOR were studied systematically.

Rabbits were exposed to posteroanterior, optokinetic stimulation which was presented at a fixed velocity of 10 to 20 deg/sec, for 20 to 60 min. Following this stimulation, the rabbit was placed in total darkness and the spontaneous eye movements as well as the HVOR were recorded. The influence of prolonged optokinetic stimulation was impressive. Such stimulation frequently evoked an optokinetic after-nystagmus (Fig. 29B). This after-nystagmus was potentiated in frequency of beats and in its slow phase velocity by HVOR testing. The duration of this optokinetically induced after-nystagmus, and the consequent bias of the HVOR, was studied during alternate "post-conditioning" periods. The after-nystagmus often lasted for 10 to 100 min (Fig. 30), and it always disappeared before the bias of the HVOR returned to normal values (Fig. 30). This prolonged after-nystagmus could not be induced in all rabbits which were tested. However, in instances where the effect could not be induced, the optokinetic gain was lower than normal.

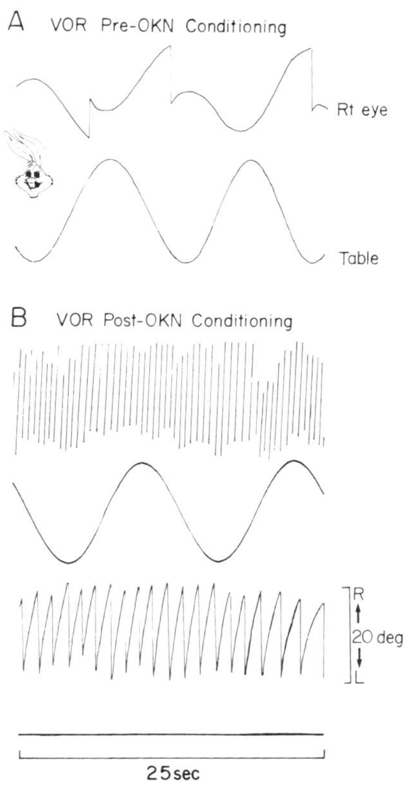

FIG. 29. Influence of sustained, unidirectional optokinetic stimulation on the vestibuloocular reflex. **A:** VOR recorded during stimulation of .08 Hz, ± 10 deg. The rabbit was then exposed to unidirectional (posteroanterior) optokinetic stimulation of the left eye for 60 min, at a stimulus velocity of 8 deg/sec. This stimulus evoked optokinetic eye movements with a gain of .8, creating an average retinal slip velocity of about 1.6 deg/sec. **B:** Spontaneous, optokinetically induced eye movements persisted following the cessation of optokinetic stimulation *(lower traces).* The velocity of these optokinetically induced eye movements was modulated by HVOR stimulation *(upper traces).*

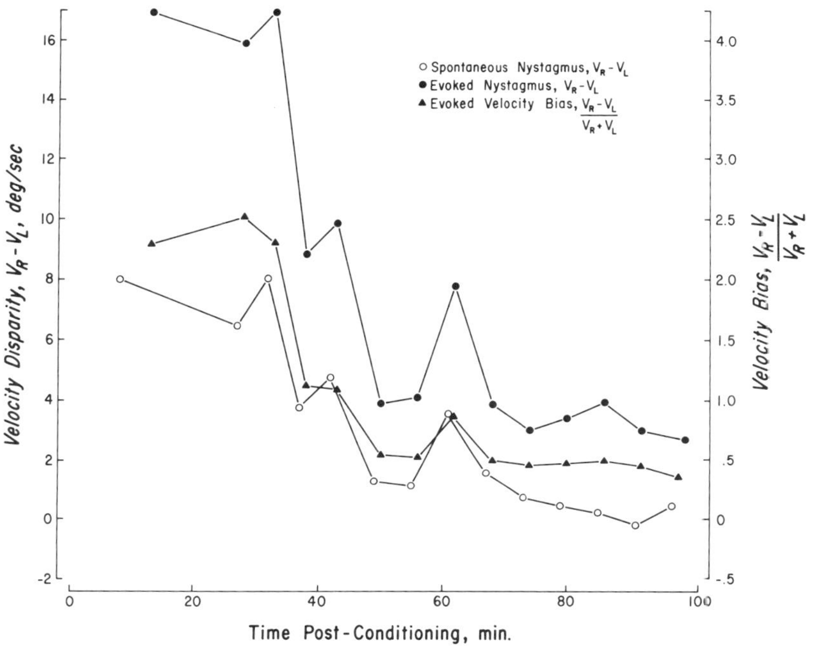

FIG. 30. The time course of the effect of constant-velocity optokinetic stimulation on spontaneous and vestibularly evoked eye movements. As in Fig. 29 the "conditioning optokinetic stimulus" was presented to the left eye moving in the posteroanterior direction at 8 deg/sec, with reference to the "viewing" left eye. The position of the occluded right eye was recorded. A spontaneous nystagmus to the right was observed when the rabbit was placed in total darkness following 60 min of optokinetic "conditioning." The VOR (sinusoidal, .1 Hz ± 10 deg) was tested at 5-min intervals following the "conditioning" period. The *velocity disparity, velocity bias* of the VOR, and the *spontaneous nystagmus* velocity are plotted as a function of time. Note the slow time course with which these measurements return toward the normal range.

Nystagmus Induced by Microstimulation of the Dorsal Cap

Earlier experiments on microstimulation of the dorsal cap suggested that such stimulation might produce a prolonged drift of the eyes. In fact, in many of these earlier experiments it was necessary for a full minute to elapse between trains of microstimulation to allow the evoked drift of the eyes to abate. Consequently, an attempt was made to induce a nystagmus similar to that observed following prolonged optokinetic stimulation, by delivering multiple trains of microstimulation to the right dorsal cap of unanesthetized rabbits. In the experiment illustrated in Fig. 31, the placement of the microstimulation electrode was guided by physiologically monitoring the multiple-unit activity evoked in the right dorsal cap by optokinetic stimulation of the left eye. Multiple trains of microstimulation to the dorsal cap were delivered over a 15-min period. A rightward nystagmus was produced which was temporarily suppressed when

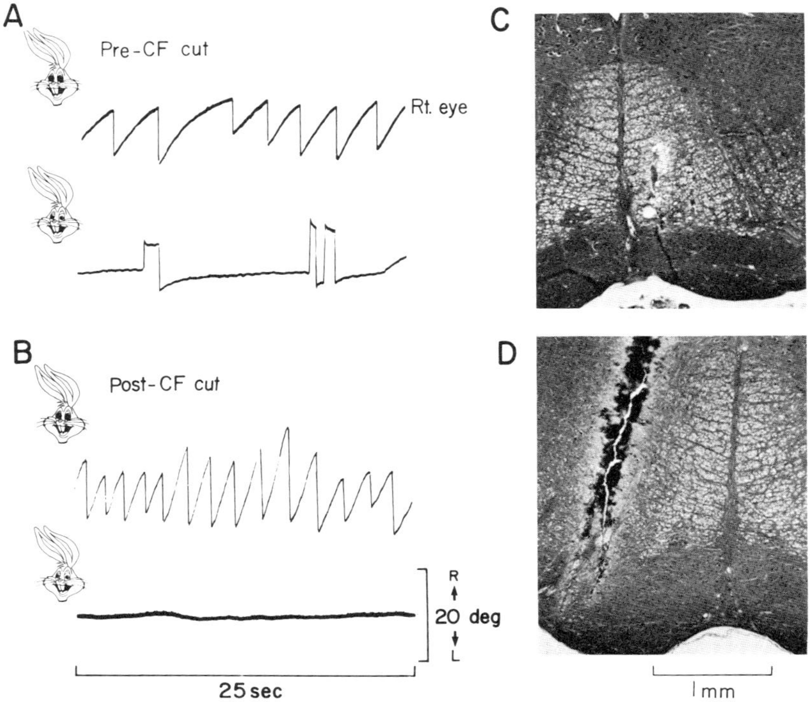

FIG. 31. The effect of transection of the olivocerebellar pathway on nystagmus induced by electrical stimulation of the right dorsal cap. The right dorsal cap was stimulated with 2-sec trains, 30 pps, 200-μ sec pulses, 20-50 μa. The animal received 50 trains of stimulation over a 15-min period. This electrical stimulation evoked spontaneous eye movements which were directed toward the right. These eye movements were suppressed when the rabbit was exposed to visual contrast. **B:** The rabbit was placed under halothane anesthesia for 10 min during which the olivocerebellar pathway originating from the right dorsal cap was severed by a transection of the left brainstem for 4 mm in the longitudinal plane. This transection did not alter the spontaneous nystagmus previously evoked by electrical stimulation of the right dorsal cap. **C:** Histological verification of the location of the stimulating electrode within the right dorsal cap. **D:** Histological verification of the transection of the left brainstem at a more rostral level of the dorsal cap.

the rabbit was exposed to visual contrast (Fig. 31A). This finding suggested that a sustained rightward drift of the eyes could have been caused by an electrically induced and endogenously maintained increase in CF discharge. Alternatively, it was possible that the rightward drift was induced by a CF-evoked change in the excitability of neurons at a more central location. In order to choose between these two alternatives, the rabbit was temporarily anesthetized with halothane, and the crossed CF projection from the right dorsal cap to the left flocculus was transected as it passed through the left brainstem (Fig. 31D). After the rabbit recovered from the halothane, it was apparent that neither the anesthetic, nor the transection of the CF projection, had caused the nystagmus to abate (Fig. 31B). These data suggest that electrical stimulation of the visual olivocerebellar pathway can induce a long-term drift of the eyes, but that once this drift is induced, it is not necessary that the olivocerebellar pathway remain intact for the drift to be maintained.

SUMMARY AND CONCLUSIONS

The visual olivocerebellar pathway appears to participate in the control of eye movement. It seems likely that the same circuitry would be used to regulate other kinds of postural adjustment and might also be useful in providing a relatively fixed spatial reference. However, for the sake of technical and theoretical simplicity, these other kinds of postural adjustments have not been examined in these experiments. The CFs which originate from the dorsal cap encode information concerning the horizontal retinal slip, at low velocities, of contour-rich stimuli of large angular subtense. This information modulates the activity of floccular Purkinje cells, which in turn modulate the activity of the vestibular nuclei neurons which project onto oculomotor neurons. One immediate behavioral consequence of activity in this circuitry is the reduction of slip of the visual world on the retina. However, it seems unlikely that an immediate reduction of retinal slip velocity is the only behavioral function of this circuitry, considering the rich possibilities for interactions with other classes of sensory information within the cerebellum and at the level of the brainstem. Damage to either the dorsal cap or the flocculus reduces the *gain* of optokinetically induced eye movements, but does not abolish optokinetic reflexes. These observations are in agreement with previous reports which suggested that damage to the flocculus would reduce the efficacy of vision in suppressing vestibular nystagmus induced calorically (89), or in modifying the phase and gain of the HVOR (54). The present series of experiments have demonstrated that the influence of the visual olivocerebellar pathway on eye movements is conjugate. These observations do not agree with a previous suggestion that the functional influence of the flocculus is restricted to movements of the ipsilateral eye (54). Even when the dorsal cap or the flocculus is destroyed, the activity of neurons of the medial and superior vestibular nuclei can still be modified by optokinetic stimulation (Barmack and Pettorossi, *unpublished observations*). Therefore, the unique properties of the visual olivocerebellar circuitry must include functions upon which our experiments have only touched. The sustained bias of the HVOR caused by prolonged activation of this circuitry might represent a unique function of the visual olivocerebellar pathway, not shared by extracerebellar optokinetic circuits. But, at present, the evidence for attributing this long-term modulatory property solely to the visual olivocerebellar pathway is not sufficiently decisive. Nor do we understand where or how the sustained changes evoked by optokinetic stimulation and electrical stimulation of the inferior olive occur. Because equivalent modifications of horizontal eye movements have been induced by optokinetic stimulation and electrical stimulation, it is tempting to assume that the modifications are induced in a common set of elements which are activated by both modes of stimulation. However, this may not be the case. In fact, this perverse property of modifiability may be shared by a number of different optokinetic circuits. On the basis of the present experiments, it *is* possible to exclude the inferior olive as a locus of the sustained change in excitability evoked by optokinetic stimulation or electrical stimulation. It seems reasonable to surmise that the cerebellum or the vestibular nuclei might be anatomical loci where optokinetically or electrically induced changes in the HVOR could be produced, but the mechanism by which the sustained changes are induced remains obscure. One is tempted to examine the trigger features of floccular Purkinje cells during optokinetic conditioning. However, assuming that the technical problems associated with the necessity of monitoring unit activity for periods of hours could be solved, the interpretation of a positive result might be confounded by the reciprocal connections of the flocculus with the vestibular nuclei. Similarly, a negative result would not necessarily discount the flocculus as a site of sustained modification. For example, if there is convergence of a subpopulation of Purkinje cells onto the vestibular nuclei, a microelectrode might not sample, without additional criteria for selection of cells, members of that subpopulation. And if there is convergence of Purkinje cells onto individual vestibular neurons, then very small changes in Purkinje cell discharge could be translated into large changes in vestibular excitability.

If one examines the circuitry of the cerebellar cortex, one is impressed with the strategic location of Golgi cells for limiting the activity of the mossy fiber–granule cell inputs to Purkinje cells (Fig. 23). The fact that Golgi cells receive a CF input suggests that these cells may participate in the CF-evoked pause in SS activity. Future attempts to develop criteria for identifying microelectrode recordings from Golgi cells in unanesthetized animals might lead to a better understanding of the role of this particular inhibitory interneuron in the immediate and sustained modification of Purkinje cell activity.

ACKNOWLEDGMENTS

This research was supported by N.I.H. Grant EY-00848 and the Oregon Lions Sight Foundation. The collaboration of D. T. Hess and J. I. Simpson on some of the experiments reported herein is gratefully acknowledged.

REFERENCES

1. Albus, J. S. (1971): A theory of cerebellar function. *Math. Biosci.,* 10:25–61.
2. Alley, K., Baker, R., and Simpson, J. I. (1975): Afferents to the vestibulo-cerebellum and the origin of the visual climbing fibers in the rabbit. *Brain Res.,* 98:582–589.
3. Angaut, P., and Brodal, A. (1967): The projection of the "vestibulocerebellum" onto the vestibular nuclei in the cat. *Arch. Ital. Biol.,* 105:441–479.
4. Armstrong, D. M. (1974): Functional significance of connections of the inferior olive. *Physiol. Rev.,* 54:358–417.

5. Baker, R., and Berthoz, A. (1975): Is the prepositus hypoglossi nucleus the source of another vestibulo-ocular pathway? *Brain Res.,* 86:121–127.
6. Baker, R., and Highstein, S. M. (1975): Physiological identification of interneurons and motoneurons in the abducens nucleus. *Brain Res.,* 91:292–298.
7. Barmack, N. H. (1977): Visually evoked activity of neurons in the dorsal cap of the inferior olive and its relationship to the control of eye movements. In: *Control of Gaze by Brain Stem Neurons, Developments in Neuroscience,* edited by R. Baker and A. Berthoz. pp. 361–370. Elsevier/North-Holland Biomedical Press, Amsterdam.
8. Bell, C., and Kawasaki, T. (1972): Relations among climbing fiber responses of nearby Purkinje cells. *J. Neurophysiol.,* 35:155–169.
9. Berman, N. (1977): Connections of the pretectum in the cat. *J. Comp. Neurol.,* 174:227–254.
10. Berthoz, A., and Llinás, R. (1974): Afferent neck projection to the cat cerebellar cortex. *Exp. Brain Res.,* 20:385–401.
11. Beinfang, D. C. (1978): The course of direct projections from the abducens nucleus to the contralateral medial rectus subdivision of the oculomotor nucleus in the cat. *Brain Res.,* 145:277–289.
12. Boylls, C. C., Jr. (1975): A theory of cerebellar function with applications to locomotion. I. The physiological role of climbing fiber inputs in anterior lobe operation. *COINS Technical Report 75C-6, Computer and Information Science.* University of Massachusetts at Amherst.
13. Boylls, C. C., Jr. (1975): A theory of cerebellar function with applications to locomotion. II. The relation of anterior lobe climbing fiber function to locomotor behavior in the cat. *COINS Technical Report 76–1, Computer and Information Science.* University of Massachusetts at Amherst.
14. Bradley, P., and Berry, M. (1976): Quantitative effects of climbing fibre deafferentation on the adult Purkinje cell dendritic tree. *Brain Res.,* 112:133–140.
15. Bradley, P., and Berry, M. (1976): The effects of reduced climbing and parallel fibre input on Purkinje cell dendritic growth. *Brain Res.,* 109:133–151.
16. Brodal, A. (1967): Anatomical organization of cerebello-vestibulo-spinal pathways. In: *Ciba Foundation Symposium of Myotatic, Kinesthetic and Vestibular Mechanisms,* edited by A. V. S. deReuck and J. Knight, pp. 148–166. J. & A. Churchill Ltd., London.
17. Brodal, A., and Hoivik, B. (1964): Site and mode of termination of primary vestibulocerebellar fibres in the cat. *Arch. Ital. Biol.,* 102:1–21.
18. Collewijn, H. (1975): Direction-selective units in the rabbit's nucleus of the optic tract. *Brain Res.,* 100:489–508.
19. Courville, J. (1975): Distribution of olivocerebellar fibers demonstrated by a radioautographic tracing method. *Brain Res.,* 95:253–263.
20. Courville, J., and Diakiw, N. (1976): Cerebellar corticonuclear projection in the cat. The vermis of the anterior and posterior lobes. *Brain Res.,* 110:1–20.
21. De Montigny, C., and Lamarre, Y. (1973): Rhythmic activity induced by harmaline in the olivo-cerebello-bulbar system of the cat. *Brain Res.,* 53:81–95.
22. Desclin, J. C. (1974): Histological evidence supporting the inferior olive as the major source of cerebellar climbing fibers in the rat. *Brain Res.,* 77:365–384.
23. Desclin, J. C. (1976): Early terminal degeneration of cerebellar climbing fibers after destruction of the inferior olive in the rat. Synaptic relationships in the molecular layer. *Anat. Embryol.,* 149:87–112.
24. Dow, R. S. (1969): Cerebellar syndromes. In: *Handbook of Clinical Neurology,* edited by P. J. Vinken and G. W. Bruyn, pp. 392–431. North-Holland Publishing Company, Amsterdam.
25. Eccles, J. C., Llinás, R., and Sasaki, K. (1966): The mossy fibre-granule cell relay of the cerebellum and its inhibitory control by Golgi cells. *Exp. Brain Res.,* 1:82–101.
26. Eccles, J. C., Llinás, R., and Sasaki, K. (1966): The excitatory synaptic action of climbing fibres on the Purkinje cells of the cerebellum. *J. Physiol.,* 182:268–296.
27. Fox, C. A., Andrade, A., and Schwyn, R. C. (1969): Climbing fiber branching in the granular layer. In: *Neurobiology of Cerebellar Evolution and Development,* edited by R. Llinas, pp. 603–611. American Medical Association, Chicago.
28. Fox, C. A., Hillman, D. E., Siegesmund, K. A., and Dutta, C. R. (1967): The primate cerebellar cortex: A Golgi and electron microscopic study. In: *The Cerebellum: Progress in Brain Research,* edited by C. A. Fox and R. S. Snider, pp. 174–225. Elsevier Publishing Company, Amsterdam.
29. Freedman, S. L., Voogd, J., and Vielvoye, G. J. (1977): Experimental evidence for climbing fibers in avian cerebellum. *J. Comp. Neurol.,* 175:243–252.

30. Friedrich, V. L., Jr., and Mugnaini, E. (1978): Relative numbers of granule and Purkinje cells in cat cerebellar cortex. 91st Session Am. Assoc. Anatomists, Vancouver, B.C.
31. Gacek, R. R. (1971): Anatomical demonstration of the vestibulo-ocular projections in the cat. *Acta Otolaryngol.,* 293:1–63.
32. Gacek, R. R. (1977): Location of brain stem neurons projecting to the oculomotor nucleus in the cat. *Exp. Neurol.,* 57:735–749.
33. Ghelarducci, B., Ito, M., and Yagi, N. (1975): Impulse discharges from flocculus Purkinje cells of alert rabbits during visual stimulation combined with horizontal head rotation. *Brain Res.,* 87:66–72.
34. Gilbert, P. F. C., and Thach, W. T. (1977): Purkinje cell activity during motor learning. *Brain Res.,* 128:309–328.
35. Giolli, R. A., and Guthrie, M. D. (1969): The primary optic projections in the rabbit. An experimental degeneration study. *J. Comp. Neurol.,* 136:99–126.
36. Granit, R., and Phillips, C. G. (1956): Excitatory and inhibitory processes acting upon individual Purkinje cells of the cerebellum in cats. *J. Physiol.,* 133:520–547.
37. Graybiel, A. M. (1977): Direct and indirect preoculomotor pathways of the brainstem: An autoradiographic study of the pontine reticular formation in the cat. *J. Comp. Neurol.,* 175:37–78.
38. Graybiel, A. M., and Hartwieg, E. A. (1974): Some afferent connections of the oculomotor complex in the cat: An experimental study with tracer techniques. *Brain Res.,* 81:543–551.
39. Groenewegen, H. J., and Voogd, J. (1977): The parasagittal zonation within the olivocerebellar projection. I. Climbing fiber distribution in the vermis of cat cerebellum. *J. Comp. Neurol.,* 174:417–488.
40. Haines, D. E. (1977): Cerebellar corticonuclear and corticovestibular fibers of the flocculonodular lobe in a prosimian primate *(Galago senegalensis). J. Comp. Neurol.,* 174:607–629.
41. Hámori, J. (1973): The inductive role of presynaptic axons in the development of postsynaptic spines. *Brain Res.,* 62:337–344.
42. Hámori, J., and Szentágothai, J. (1966): Identification under the electron microscope of climbing fibers and their synaptic contacts. *Exp. Brain Res.,* 1:65–81.
43. Hendrickson, A., and Winfield, J. A. (1978): The fast visual mossy fiber pathway to flocculus originates in the medial terminal nucleus of the accessory optic tract. 91st Session Am. Assoc. Anatomists, Vancouver, B.C.
44. Highstein, S. M. (1973): The organization of the vestibulo-oculomotor and trochlear reflex pathways in the rabbit. *Exp. Brain Res.,* 17:285–300.
45. Highstein, S. M. (1973): Synaptic linkage in the vestibulo-ocular and cerebello-vestibular pathways to the VIth nucleus in the rabbit. *Exp. Brain Res.,* 17:310–314.
46. Highstein, S. M., Ito, M., and Tsuchiya, T. (1971): Synaptic linkage in the vestibulo-ocular reflex pathway of rabbit. *Exp. Brain Res.,* 13:306–326.
47. Hoddevik, G. H., and Brodal, A. (1977): The olivocerebellar projection studied with the method of retrograde axonal transport of horseradish peroxidase. V. The projections to the flocculonodular lobe and the paraflocculus in the rabbit. *J. Comp. Neurol.,* 176:269–280.
48. Hoffmann, K.-P., Behrend, K., and Schoppmann, A. (1976): A direct afferent visual pathway from the nucleus of the optic tract to the inferior olive in the cat. *Brain Res.,* 115:150–153.
49. Ito, M. (1976): Cerebellar learning control of vestibulo-ocular mechanisms. In: *Mechanisms in Transmission of Signals for Conscious Behavior,* edited by T. Desiraju, pp. 1–22. Elsevier, Amsterdam.
50. Ito, M., Nisimara, N., and Yamamoto, M. (1976): Pathways for the vestibulo-ocular reflex excitation arising from semicircular canals of rabbits. *Exp. Brain Res.,* 24:257–271.
51. Ito, M., Nisimaru, N., and Yamamoto, M. (1976): Postsynaptic inhibition of oculomotor neurons involved in vestibulo-ocular reflexes arising from semicircular canals of rabbits. *Exp. Brain Res.,* 24:273–283.
52. Ito, M., Nisimaru, N., and Yamamoto, M. (1976): Inhibitory interaction between the vestibulo-ocular reflexes arising from semicircular canals of rabbits. *Exp. Brain Res.,* 26:89–103.
53. Ito, M., Nisimaru, N., and Yamamoto, M. (1977): Specific patterns of neuronal connexions involved in the control of the rabbit's vestibuloocular reflexes by the cerebellar flocculus. *J. Physiol.,* 265:833–854.
54. Ito, M., Shida, T., Yagi, N., and Yamamoto, M. (1974): Visual influence on rabbit horizontal vestibulo-ocular reflex presumably effected via the cerebellar flocculus. *Brain Res.,* 65:170–174.

55. Kanaseki, T., and Sprague, J. M. (1974): Anatomical organization of pretectal nuclei and tectal laminae in the cat. *J. Comp. Neurol.,* 158:319–337.
56. Kawaguchi, S., Yamamoto, T., Mizuno, N., and Iwahori, N. (1975): The role of climbing fibers in the development of Purkinje cell dendrites. *Neurosci. Letters,* 1:301–304.
57. Kooy, F. H. (1916): The inferior olive in vertebrates. *Folia Neurobiol. (Leipzig),* 10:205–369.
58. Kornguth, S. E., and Scott, G. (1972): The role of climbing fibers in the formation of Purkinje cell dendrites. *J. Comp. Neurol.,* 146:61–82.
59. Lamarre, Y., and Weiss, M. (1973): Harmaline-induced rhythmic activity of alpha and gamma motoneurons in the cat. *Brain Res.,* 63:430–434.
60. Lange, W. (1975): Cell number and cell density in the cerebellar cortex of man and some other mammals. *Cell. Tiss. Res.,* 154:115–124.
61. Larramendi, L. M. H., and Victor, R. (1967): Synapses on the Purkinje cell spines in the mouse. An electron microscopic study. *Brain Res.,* 5:15–30.
62. Llinás, R. (1970): Neuronal operations in cerebellar transactions. In: *The Neurosciences: Second Study Program,* edited by F. O. Schmitt, pp. 409–426. Rockefeller Univ. Press, New York.
63. Llinás, R., and Volkind, R. A. (1973): The olivo-cerebellar system: Functional properties as revealed by harmaline-induced tremor. *Exp. Brain Res.,* 18:69–87.
64. Maciewicz, R. J., Eagen, K., Kaneko, C. R. S., and Highstein, S. M. (1977): Vestibular and medullary brain stem afferents to the abducens nucleus in the cat. *Brain Res.,* 123:229–240.
65. Maekawa, K., and Simpson, J. I. (1973): Climbing fiber responses evoked in vestibulocerebellum of rabbit from visual system. *J. Neurophysiol.,* 36:649–666.
66. Maekawa, K., and Takeda, T. (1975): Mossy fiber responses evoked in the cerebellar flocculus of rabbits by stimulation of the optic pathway. *Brain Res.,* 98:590–595.
67. Maekawa, K., and Takeda, T. (1976): Electrophysiological identification of the climbing and mossy fiber pathways from the rabbit's retina to the contralateral cerebellar flocculus. *Brain Res.,* 109:169–174.
68. Marani, E., and Voogd, J. (1977): An acetylcholinesterase band-pattern in the molecular layer of the cat cerebellum. *J. Anat.,* 124:335–345.
69. Marr, D. (1969): A theory of cerebellar cortex. *J. Physiol.,* 202:437–470.
70. McMasters, R. E., Weiss, A. H., and Carpenter, M. B. (1966): Vestibular projections to the nuclei of the extraocular muscles. Degeneration resulting from discrete partial lesions of the vestibular nuclei in the monkey. *Am. J. Anat.,* 118:163–193.
71. Mizuno, N., Mochizuki, K., Akimoto, C., and Matsushima, R. (1973): Pretectal projections to the inferior olive in the rabbit. *Exp. Neurol.,* 39:498–506.
72. Mizuno, N., Nakamura, Y., and Iwahori, N. (1974): An electron microscope study of the dorsal cap of the inferior olive in the rabbit, with special reference to the pretecto-olivary fibers. *Brain Res.,* 77:385–395.
73. Mlonyeni, M. (1973): The number of Purkinje cells and inferior olivary neurones in the cat. *J. Comp. Neurol.,* 147:1–9.
74. Murphy, J. T., MacKay, W. A., and Johnson, F. (1973): Differences between cerebellar mossy and climbing fibre responses to natural stimulation of forelimb muscle proprioceptors. *Brain Res.,* 55:263–289.
75. Oscarsson, O. (1969): The sagittal organization of the cerebellar anterior lobe as revealed by the projection patterns of the climbing fiber system. In: *Neurobiology of Cerebellar Evolution and Development,* edited by R. Llinás, pp. 525–537. American Medical Association, Chicago.
76. Palay, S. L., and Chan-Palay, V. (1974): *Cerebellar Cortex.* Springer-Verlag, New York.
77. Palkovits, M., Magyar, P., and Szentágothai, J. (1971): Quantitative histological analysis of the cerebellar cortex in the cat. I. Number and arrangement in space of the Purkinje cells. *Brain Res.,* 32:1–13.
78. Precht, W., and Llinás, R. (1969): Functional organization of the vestibular afferents to the cerebellar cortex of frog and cat. *Exp. Brain Res.,* 9:30–52.
79. Rushmer, D. S., Roberts, W. J., and Augter, G. K. (1976): Climbing fiber responses of cerebellar Purkinje cells to passive movements of the cat forepaw. *Brain Res.,* 106:1–20.
80. Scott, T. G. (1965): The specificity of 5′-nucleotidase in the brain of the mouse. *J. Histochem. Cytochem.,* 13:657–667.
81. Scott, T. G. (1967): The distribution of 5′-nucleotidase in the brain of the mouse. *J. Comp. Neurol.,* 129:97–114.
82. Shinoda, Y., and Yoshida, K. (1975): Neural pathways from the vestibular labyrinths to the flocculus in the cat. *Exp. Brain Res.,* 22:97–111.

83. Simantov, R., Snyder, S. H., and Oster-Granite, M. L. (1976): Harmaline-induced tremor in the rat: Abolition by 3-acetylpyridine destruction of cerebellar climbing fibers. *Brain Res.*, 114:144–151.
84. Simpson, J. I., and Alley, K. E. (1974): Visual climbing fiber input to rabbit vestibulocerebellum: A source of direction-specific information. *Brain Res.*, 82:302–308.
85. Sotelo, C., and Arsenio-Nunes, M. L. (1976): Development of Purkinje cells in absence of climbing fibers. *Brain Res.*, 111:389–395.
86. Sotelo, C., Hillman, D. E., Zamora, A. J., and Llinás, R. (1975): Climbing fiber deafferentation: Its action on Purkinje cell dendritic spines. *Brain Res.*, 98:574–581.
87. Stein, B. M., and Carpenter, M. B. (1967): Central projections of portions of the vestibular ganglia innervating specific parts of the labyrinth in the rhesus monkey. *Am. J. Anat.*, 120:281–317.
88. Takeda, T., and Maekawa, K. (1976): The origin of the pretecto-olivary tract. A study using the horseradish peroxidase method. *Brain Res.*, 117:319–325.
89. Takemori, S., and Cohen B. (1974): Loss of visual suppression of vestibular nystagmus after flocculus lesions. *Brain Res.*, 72:213–224.
90. Tarlov, E. (1970): Organization of vestibulo-oculomotor projections in the cat. *Brain Res.*, 20:159–179.
91. Thach, W. T., Jr. (1967): Somatosensory receptive fields of single units in cat cerebellar cortex. *J. Neurophysiol.*, 30:675–696.
92. Thorpe, P., and Herbert, J. (1976): The accessory optic system of the ferret. *J. Comp. Neurol.*, 170:295–309.
93. Wilson, V. J., and Maeda, M. (1973): Connections between semicircular canals and neck motoneurons in the cat. *J. Neurophysiol.*, 37:346–357.
94. Wilson, V. J., Maeda, M., Franck, J. I., and Shimazu, H. (1976): Mossy fiber neck and second-order labyrinthine projections to cat flocculus. *J. Neurophysiol.*, 39:301–310.
95. Yamamoto, M. (1978): Localization of rabbit's flocculus Purkinje cells projecting to the cerebellar lateral nucleus and the nucleus prepositus hypoglossi investigated by means of the horseradish peroxidase retrograde axonal transport. *Neurosci. Letters*, 7:197–202.

Posture and Movement, edited by
R. E. Talbott and D. R. Humphrey.
Raven Press, New York © 1979.

Characteristics of the Motor Programs Underlying Visually Evoked Movements

Emilio Bizzi and Andres Polit

Department of Psychology, Massachusetts Institute of Technology, Cambridge, Massachusetts 02139

The experiments described here are addressed at identifying some of the processes underlying arm and head movements in monkeys. We have indicated that these movements depend, at least partially, upon a process that specifies an equilibrium point between agonist and antagonist muscles rather than amplitude and duration of movement. In addition, our findings suggest that the major role of the feedback system is in updating and adjusting the execution of learned motor patterns when posture is changed.

INTRODUCTION

In this chapter we will touch upon two interrelated themes. First, we will report on recent findings in monkeys that have revealed some characteristics of the process whereby head and arm movements reach an intended posture. Second, we will describe experiments designed to elucidate the interactions between the motor programs underlying these head and arm movements and proprioceptive feedback originating from neck and arm muscles.

Although proprioceptive feedback has been studied by many physiologists, there is still a great deal of uncertainty about its role in the execution of voluntary movements. It has been proposed, for instance, that the termination of a voluntary movement and the subsequent maintenance of a posture depend on a readout of proprioceptive afferent signals generated during the movement. Some of the proponents of this idea have suggested that there is a "comparator" in some part of the central nervous system which, upon receiving inputs from motor commands and afferent feedback, provides a signal leading to the cessation of an ongoing motor pattern (4,5). Other hypotheses have been proposed. Among these are that the proprioceptive apparatus plays a role in providing compensation for unexpected changes in load (7), in linearizing muscle properties (8) and, more generally, in reprogramming. Naturally, the proprioceptive apparatus might well subserve all of these functions, since they are not mutually exclusive. It is important, however, in order to gain perspective on the proprioceptive apparatus, to determine which of its functions are of major importance and which are of only minor importance. Accordingly, we will attempt to evaluate and hierarchically order the functions of proprioceptive feedback in a well-defined category of voluntary movement: head and arm movement elicited by the presentation of a visual target.

Before presenting our ideas on the contribution of proprioception to these movements, we will consider some of the properties of the central process or command by which they are specified. We feel that there is a good reason for considering the two aspects of these movements in this order; obviously the afferent feedback, particularly from muscle spindles, depends on the specifications contained in the initial motor command. Accordingly, we will first describe the process whereby the monkey's head moves in response to the presentation of a visual target. Second, we will present evidence indicating that visually triggered arm movements are controlled, to a first approximation, by a similar process. We will conclude by discussing the role of afferent feedback in visually triggered head and arm movements.

HEAD MOVEMENT AND POSTURE IN INTACT AND IN DEAFFERENTED MONKEYS

In our experiments on the mechanism of movement termination, we examined the head movements which are part of the monkey's coordinated eye-head response to the presentation of a visual stimulus (3). We disturbed centrally initiated head movements by applying loads; our goal was to observe the effect of the resulting proprioceptive response on the final position of the head. When we applied a constant-torque load whose effect extended beyond the dynamic phase, we observed a constant degree of head undershoot (Fig. 1). While the constant load was being applied, there was an increase in muscle-spindle discharge, indicated by an increase in electromyographic (EMG) activity. Presumably, tendon organ activity also increased, and there was a modification of postural information from joint receptors. However, in spite of these changes in the flow of proprioceptive activity, the head reached its "intended" final position after the constant load was removed (Fig. 1). This final head position was equal (on average) to that reached when the load had not been applied, suggesting that the program for final position was maintained during load application and was not readjusted by proprioceptive signals acting at segmental and suprasegmental levels. We concluded that the central program establishing final head position is not dependent on a readout of proprioceptive afferents generated during the movement, but is preprogrammed.

To test this hypothesis further, we investigated the way in which our monkeys reached final head position when they were deprived of neck proprioceptive feedback in addition to visual feedback. The goal here was to observe how monkeys moving their heads in an "open-loop" mode dealt with a constant torque applied during centrally initiated movements. Figure 2 shows that follow-

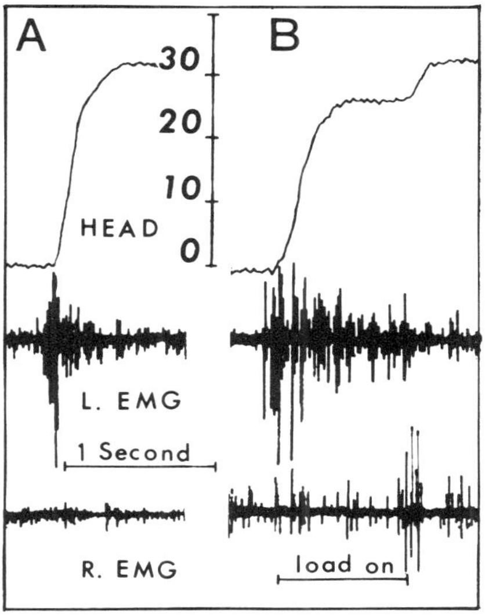

FIG. 1. Typical visually triggered head movement in chronically vestibulectomized monkey to appearance of target at 40° but performed in total darkness. **A:** An unloaded movement. **B:** A constant-force load was applied at the start of the movement resulting in an undershoot of final position relative to **A,** despite the increase in EMG activity. Note head returns to the same final position after removal of the load. Vertical calibration in degrees; time marker is 1 sec. EMG recorded from left and right splenium capitis.

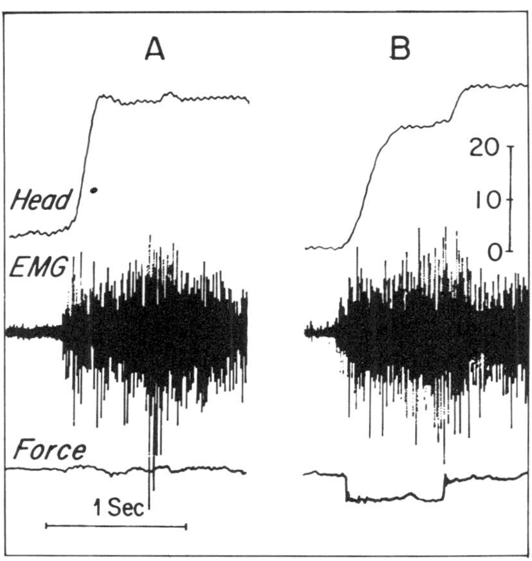

FIG. 2. Typical movements of a chronically vestibulectomized monkey with sectioned dorsal roots (C1-T3) made open-loop (in total darkness). **B:** A constant force load was applied at the start of the movement, resulting in an undershoot while the load was on. Similarity of EMG pattern in **A** and **B** shows lack of stretch reflex. Vertical calibration in degrees. [From ref. 3, E. Bizzi, A. Polit, and P. Morasso, *J. Neurophysiol.,* 39:435–443 (1976).]

ing the unexpected application of a constant-torque load at the beginning of a visually triggered movement, the head attained a posture short of its intended final position. It should be emphasized that although the head movement was elicited by the target, it was not guided by a target light. It should also be noted that the animals were chronically vestibulectomized. Again, the position that the head attained after removal of the constant torque was found to be statistically equal to the position that the head reached when the load was not applied (3). These results indicate that the behavior of the motor system with respect to head posture is the same before and after deafferentation.

We believe that the results of this series of experiments contribute to our understanding of the mechanism whereby movement is terminated and a newly acquired position is maintained. If we assume that the "program" for head movements and posture specifies a given level of alpha motoneuron activity to both agonist and antagonist muscles, and that the firing of these neurons will determine a particular length-tension curve in each muscle, then we must conclude that the final resting position of the head is determined by the length-tension properties of all of the muscles involved. This hypothesis explains both the head undershoot when a constant load is applied, and the attainment of the intended final head position following the removal of the load, shown in Figs. 1 and 2, respectively. Although the process of selecting a new equilibrium between the length-tension properties of agonists and antagonists should result in *movement* and the attainment of a new head position, it should be clear that our experiments do not rule out the presence of other, parallel processes. It is possible, for instance, that the dynamic changes are controlled through separate mechanisms. Furthermore, as is clearly shown in Fig. 1, the process specifying agonist-antagonist equilibrium must take into account the activation of the proprioceptors during both loaded and unloaded movements.

VISUALLY TRIGGERED ARM MOVEMENT IN INTACT AND DEAFFERENTED MONKEYS

In a complementary set of experiments involving arm movement, we extended the previously described findings on the final position of the head. Adult rhesus monkeys were trained to point to a target light with the forearm and to hold the arm in that position for about 1 sec in order to obtain a reward. The monkey was seated in a primate chair, and its forearm was fastened to an apparatus which permitted flexion and extension of the forearm about the elbow in the horizontal plane (9). A torque motor in series with the shaft of this apparatus was used to load the arm. The experiments were conducted in a dark room to minimize visual cues; at no time during an experiment was the animal able to see its forearm. At random times, we displaced the initial position of the forearm. In most cases, the positional disturbance was applied immediately after the appearance of the target light and was stopped just prior to the activation of the motor units in the agonist muscle. Hence, when the motor command specifying a given forearm movement occurred, the positional disturbance had altered the length of the agonist and antagonist muscles, and the proprioceptive stimulation resulting from this disturbance had altered their state of activation (Fig. 3). In spite of these changes, the intended final arm position was always reached; this was true whether the torque motor had displaced the forearm further away from, closer to, or even beyond the intended final position. There were no significant differences among the final positions achieved in these three conditions. Naturally, the attainment of the intended arm position in this experiment could be explained by assuming that afferent proprioceptive information modified the original motor command (note the large reflex response following arm displacement). However, the results of previous work on final head position suggest an alternative hypothesis: that the motor program underlying arm movement specifies, through the selection of a new set of length-tension curves, an equilibrium point between agonists and antagonists that correctly positions the arm in relation to the visual target. To investigate this hypothesis, we retested the monkey's pointing performance after it had undergone a bilateral C1-T3 dorsal rhizotomy. Remarkably, we could elicit the pointing response very soon after the surgery (within 2 days in some of the animals), even though we observed no spontaneous arm movements other than an occasional quick, flinging movement in response to a threat. The forearm was again displaced (at random times) immediately after the appearance of the target light and released just prior to the activation of motor units in the agonist muscles. Because the arm

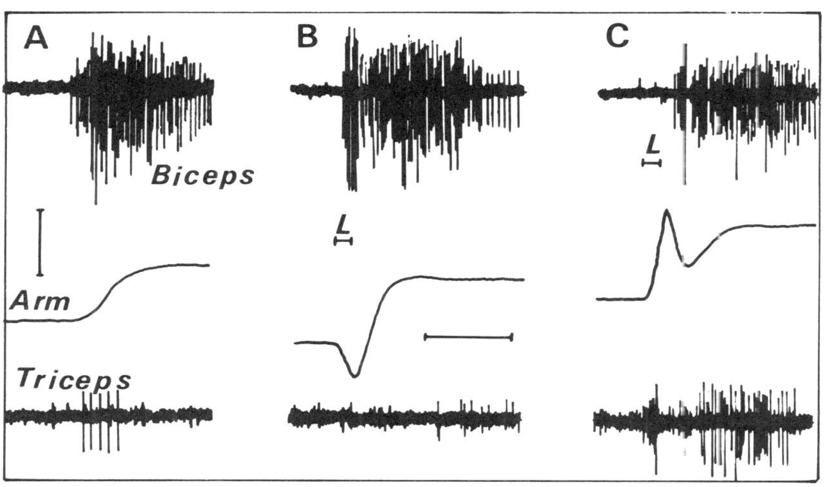

FIG. 3. Flexion movements made by intact monkeys to visual targets without sight of their limbs. **A:** A normal movement and the pattern of EMG activity recorded from the biceps and tricpes muscles. **B:** A movement in which the torque motor has displaced the arm away from the target. **C:** An instance in which the motor moved the limb toward the target zone. Note in **B** and **C** an increase in activity of stretched muscles. Time calibration = 1 sec; vertical bar = 15°.

was not visible to the animal and the proprioceptive activity could not reach the spinal cord, the arm reached its intended final position "open-loop." The fact that we never observed any sign of reflex response or reprogramming in the EMG activity corroborates this supposition. We found that the final arm position was reached even when the initial position was displayed. For each target position, *t*-tests were performed to test for differences between the unloaded and the loaded average final positions. No significant differences were found. This finding suggests that what is programmed is an intended equilibrium point, resulting from the interaction of agonist and antagonist muscles. The results when a constant force was applied with the torque motor throughout the arm movement add further support to this idea. The arm undershoot which we observed in this case suggests that the intended equilibrium point of the arm muscles was modified, in a predictable way, by the application of an external force.

If we assume that no radical change in motor programming occurred during the days immediately following the rhizotomy, we can conclude that visually triggered arm and head movements are subserved by a process that specifies a new intended equilibrium point between agonists and antagonists rather than amplitude or duration of movement. The fact that immediately after undergoing deafferentation, the monkey was able to execute accurate forearm movements in response to any one of 12 visual targets to which it had been trained to respond before the operation strongly implies that "programs" were released when the targets were presented. However, we do not know where these programs are located in the central nervous system, what they encode, or how they are implemented.

These findings ought to be qualified. First, although we may have detected a process underlying arm and head movement, it is certainly not the only process that occurs during the movement. It is quite clear, for instance, that the head (or arm) movements that monkeys use to reach a given position can vary in velocity. Consequently, the mechanism by which an intended posture is achieved must coexist with a mechanism specifying intended head (or arm) velocity. Second, the successful execution of the hypothesized "programs" in the deafferented animal is contingent upon the animal's knowing the position of the arm relative to the body. Whenever we changed the usual spatial relationship between the animal and the arm apparatus, the monkey's pointing response to the target was inaccurate. All of our intact monkeys, in contrast, were able to compensate quickly for any variations in their accustomed position with respect to the arm apparatus. The dramatic inability of the deafferented monkey to execute accurate pointing responses in an unusual postural setting underscores the great importance of afferent feedback in the control of movement.

CONCLUSIONS

We feel that in the experiments discussed here we have identified one of the processes underlying arm and head movement. Specifically, we have indicated that these movements depend, at least partially, upon a process that specifies an equilibrium point between agonists and antagonists rather than amplitude or duration of movement. We have also demonstrated that this process and the mechanism by which changes in load are offset develop independently and in parallel (2). Furthermore, the latter's contribution to the type of movement we have analyzed is rather modest (2). Because the velocity of both arm and head movements used to reach a given position could be programmed separately, we have postulated an independent process to account for this aspect of the movements. Taken together, our experiments, then, strongly suggest that at least three parallel processes underlie arm and head movement and that motor control is organized in a modular fashion. It should be stressed that we have been concerned with the logical organization of the motor system, and we have not tackled the demanding and difficult task of relating these processes underlying arm and head movement to brain circuits.

The second goal of our studies was to develop some perspectives on the role of afferent feedback during voluntary movement. Our main finding suggests that the major role of the feedback system is in updating and adjusting the execution of learned motor patterns when posture is changed. Compensation for unexpected disturbances occurring during a preprogrammed (learned) movement was not found to be a major function of afferent feedback (2).

Finally, our results, particularly the demonstration that visually elicited movements might be represented as commands for selecting and shifting equilibria between agonist and antagonist length-tension curves, are in line with the hypothesis that has been put forward to account for speech production (6). In addition, they are consistent with the work of Asatryan and Feldman (1), who have suggested that arm movements in man may result from a shift in the equilibrium point of the muscle-load system.

ACKNOWLEDGMENTS

These studies have been supported by the National Institutes of Health Research Grant NS 09343, awarded by the National Institute of Neurological Diseases and Stroke, and by the National Aeronautics and Space Administration Grant NGR 22–009–798.

REFERENCES

1. Asatryan, D. G., and Feldman, A. G. (1965): Biophysics of complex systems and mathematical models. Functional tuning of nervous system and control of movement of maintenance of a steady posture. I. Mechanographic analysis of the work of the joint on execution of a postural task. *Biophysics,* 10:925–933.
2. Bizzi, E., Dev, P., Morasso, P., and Polit, A. (1978): The effect of load disturbances during centrally initiated movements. *J. Neurophysiol.,* 41:542–556.
3. Bizzi, E., Polit, A., and Morasso, P. (1976): Mechanisms underlying achievement of final head position. *J. Neurophysiol.,* 39:435–443.
4. Eccles, J. C., Sabah, N. H., Schmidt, R. F., and Taborikova, H. (1972): Modes of operation of the cerebellum in the dynamic loop control of movement. *Brain Res.,* 40:73–80.

5. Gibbs, C. B. (1954): The continuous regulation of skilled responses by kinaesthetic feedback. *Br. J. Psychol.*, 45:24–39.
6. MacNeilage, P. F. (1970): Motor control of serial ordering of speech. *Psychol. Rev.*, 77:182–196.
7. Merton, P. A. (1972): How we control the contraction of our muscles. *Sci. Am.*, 226:30–37.
8. Nichols, T. R., and Houk, J. C. (1976): The improvement in linearity and the regulation of stiffness that results from the actions of the stretch reflex. *J. Neurophysiol.*, 39:119–142.
9. Polit, A., and Bizzi, E. (1979): Characteristics of the motor programs underlying arm movements in monkeys. *J. Neurophysiol.*, 42:183–194.

Posture and Movement, edited by
R. E. Talbott and D. R. Humphrey.
Raven Press, New York © 1979.

Visual-Vestibular Interaction

Laurence R. Young

Man Vehicle Laboratory, Department of Aeronautics and Astronautics, Massachusetts Institute of Technology, Cambridge, Massachusetts 02139

Following the pioneering observations of Ernst Mach (14) in the late 19th century, there was a relative hiatus in work on what we now term "visual-vestibular interaction." Serious interest in the roles played by visual scenes and vestibular inputs to spatial orientation began again with the earliest days of night flying and instrument flight. It soon became clear that, in the absence of a quasi-static visual or instrument indication of vertical and direction, attempts to "fly by the seat of the pants" led to disorientation at best. On the other hand, the motion cues provided by flight or in a moving-base simulator serve a vital role, not fulfilled by vision, in providing high-frequency acceleration information which allows the pilot to maintain stability.

Quantitative assessment of the relative importance of visual and nonvisual cues was investigated in the tilted-room experiments by Witkin and his associates (17) in the 1940s. By controlling the tilt of the observer's chair and the orientation of the room being observed, the experimenters could demonstrate the influence of the visual and nonvisual cues on the subjective indication of the vertical. Normally, the subject set his perceived vertical between the "visual vertical" and the true direction of down. Various sex-linked and personality traits were explored in terms of the relative strength of the "external" (visual) and "internal" (vestibular) cues.

In recent years, considerable attention has been directed to the phenomenon of an induced self-motion resulting from the observation of a wide, continuously moving field. Some well-known experiences of self-motion are commonly found when looking at a flowing river from a bridge, or having the adjacent train begin to move while seated in a stationary train. Mach first experimented with a wood and paper drum which enclosed his subject and could rotate independently of the subject about a vertical axis, inducing an alternating sensation of drum motion or self-motion opposite to a stationary drum. The phenomenon, later termed *circularvection,* could not be fully explained by nystagmic eye move-

This work was first presented at the Sixth International Symposium on Biocybernetics, "Control Mechanisms in Bio- and Ecosystems," International Federation of Automatic Control, Leipzig, East Germany, September, 1977.

ments, since it was present, and even strengthened, when eye movements were minimized by means of a fixation point. Several laboratories are currently active in the exploration of the phenomenon of visually induced motion, both the physiological description and the underlying neurophysiological mechanisms. In order to generate a strong and compelling illusion of motion (circularvection or its translatory equivalent, linearvection), one must provide a large, moving, peripheral visual field, up to 180°, containing sufficient high-contrast borders, and eliminate any stationary background features. Some of the most effective displays for circularvection have been in 360° or wide-screen motion pictures, amusement parks, and in wide-field-of-view ship or flight simulators.

Two basic tools of psychophysics are applied to the quantitative determination of visually induced motion effects. In the first of these, magnitude estimation, the subject is simply taught a reference scale against which to rate his instantaneous sensation of motion during the test series. Suppose for example, that a subject is rotated about a vertical axis, in the light, at 15°/sec and at 30°/sec, and instructed to call these velocities 15 and 30 respectively. When he is then either rotated in the dark (pure vestibular stimulation) or exposed to a rotating visual field (circularvection stimulation), he is able to estimate his apparent speed relative to that reference scale, either by use of a dial and meter as is practiced in our laboratory, or by calling out numbers. Despite the apparent roughness of the method, used so successfully by Stevens (15) and his followers, the standard error of the estimates is normally less than 10% of the estimate. The other principal technique used in measurement of circularvection is the nulling method, in which a subject attempts to balance the visually induced motion by a controlled and measured true body motion until he feels that he is returned to the reference null conditions. For both magnitude estimation and nulling procedures, it must be recognized that the measured output includes not only the dynamic response of the perceptual processes but also that of the motor control programs involved in setting an indicator. Fortunately, the manual-control literature provides us with adequate estimates of the portions of the response attributable to the tracking task.

The response to a full-field surround which suddenly begins to move at constant velocity is rather startling, although quite repeatable. At first, the veridical motion is sensed—the surround appears to be moving and the subject feels himself stationary. However, if the surround is large enough and placed in the peripheral field, and if the experiment is so arranged that the subject's mental set would permit self-rotation, a remarkable new sensation occurs. After a period of typically 2 to 5 sec, the visual field appears to slow down, often to a stop, and the subject perceives himself as rotating in the opposite direction. The postural reactions to this illusory motion can cause some observers to fall over. To prevent this, most 360° exhibits, like the one in Disneyland, provide a support to hang onto. The sensation of rotation builds to a maximum over a period of 3 to 10 sec, rising approximately as an exponential. The exact relationship between the final sensation of self-rotation and the field velocity depends to a great extent upon the details of the experimental situation. In order to achieve "saturation" of the effect, in which the perception of field motion is entirely replaced by one of self-motion with respect to a stationary field, it is useful to have a wide, rich field of view in the periphery, moving uniformly at velocities less than 60°/sec (5). Mach's paper drum, which rotated around a seated subject, is a particularly effective stimulus. If the visual surround is allowed to accelerate smoothly to its final velocity, at accelerations of only a few degrees/sec^2, comparable to the acceleration thresholds of the semicircular canals, then the phenomenon of self-motion is more likely to be a smooth, continuous development of circularvection. A model of conflict resolution in multisensory integration is useful (Fig. 1). When the self-motion implied by visual-field motion is slowly accelerating, and processed by some internal model of the labyrinth, it is in general agreement with the actual semicircular canal signals, which are indicating no change in velocity (18). However, when the visual field changes its velocity suddenly, the implied acceleration pulse is clearly not confirmed by any vestibular activity. The null semicircular canal signals appear to dominate over the visual in the short term, and leave the subject with an initial sensation of remaining stationary in the presence of a rotating visual field, only gradually changing to the steady state sense of rotation. The importance of the semicircular signals in modifying the interpretation of moving fields is clearly demonstrated by simply giving the subject a small velocity pulse opposite to the field motion. Even this brief vestibular stimulus in the correct direction, reducing the initial conflict between vestibular and visual signals, is sufficient to hasten the onset of circularvection.

Although it is tempting to explain circularvection on the basis of the optokinetic nystagmus pattern of eye movements induced by this rotating field, this is in fact inadequate. Circularvection is facilitated by the presence of a fixation point in the foreground, thereby minimizing nystagmus. Fischer and Kornmuller (8), working in Germany in the 1930s, and recently Dichgans and his colleagues

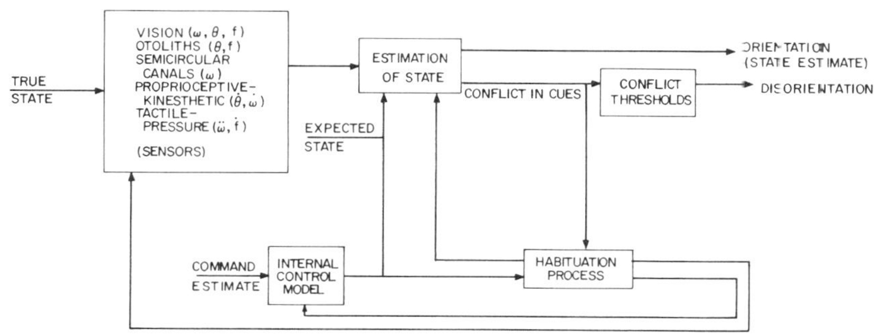

FIG. 1. Schematic representation of the human spatial orientation system, with each sensory channel assigned the geometric variables with which it is chiefly stimulated (ω = angular velocity, θ = orientation angle with respect to the vertical, f = gravitoinertial acceleration).

at Freiburg (3,5), showed that the generation of nystagmus from a central field stimulus did nothing to inhibit the development of circularvection based upon the oppositely moving peripheral field. It is conceivable that the facilitation of circularvection, when the eyes are fixed on a stationary spot and the field sweeps past, is related to the Aubert-Fleischel phenomenon, wherein a moving target appears faster when it sweeps across the retina of a stationary eye than when the eye follows it in pursuit movement.

Once circularvection is established, even though the visual field may be rotating about a vertical axis at a constant velocity, there is frequent waxing and waning of the phenomenon, with occasional precipitous losses of circularvection in which the subject feels himself to be stopped and the surround moving once more, followed by the characteristic slow reestablishment of circularvection sensation. The sensation gradually habituates, and on occasion even reverses direction. Similar findings hold for "linearvection," resulting from translating the entire surround. Brandt and his colleagues have recently shown that the occurrence of these sudden reversals of circularvection is highly correlated with eye movements into the direction that the field is moving, although voluntary eye movements do not either produce or inhibit such reversals (4).

Some simple experiments using simultaneous visual and vestibular stimuli are useful in revealing the manner in which our orientation system deals with diverse, often incompatible, inputs (Fig. 2). As described earlier, a constant rotation of the visual surround produces, after some seconds, a nearly constant circularvection in the opposite direction. A series of brief, constant-yaw-rate

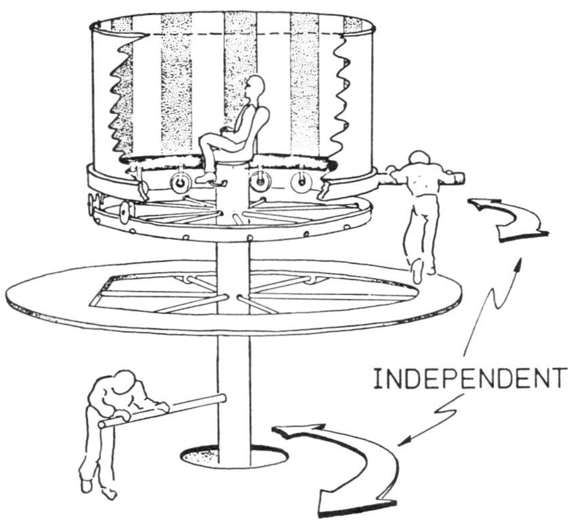

FIG. 2. A whimsical view of our visual-vestibular interaction experiments. The subject may be stimulated by independent drives of the base or of the surround relative to the base. The visual surround may also be fixed, either to the base or to the laboratory. (By Harkness, from Zacharias, ref. 23.)

turns of the subject's cabin also produces a turning sensation which is well predicted on the basis of models of semicircular canal function. What happens when both inputs are applied simultaneously? None of the obvious guesses is correct. The subject does not merely add the independent visual and vestibular response. Nor does he adopt just one or the other. Rather, he switches between the two—adopting largely the visual input when the vestibular signal is in agreement with it, and rapidly losing circularvection and adopting the vestibular output when visual and vestibular information are in significant conflict as to sign or large rates of change of rotation (19). In other words, the fast vestibular filter must gate the visual signal to permit circularvection to be adopted or maintained. Similarly, the presence of circularvection influences the processing of subsequent vestibular inputs, delaying the perception of true motion in the direction opposite to the visually induced turn. This simple effect may be of significant value for economizing on flight simulators for pilot training, where every foot of base motion adds to the simulator cost. By maintaining sustained motion sense with vection, the true base motion may be "washed out" or returned to neutral more quickly without alerting the pilot to the false cues.

Although the nonlinear gating or switching characteristic is revealed in some of the conflict-producing experiments, it was still of considerable interest to see how the human orientation system uses visual and vestibular cues in a more general, random-input situation. This was achieved in our modified Link Trainer by driving the rotating visual field randomly (as a sum of many sinusoids) and simultaneously turning the cabin according to another, independent random signal (Fig. 3). The subject was required to attempt to steer the trainer to keep it stationary in the laboratory. By using spectral analysis to calculate the subject's transfer function, from each of the input signals to his control-stick output, we were able to derive frequency responses for the visual and vestibular channels separately (22,23). As we had expected, the orientation system uses the different inputs appropriately, according to their role in the frequency spectrum, just as an optimal control system would do. At lower frequencies, where semicircular canal information contains excessive phase lead, and is unreliable because of drift, the visual input is dominant. Conversely, at higher frequencies, where visual rate information is noisier and the vestibular transducers faithfully measure velocity, the manual control is based almost entirely on the vestibular channel (Fig. 4). Linear approximations have their limitations, however, and more satisfactory models incorporate some of the major nonlinear "switching" aspects of visual-vestibular interaction. Zacharias (23) has simulated a "soft switch" model, in which the gain of the vestibular channel rises when the "conflict" between the channels becomes large (Fig. 5). This schema, which implies an internal model of the sensory dynamics, yields good quantitative predictions of transient and frequency-response data.

It is clear then that moving visual fields not only influence the perception of motion and orientation, but also in many ways the effect is that of mimicking pure vestibular responses. Inevitably, we are led to search for the underlying

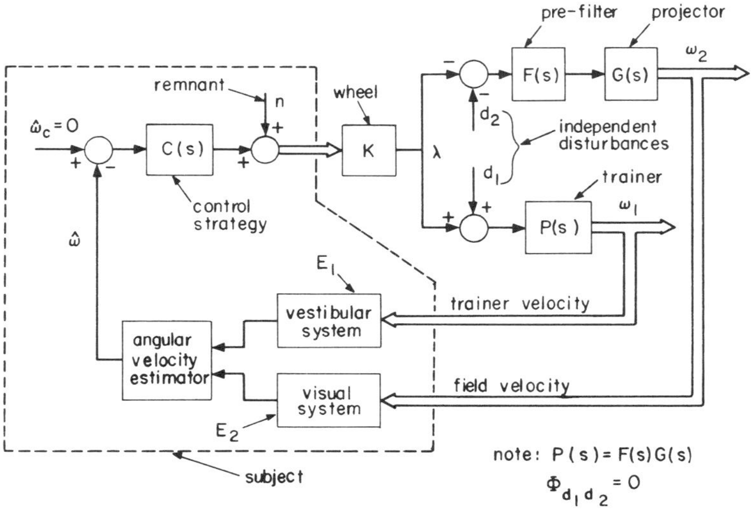

FIG. 3. Block diagram of the experiment to determine visual and vestibular influences on perceived motion as reflected in the task of stabilizing a Link trainer in yaw. (From Zacharias, ref. 23.)

neurophysiological mechanisms and especially for sites in the brain in which visual-vestibular interaction can be observed at the single-unit level of neural response. Since no direct anatomical pathways from the visual centers to the vestibular system had been established, it was natural to begin looking for such

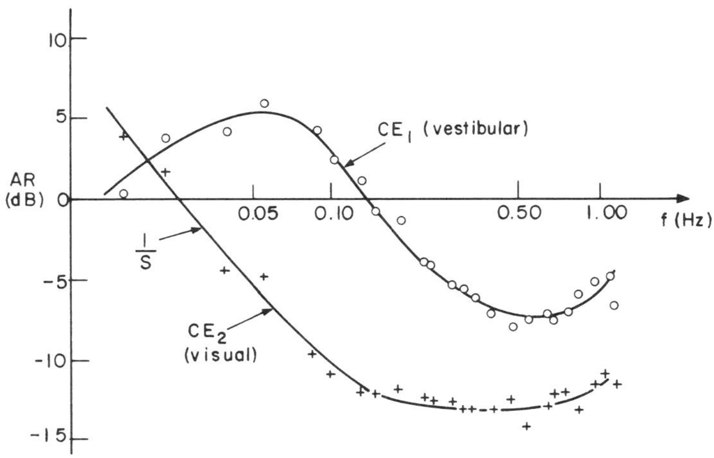

FIG. 4. Frequency response of manual control gains relative to visual and vestibular random inputs. (From Zacharias, ref. 23.)

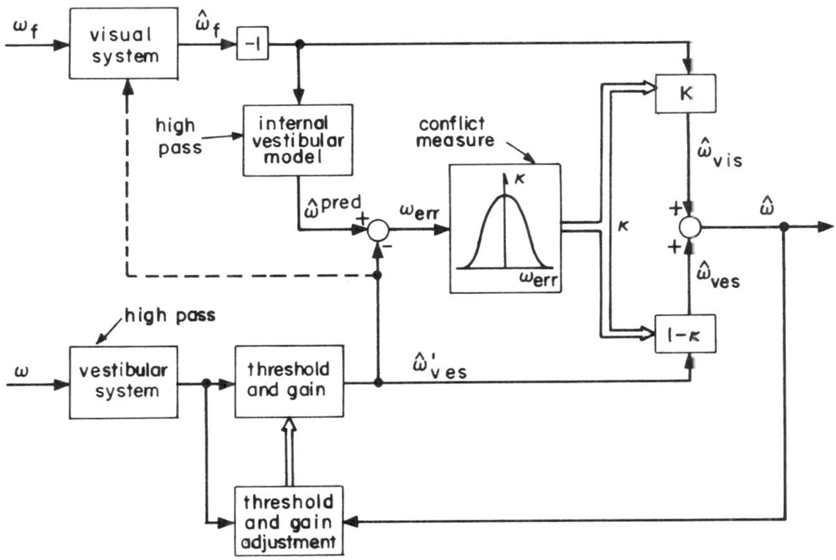

FIG. 5. A "soft switch" conflict model, in which visual information is used only when the vestibular signal it implies ($\hat{\omega}_{pred}$) is in good agreement with the actual vestibular signal ($\hat{\omega}'_{ves}$). (From Zacharias and Young, ref. 22.)

interactions at the earliest possible junction in the vestibular processing network—the eighth nerve leading directly from the labyrinth and the vestibular nucleus where the nerve first synapses. (As will be clear from what follows, the terminology "vestibular nucleus," consisting of four nuclei on each side of the brain stem, is a misleading misnomer, since it is so heavily influenced by other sensory input, particularly the visual and proprioceptive.)

For some time, the influence of head orientation on visual cortex units had been demonstrated, most convincingly at first by Grüsser (9), then working in Jung's laboratory in Freiburg, Germany. The first direct demonstration of the opposite influence (visual effect on vestibular units) was the result of an ingenious experiment by Klinke and Schmidt (13) in Berlin. Having first identified single units as most probably first-order vestibular afferents in the goldfish, they then exposed the paralyzed fish to a surround of horizontally moving vertical stripes. When the field moved to the left, with the fish held stationary, the semicircular canal afferent signal was modulated just as though the animal were actually rotated to the right. One interpretation of these results is that the visual-vestibular integration takes place at the end organ in fish. On the other hand, the field rotation presumably provokes an attempt to swim with the moving field, and the motor command, even though not carried out, could modulate vestibular efferents and consequently the afferents in anticipation of the expected gross vestibular stimulation. In monkeys, however, Keller (12) recently recorded from the eighth nerve, but failed to find any modulation of the first-order vestibular afferents by oscillation of the surround. It remains to be seen whether or not this negative result could be attributed to his use of a relatively short-period oscillation, not allowing time enough for circularvection to develop.

The evidence for visual-vestibular interaction is much clearer at the level of the vestibular nucleus. Dichgans et al., recording from goldfish vestibular nuclei, found that most units responded to vestibular stimulation and to visual-field motion (6). When the animal was rotated in the light, the unit activity remained elevated during constant turning, reflecting the role of visual influences in sustaining the turning sensation. Dichgans referred to these neurons as "speedometer units." Meanwhile, Henn, Finley, and I, working in von Baumgartner's neurophysiology lab in Zurich, were also searching for evidence of visual-vestibular integration in vestibular nucleus of awake, unanesthetized rhesus monkeys. By using an awake primate and stimulus conditions similar to those used for human rotation tests, we could search for neurophysiological correlates of circularvection and more general spatial-orientation behavior. We found that all the vestibular nucleus firing rates were not only modulated in the expected direction by moving visual fields, but that there was excellent qualitative agreement between the dynamics of unit response to visual and vestibular stimuli and the dynamics of motion sensation in human subjects (11). We wrote, "The slow gain of activity when visual and vestibular responses agree in direction, the sudden drop in activity when they disagree, the delay in onset of activity following optokinetic stimulation, and the outlasting of activity after the end of a moving visual stimulus are all qualitatively similar to the pattern of circularvection sensation in equivalent psychophysical experiments on humans." Further work on the same preparation by Waespe and Henn has demonstrated qualitatively the presence of a switching behavior between visual and vestibular influences on vestibular nuclei units—exactly mirroring the human circularvection results (16) (Fig. 6). Similar work on the paralyzed goldfish by Allum shows general agreement, although there are differences in detail which could be attributed to species difference (1).

Quite similar self-motion phenomena occur in response to linear translations of the visual field, in which the sensation is referred to as *linearvection.* In our modified Link Trainer at MIT and in a "visual tunnel" device at the CRNS Laboratory of Work Physiology in Paris, Alain Berthoz and I have explored linearvection in the fore-aft direction, creating a sensation of driving through a long tunnel (2). The sensation of velocity increases with field velocity up to the saturation level of approximately 1 meter per second. During prolonged exposure, the sensation of speed slowly decays in an adaptation phenomenon also seen in circularvection. This adaptation to prolonged visually induced velocity is well known to those of us who have found our automobile speed creeping up slowly during turnpike driving, or who have noticed how much slower than the speedometer reading we appear to be traveling after exiting from a high-speed highway. One somewhat different characteristic of linearvection in the

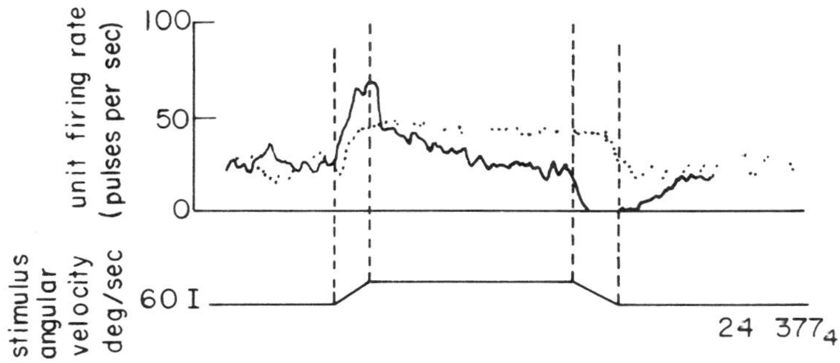

Fig. 6. Unit activity in the monkey vestibular nucleus showing responses to yaw rotation and circularvection stimuli which are similar to human motion perception time course. For all test cases, the stimulus is a velocity trapezoid as shown. The dots are the unit running-average firing rate for visual stimuli alone, showing a gradual buildup, maintained steady response with minimal adaptation during the minute of constant angular velocity, and then a sudden drop with slight undershoot when the field velocity returns to zero. The solid line is the visual response to a vestibular stimulus—a strong acceleration response, returning to rest level during constant-velocity rotation in the dark, and a negative (inhibitory) response during deceleration back to a stop. The combined visual and vestibular response *(dashed line),* corresponding to rotation in the light with the field fixed in space, reveals the switching behavior wherein phasic fast vestibular responses are initially adapted as the vestibular signal decays, and finally, the strong inhibitory vestibular signal during deceleration brings the combined response rapidly back to rest. (From Waespe and Henn, ref. 16.)

horizontal plane is a definite fore-aft asymmetry, in which the development of a rearward-moving linearvection is stronger, more rapid, and more easily sustained than is the sensation of moving in the forward direction (20). Furthermore, vertical movement of the peripheral field produces a sustained sensation of vertical linearvection, similar to the feeling of going up or down in a European-style, open-cage elevator. In fact, a New York restaurant has been reported to use this phenomenon to create the sensation of descending deep into a coal mine as part of its ambience. The frequency response of the subjective sensation of linearvection has been measured for both horizontal and vertical linearvection, using single sinusoid inputs and, to avoid the complications of predictability of the input, a quasi-random signal composed of many sinusoids. The results indicate, as expected, a strong low-frequency linearvection sensation which drops rapidly in strength at frequencies above approximately 0.1 Hz.

Although the phenomena of circularvection and linearvection were reasonably well-known based upon common experience, the finding of visually induced tilt in response to a visual field rotating about a horizontal axis was somewhat of a surprise. When the entire visual field rolls about the horizontal saggital axis, after a period from a few seconds up to 30 sec, the subject begins to report a paradoxical phenomenon (7). Just as in the case of circularvection, the moving pattern appears to slow down or stop in space and the subject senses himself rolling in the direction opposite to the vertical and when asked to set an indicator to the vertical, indicates it as displaced in the direction opposite to the moving field, frequently at angles from 15 to 45°. Similarly, his postural reaction is one of balancing an apparent tendency to fall in the direction of the induced tilt. This is, of course, a paradoxical sensation since constant roll velocity and a steady roll angle relative to the vertical are internally inconsistent. However, such phenomena of motion without displacement are by no means unknown in psychology. Ocular countertorsion is induced by the rotating field, but it is limited to only a few degrees and is insufficient to account for the tilt sensation. Held and his co-workers (10) have since quantified the necessary visual-field border-density placement and velocity to achieve a given induced tilt. Utilizing the 360° field-of-view capability of the point light-source system of the "Dual Maneuvering Spheres" at NASA Langley Research Center in Virginia, Oman, Dichgans, and I further explored the matter of visually induced tilt with fields of different density and realism; and we demonstrated that pitching phenomena were produced in a manner analogous to that of roll, but with some important differences (21). Realism in the moving scene was unimportant. Another asymmetry appeared in pitch. When the entire visual field was moving upward, the sensation was a strong one of pitching down or falling forward. However, when the field was moving downward, there was much less of an upward-pitch phenomenon. Rather, the subjects more often reported a strong component of vertical linearvection or moving straight up. In addition, it was found that movement of the upper visual field was of prime importance in visually induced pitch, whereas the lower-visual-field movement was more often interpreted in terms of linearvection. Some teleological arguments, admittedly speculative, can be used to explain this phenomenon. Although we are accustomed to rapid linear motion of the lower near field in locomotion, our usual experience of uniform motion in the upper field occurs when we are falling forward or backward and see the stars or clouds moving about us.

The phenomenon of visually induced tilt provided a particularly attractive means of testing our theories of visual-vestibular interaction. Presumably, the extent of the static visually induced tilt, generated by a field rotating about an earth horizontal axis, is limited only by the information from the nonvisual graviceptors which deny the presence of any tilt. By manipulating the head into different orientations, in which the otolith organs are presumed to be less effective or to have "blind spots," we would expect the strength of the visual effect to increase, which indeed it does. Finally, if the same field is rotated about a vertical axis and viewed from below, while lying supine, there is no conflict between visually induced roll and the lack of confirming nonvisual cues. The results are in conformity with expectation, and the sensation of roll rate continues throughout the stimulus duration except for occasional losses as in the case of circularvection.

In all experimental situations thus far examined, the visually induced motion is limited in rate of onset by the expected response of the semicircular canals, and limited in static tilt by the expected response of the otolith organs. Visual-vestibular conflicts can induce motion sickness, even when there is no true body motion to excite the vestibular apparatus. The elucidation of such mechanisms, and attempts to cure the malady or speed habituation, are at the center of research on "space motion sickness" which affects astronauts and was so prominent during the Skylab flights. A series of experiments concerning visually induced motion and visual-vestibular interaction is scheduled to begin on board the Space Shuttle, in Spacelab-1, scheduled for launch in 1981. It remains to be seen whether or not adaptation to the weightlessness of space will modify the way in which self-motion is generated from visual cues.

SUMMARY

Human subjective orientation in space is treated as a multi-input problem with characteristic sensor dynamics and noise. The evidence for effects of moving visual fields on orientation illusions is reviewed and the interactions with vestibular cues discussed. A "soft switch" model based upon intersensory conflict-resolution is considered, along with some of the existing relevant neurophysiological findings.

REFERENCES

1. Allum, J. H. J., Graf, W., Dichgans, J., and Schmidt, C. L. (1976): Visual-vestibular interactions in the vestibular nuclei of the goldfish. *Exp. Brain Res.*, 25:463–485.
2. Berthoz, A., Pavard, B., and Young, L. R.(1975): Perception of linear horizontal self-motion induced by peripheral vision (linearvection): Basic characteristics and visual-vestibular interactions. *Exp. Brain Res.*, 23:471–489.
3. Brandt, Th., Dichgans, J., and Koenig, E. (1973): Differential effects of central versus peripheral vision on egocentric and exocentric motion perception. *Exp. Brain Res.*, 16:476–491.
4. Brandt, Th., Dichgans, J., and Buchele, W. (1974): Motion habituation: Inverted self-motion perception and optokinetic after-nystagmus. *Exp. Brain Res.*, 21:337–352.
5. Dichgans, J., Diener, H. C., and Brandt, Th. (1974): Optokinetic-graviceptive interaction in different head positions. *Acta Otolaryngol.*, 78:391–388.
6. Dichgans, J., Schmidt, C. L., and Graf, W. (1973): Visual input improves the speedometer function of the vestibular nuclei in the goldfish. *Exp. Brain Res.*, 18:319–322.
7. Dichgans, J., Held, R., Young, L. R., and Brandt, Th. (1972): Moving visual scenes influence the apparent direction of gravity. *Science*, 178:1217–1219.
8. Fischer, M. H., and Kornmuller, A. E. (1930): Optokinetisch ausgelöste bewegungswahrnehmungen und optokinetischer nystagmus. *J. Psychol. Neurol.*, 41:273–308.
9. Grüsser, O. J., and Grüsser-Cornehls, U. (1972): Interaction of vestibular and visual inputs in the visual system. In: *Progress in Brain Research, Vol. 37: Basic Aspects of Central Vestibular Mechanisms*, edited by A. Brodal and O. Pomeiano, pp. 573–583. Elsevier, Amsterdam.
10. Held, R., Dichgans, J., and Bauer, J. (1975): Characteristics of moving visual scenes influencing spatial orientation. *Vision Research*, 15:357–365.
11. Henn, V., Young, L. R., and Finley, C. (1974): Vestibular nucleus units in alert monkeys are also influenced by moving visual fields. *Brain Research*, 71:144–149.
12. Keller, E., and Daniels, P. (1975): Oculomotor related interaction of vestibular and visual stimulation in vestibular nucleus cells in alert monkeys. *Exp. Neurol.*, 46:187–198.
13. Klinke, R., and Schmidt, C. L. (1970): Efferent influence on the vestibular organ during active movement of the body. *Pfluegers Arch. Ges. Physiol.*, 318:325–332.
14. Mach, E. (1875): *Grundlinien der Lehre von den Bewegungsempfinden.* Engelmann, Leipzig (Bonset, Amsterdam, 1967).
15. Stevens, S. S. (1968): Measurement, statistics, and the schemapiric view. *Science*, 161:849–856.

16. Waespe, W., and Henn, V. (1977): Neuronal activity in the vestibular nuclei of the alert monkey during vestibular and optokinetic stimulation. *Exp. Brain Res.*, 27:523–538.
17. Witkin, H. A., and Asch, S. E. (1948): Studies on space orientation. III. Perception of the upright in the absence of a visual field. *J. Exp. Psychol.*, 38:603.
18. Young, L. R. (1970): On visual vestibular interaction. *Fifth NASA Symposium on the Role of the Vestibular Organs in Space, NASA SP-314.*
19. Young, L. R., Dichgans, J., Murphy, R., and Brandt, Th. (1974): Interaction of optokinetic and vestibular stimuli in motion perception. *Acta Otolaryngol.*, 77:159–166.
20. Young, L. R., and Chu, W.: Dynamic response of linearvection. *(In preparation.)*
21. Young, L. R., Oman, C. M., and Dichgans, J. (1975): Influence of head orientation on visually induced pitch and roll sensation. *Aviat. Space Environ. Med.*, 46:264–268.
22. Zacharias, G. L., and Young, L. R. (1977): Manual control of yaw motion with combined visual and vestibular cues. *Proceedings of the 13th Annual Conference on Manual Control,* MIT.
23. Zacharias, G. L. (1977): *Motion Sensation Dependence on Visual and Vestibular Cues.* Ph.D. Thesis, MIT.

Posture and Movement, edited by
R. E. Talbott and D. R. Humphrey.
Raven Press, New York © 1979.

Muscle Properties: A Problem for the Motor Controller Physiologist

Lloyd D. Partridge

Department of Physiology and Biophysics, University of Tennessee Center for the Health Sciences, Memphis, Tennessee 38163

INTRODUCTION

Muscle is the filter through which all motor output must pass, regardless of how those signals are generated—by reflex, voluntary (whatever that is), or any mixture of reflex and voluntary activity. That filter sets limits on the accomplishment of movement, and these limits cannot be avoided by neural processes. In some cases, tricks are used to accomplish motor tasks which, by straightforward control, would exceed the limits. A bagpipe player, for example, slides a finger past a hole and closes and reopens it faster than his muscles can make a reversing movement. Another common trick is seen in the individual who has a pathological loss of the use of a particular muscle, but learns by manipulating with other muscles to approximate the movement for which the lost muscle normally would have been used. However, there is no way for the nervous system to use a muscle beyond the limits of what that muscle can accomplish. Even within the range of what the muscle can accomplish, the nervous system must use appropriate tactics in generating signals to drive the muscle, tactics that will compensate for the properties found within the muscle. The complexity of this required compensation has attracted little attention. Nevertheless, when a desired response has been accomplished, those compensations must have been built-in to the signals which drove the muscles.

We tend to be even less cognizant of the opposite side of the coin, the possibility that the nervous system may be able to take advantage of certain muscle properties in order to use more easily generated nerve signals to accomplish the desired result. It is not hard to demonstrate with artificial stimulus that the same muscle response can be produced by different stimulus patterns. In such cases, it is possible that the use of more than one control option is available to the nervous system, such that a variety of signals from simple to complex can be used to produce the same results from muscle. Perhaps some of the advantages acquired by athletic training lie in the development of a selection of the most advantageous of the neural signal options from such a range.

Most of the muscle properties and the rules of response to stimulus which I intend to discuss are well known to you (e.g., see 5,8,14,19,27,43,44,60,61,62); but I will briefly review them before attempting to put them together, in order to suggest limitations on possible strategies available for motor control. I will first review factors which affect the response of a single muscle to nerve signals. Then I will briefly look at the mechanical relationship between the motor act which is ultimately controlled and the muscle activity which contributes to that control. In this context, it would appear important to ask how accurate the motor control system must be to define a response, and the related question, how precisely successive attempts at the same action reproduce some standard form. Finally, after reviewing the muscle properties and the mechanics, I would like to question whether what we already know about peripheral systems may provide us with some hints about the possible ways that are available to accomplish the central control of motor activity.

At this point, let me introduce another version of the flow diagram for the motor response system (Fig. 1). Although it looks quite different from Houk's diagrams *(this volume),* it is not intended to disagree about any of the signal pathways. By accident, this diagram even duplicates the "command signal" term from Humphrey's discussion *(this volume).* The broad arrows representing multiple signals are intended to include Brookhart's and Young's descriptions *(this volume)* of multiple signals acting to determine either posture or movement for vehicle control, and even would include multifactored inputs to respiratory control found in Brookhart's first papers (6). The general term *motor output* includes those in which the end result involves coordinated eye, head, and trunk movement, as well as the more typically described postural, walking, and jumping activities with their controlled joint angles. For good measure, let's throw in the important control of the two ends of the GI tract, in the tongue and the anal sphincter. The points of convergence of the signals in this figure are not restricted to simple summation, but should accommodate inclusion of, for example, both transfer functions familiar to Carlos Terzuolo (54) and linear and nonlinear combinations. To give the diagram maximum flexibility, let us consider each point of convergence to have an output that is in an unspecified functional relationship to its total input. I would point out I really haven't forgotten about the cortex; but it and some of the other ill-defined parts that form the marrow of the skull are in that unknown territory off the map to which so many signals are shown to be going and from which the command signal is coming. In fact, cortex also is probably to be found as part of the two converging points that transform the command signal and are marked CNS. There is a bias in emphasis here, but that is intentional.

The diagram shows, in the heavy arrow down the center, the flow and modification of the command signal until it becomes a final motor response. At this point, the representations seem so nonspecific that they could offend no one and would fit anything. I propose to show, however, that even this very general diagram with slippery definitions can have enough teeth to damage some concepts

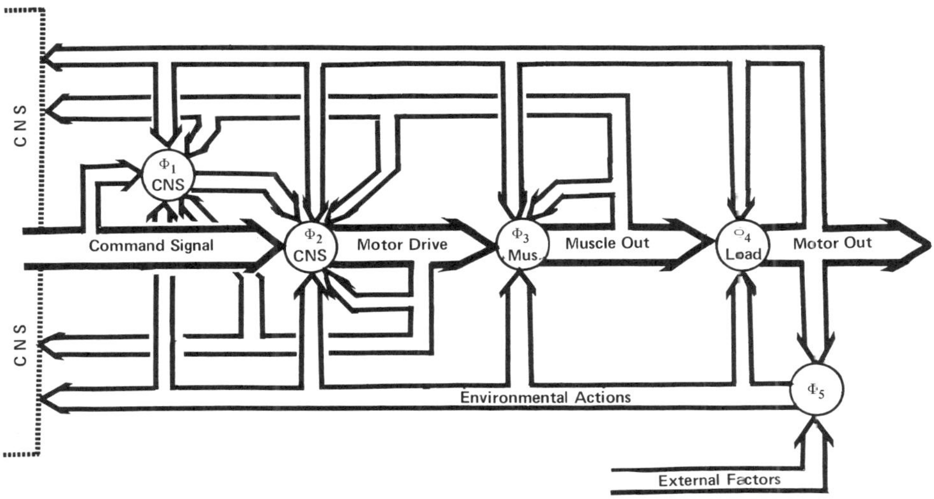

FIG. 1. One view of the motor system. *Broad arrows* are used to represent signals with multiple components. Total muscle response is distinguished from the total motor response which is the result of the combined actions of muscles on the mechanical load (ϕ_4). The rules of this combination are affected by the interaction between external and internal mechanical factors. This scheme shows the command signal, the initial representation of the desired motor response, to be modified by interaction with various neural signals (ϕ_2) before it becomes the pattern of nerve impulses that drive the involved muscles. The muscle response is a function of both the drive from motoneurons and the effect of the mechanical factors which act on the muscle (ϕ_3). The possibility is also represented (ϕ_1) that the rules for modification of the command signal can be dependent in part on characteristics of the command signal itself. The internal effect of external factors (labeled environment actions) on the system are dependent on the state due to its own previous activity (ϕ_5). Each point of interaction of signals (ϕ) is represented without designating the nature of the particular function by which the converging signals are combined; i.e., each output is defined only as some function of the various inputs. Both the site of origin and whether various factors influence the generation of the command signal are symbolically omitted from this diagram.

of motor control and to expose a different emphasis on the role of sensory feedback.

MUSCLE RESPONSE TO NERVE SIGNALS

The transformation which converts "motor drive" to "muscle output" involves the activity of many motor units and need not, and usually cannot, be confined to a single muscle. However, although well studied, even the relationship in single units between nerve signals and response is complex. First, consider the temporal relationship between impulses on the motor nerve and the mechanical response of the muscle driven. A single nerve impulse on a single motor nerve fiber produces a brief and rather trivial rise and fall of tension in the driven motor unit. If additional impulses come along the same nerve fiber before the effects of the first one have died away, a cumulative response occurs which looks quite like that response to a train of pulses which could be described by a particular type of second order differential equation, or, as it is sometimes named, "a second order lag" (4,13). In physiology, the pattern has been more commonly described as a partial or complete tetanic response.

Let us consider for a moment what any system with a second order lag does to signals that it processes. First, an impulse passed through a second order lag system will produce a rounded response appreciably longer than the pulse that was the input. This response looks very much like the twitch response of a skeletal muscle. With the initiation of a train of impulses into the lag system, the system's response is simply the sum of a series of individual responses. The resulting pattern suggests partial or complete tetanus in muscle (Fig. 3A). A sinusoidal signal passed through a second order lag results in a response that is conveniently described in terms of amplitude ratio and phase shift for the system. At low frequencies, the cyclic output of a second order lag system will be almost in phase with the input; but, as frequency increases, the part of the cycle by which the response lags the input increases. At the same time, the actual time represented by the lag between the output and the input cycle decreases. For a second order lag system with time constants similar to muscle, the lag will amount to only a few degrees of phase shift at the lower end of physiological movement frequencies. However, these few degrees may be seconds in time. This same system at frequencies that are high for physiological movement will show a phase shift approaching half a cycle, 180°, but only about 50 msec when expressed as lag time. Even this small lag would appear to be more important than conduction delay in either physiological effect or as the basis for the delay of the peak in a cross-correlation between EMG and isometric force of the muscle. The second order lag system also shows a frequency dependence of its amplitude responsiveness to input signals. At low frequencies, the system tracks the input signal with maximum amplitude; but, as frequency increases, the cyclic component of the input is less and less effectively followed. In a signal that could be considered a mixture of a cyclic and a constant value, the constant value then would be represented fully, but the fast cyclic components would be very poorly followed by such a lag system. This corresponds to the tetanic fusion of classical physiology. By using pulses delivered at a rate that is sinusoidally modulated as a stimulus to test muscle properties, we find the muscle and muscle systems do indeed act much like a second order lag system (18,44) (Fig. 2).

Closer inspection, however, reveals that while the second order lag description fits a considerable part of the muscle response (4), it is still only an approximation; and several types of deviation occur. Several years ago, Carlos Terzuolo and associates at Minnesota demonstrated that a higher order description fit the muscle response better than the second order description (54). This better fit included a time delay, as well as other linear terms.

There are also nonlinear factors that cause the response of muscle to deviate

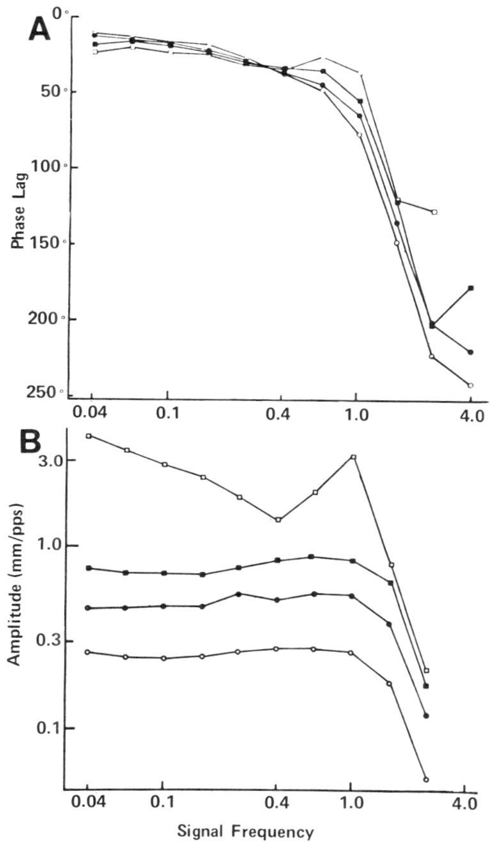

FIG. 2. Response of cat triceps surae to different amplitudes of sinusoidally modulated stimulus rate [reprinted from ref. 44, L. Partridge, *Am. J. Physiol.,* 210:1178–1191 (1966)]. Each line describes response to cycles of modulated stimulus of a particular amplitude. **A:** The lag of the response as a function of the (360°) modulation cycles. (A second order lag is similar to these curves, but would involve no lag more than 180°.) **B:** Amplitude ratio, comparing amplitude of movement with amplitude of stimulus modulation. (For a linear system with second order lag, the curves could look similar to these, but for tests with different input amplitude, the curves would all be superimposed.)

from simple second order lag predictions. Threshold-like nonlinearities are expressed by the facts that neither negative stimulus rates nor muscles that push exist. Also, a linear second order system driven by a series of pulses at a constant rate would level off at a final response amplitude that was directly proportional to the input pulse rate. This level would increase with pulse rate without limit (Fig. 3A). We all know that with tetanic experiments in frog the effectiveness of an incremental increase of stimulus rate decreases as the starting stimulus rate increases (Fig. 3B). In mammalian limb muscles, stimulus rates above 50

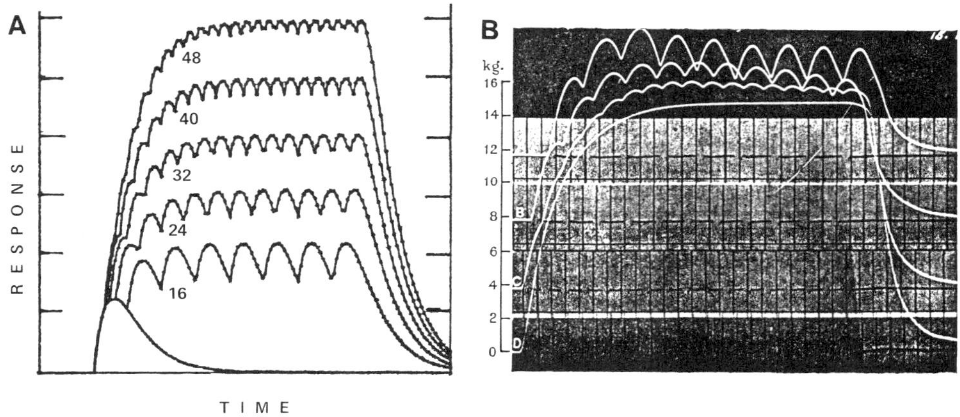

FIG. 3. Comparison of responses of a muscle with response of a second order lag when both are driven by a train of pulses delivered at different constant rates. **A:** Response of theoretical system to a single pulse and five different stimulus rates. Rate increments between lines are equal. Each response is calculated as a sum of identical pulse responses. Peak response increases without limit and proportionally to pulse rates in each train. Numbers under each curve identify pulse rate used. **B:** Response in cat gastrocnemius to trains of pulse stimuli [reprinted from ref. 14, S. Cooper and J. C. Eccles, *J. Physiol. (Lond.),* 69:377–385 (1930)]. Each response is recorded from a separate base line. Stimulus rates are: A, 19 pps; B, 23.5 pps; C, 35 pps; and D, 115 pps. These responses are not the sum of identical pulse responses. As stimulus rate increases, the contributions of individual pulses decrease. Consequently, peak responses increase less than proportionally to the stimulus rates.

pulses/sec produce relatively little change in the response with further rate increases, at least up to a few hundred pulses/sec. Burke has recently looked at this saturating type of nonlinear response as a failure of successive stimulus pulses to contribute as much response as the initial impulse does, and he plots the failure as a function of interval since the previous pulse (9).

Another deviation from simple linear dynamics in muscle might be described as "hysteresis," one term which seems to combine what has been described as several different operationally defined phenomena of muscles and may involve different internal processes. The various observations all relate to a residual effect of previous activity that does not disappear as soon as the principal response of that excitation decays. In post-tetanic potentiation, after the response to the tetanus has disappeared, something persists which causes the response to additional single stimuli to be enhanced (26). Also, after muscle has been stimulated at a relatively high rate, a subsequent reduction of stimulus rate does not correspondingly reduce the response (Fig. 4). If the stimulus is then sustained at a moderate rate, the residual effect will last at least for minutes (44), producing, in effect, an increased response to each subsequent stimulus. After this phenomenon was described in mammalian muscle, Wilson went back to invertebrates, where a catch or latch phenomenon had previously been described and there showed that additional invertebrate muscle had properties similar to the catch

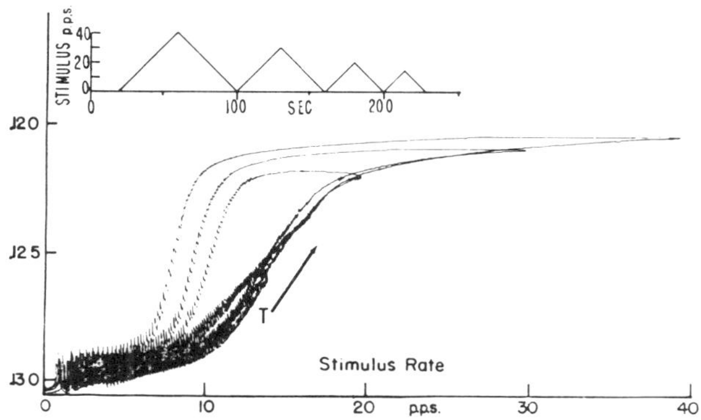

FIG. 4. Muscle contraction plotted against stimulus rate for a series of stimuli delivered with triangularly modulated rates. In successive cycles of stimulus modulation, different maximum rates were reached. The responses increased nonlinearly with increasing stimulus rates. However, when the stimulus rates decreased from levels above about 15 pps, the response did not decrease until the stimulus rate fell to a level near 10 pps. (This pattern was not related to friction in the recording system.) *Arrow* shows direction of increasing time. Insert graph shows time sequence of stimulus rates used for stimulation for this record [reprinted from ref. 45, L. Partridge, *Am. J. Physiol.,* 223:148 (1972)].

phenomena when subject to the same test procedure as those used in the mammalian muscle (63). It has also been shown that with the displacement in time of a single pulse within a train of uniformly spaced stimuli to a muscle, a response shift can be produced which may last long after the brief stimulus shift (9,63). As a result of this hysteresis in muscle, the response of a muscle to any particular stimulus pattern may be greatly dependent on details of the history of the stimulus before that pattern was applied.

Under those circumstances in which this hysteretic response occurs, it might be either an advantage or a problem to the motor controller. It might allow control with a less costly signal because, once "latched," a stable muscle response might be held with a lower level of even irregular neural drive. On the other hand, when the muscle response is "latched," the central nervous system has no continuous control to reduce the muscle response until this "latching" effect is released. These are important issues for the system which would transform a command signal into a muscle drive signal. Either to use or to get around this muscle effect, the neural signal to the muscle would need to be specially adapted to accommodate these muscle properties. Incidentally, it is my impression that part of the fine structure of EMG which tends to be looked upon as noise is, in fact, related to these factors in muscle control. That part then is necessary detail, not noise.

With the interest that has developed recently in the subject of recruiting order, there is no point for me to dwell at length on recruitment. I shall just repeat for the record that while the size of the unit recruited increases as recruitment progresses, simultaneously the shift from slow to fast units (10) would affect the dynamic properties of that system which is responding in the muscle. Such a change has been seen earlier with graded intensity of nerve stimulus (27). Does this recruitment order produce an automatic compensation for the lag in muscle? Thus, when one attempts to make a movement which approaches a step change in position, a compensation is needed for muscle lag. If that large initial burst of activity, normally seen, includes recruitment to above threshold for the faster units, those units may serve to increase the rise speed of the response as well as add to the total response, doubly compensating for the lag inherent in muscle [without change in the recruiting order (17)]. A related problem in a different context: Is it known whether the increase in motor activity associated with an increase of muscle thermogenesis crosses the same line from slow to fast units near the point of the initiation of shivering? This shift would change the dynamics and consequently affect reflex stability. The development of a periodic firing in shivering could result from a change in stability secondary to such a dynamic change. In either case, if both the size of the increment of response and also the dynamic properties of the response of a muscle to a small increment of excitation depend on the level of activity at the moment that increment is added, it complicates the rules for organizing an appropriate neural signal.

The mechanical response to neural signal further depends on the mechanical conditions at the muscle at the moment of arrival of the stimulus. For over a century, it has been known that the force delivered by a muscle is dependent on the length at which that force is delivered (5,48,49,61). For a muscle, both the maximal force which can be produced on stimulation and also the minimum force—that is, the force delivered when the muscle is unstimulated—are dependent on length. The combination of active and passive length-tension curves represents the bounds of the output force which the nervous system can command from any particular muscle in isometric conditions, assuming that the active curve referred to is measured from maximum responses (Fig. 5).

It has long been indicated and well known for almost half a century that force delivered by a muscle is also dependent on the velocity of movement of that muscle (5,19,62). Thus, for a particular muscle the maximum force that can be delivered is dependent on both length and velocity. Because of the way many early experiments were done, the force-velocity curves did not include a representation of forces in periods of stretching, even though it is obvious that muscles in physiological activity must be stretched as much as shortened if they are not to disappear. More recently, this other side of the force-velocity curve has been carefully examined (32). The force delivered by an unstimulated muscle obviously can be graphed also as a function of velocity, but this passive force-velocity curve has had relatively little study (43). Some experiments on the passive muscle done by Laurel Benton in my laboratory suggest that it may have some functionally interesting dynamic properties, but these have not been fully established. In any case, the force delivered by a muscle as a function of length is the same force as that which muscle delivers as a function of velocity,

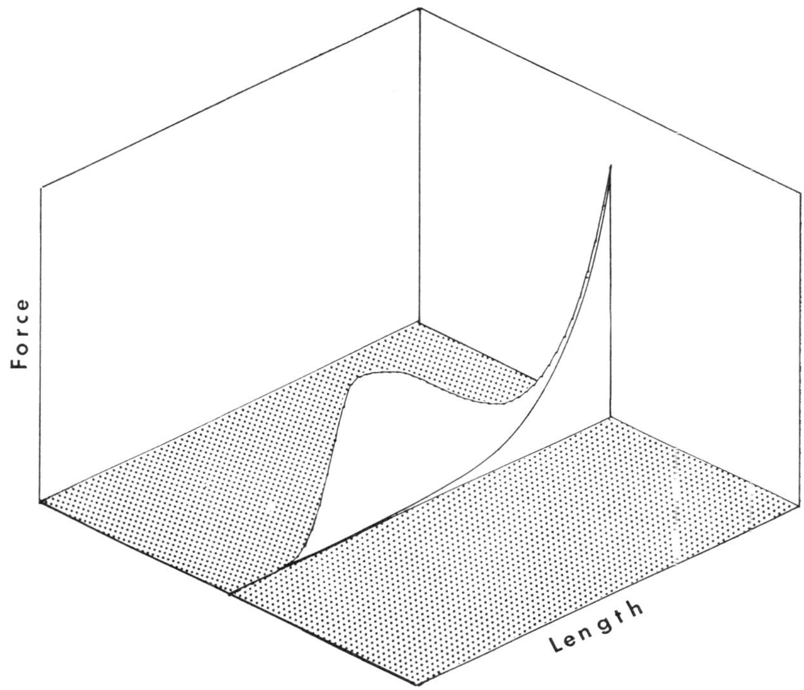

FIG. 5. Typical relationship between muscle length and muscle force plotted in an atypical coordinate system. The experimental points which are joined to make up such curves are measured isometrically. *Upper graph line* is the type of relationship frequently found with tetanically stimulated muscle, while the *lower graph* is found with unstimulated muscle. The quantitative values of length and force on the graph vary with muscle size, whereas the prominence of the left-hand peak of force in the active muscle varies among muscles and also within one muscle with intensity of muscle excitation.

so it should be possible to orient a force-velocity curve and a length-tension curve in a three-dimensional space such that this single force is represented by the point of crossing of these two graphs (Fig. 6). At other moments, when the muscle is operating at a different length and/or velocity, the maximum force that could be delivered similarly would be representing a crossing of other pairs of graphs. (Incidentally, the base plane on which this three-dimensional graph is built is a plane with velocity plotted against length, which is often used for study of mechanical systems and is called the "phase plane.") If a large number of intersections of length-tension and force-velocity graphs are plotted over the phase plane, a force surface begins to emerge (Fig. 7A). The passive length-velocity-force surface can be represented similarly (Fig. 7B). For other muscles, the effect of length on the force surface may vary (Fig. 7C) (11,60).

Such a force surface can be related to rather simple experiments. For example,

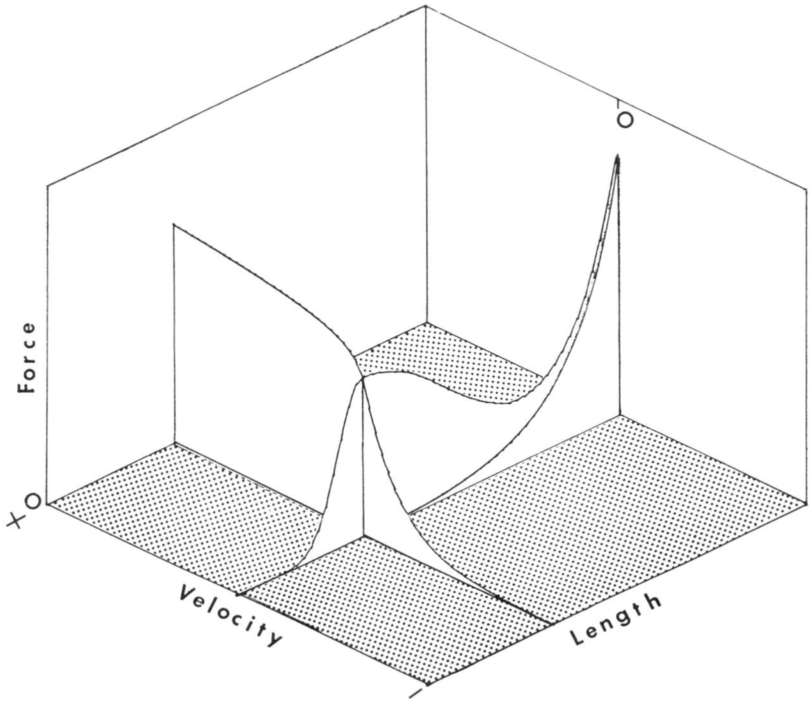

FIG. 6. The force at a particular moment, represented as a point on both a length-tension curve and a force-velocity curve. This graph shows two points in a four-dimensional relationship among length, velocity, excitation, and force for a muscle. The excitation dimension would select an intersection of a particular pair of curves, falling between the extremes of the passive curve (minimal excitation) and the active curve (maximal excitation). Both lengthening (+) and shortening (−) velocities are represented in this figure, with the force-velocity graph drawn parallel to the velocity axis with coordinates reversed from the most familiar orientation. In conformity to usual discussions, no velocity effect is shown for the unexcited muscle. Increasing length is plotted along the axis going to the right.

if a maximally excited muscle were driven by external forces to make a movement, that movement would define some trajectory on the phase plane. For any point on that phase plane, the maximum force which the muscle could deliver at that length and velocity would be shown as the point at which a line projecting up from the trajectory intersects the surface of maximal force. Projecting the

→

FIG. 7. Intersection of 26 length-tension curves with 26 force-velocity curves. (These figures differ from some previously published graphs in one or more ways: Total force, not just active force, is shown; lengthening as well as shortening velocities are included. The force is plotted on the vertical axis, over the phase plane.) **A:** Curves for maximally excited muscle, as seen in some muscles. **B:** Curves for unexcited muscle (effect of velocity on passive force is not shown). **C:** Variation of length-tension curve, as seen in other muscles. Most reported muscle tests would produce surfaces with characteristics between those shown in **A** and **C.**

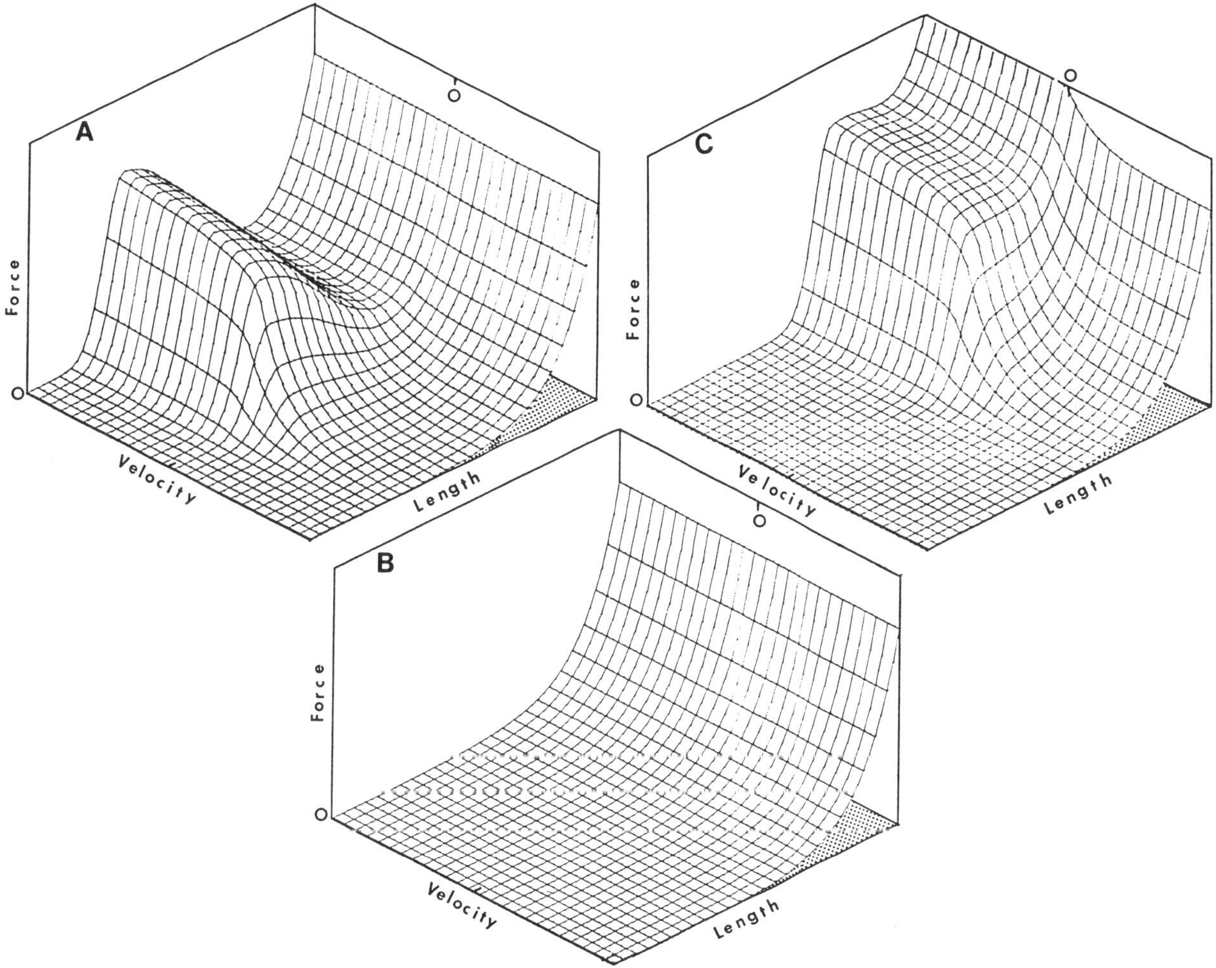
A
Force
O
Velocity
Length
O
C
Force
O
Velocity
Length
O
B
Force
O
Velocity
Length
O

whole trajectory up to intersect the force surface results in a representation of the variation of the maximum force which the muscle could produce through the sequence of different operating conditions in the driven movement (Fig. 8). In this case, since the force pattern represents maximal force and presumably maximal stimulation, the variation in force through the cycle represents only the effect of movement, not of changes in stimulus.

On the other hand, for a muscle itself to move a load along the length pattern illustrated in Fig. 8 requires the muscle to deliver a force output that changes in the specific way needed to impart that movement to a load with those particular mechanical properties. For the movement to be accomplished, that force requirement must involve no force higher than the maximum force or lower than the passive force surface. Most of the time, the required force for a movement would be not at either of the extremes, but between them, and produced by neural excitations greater than zero, but less than maximal.

For convenience in discussing the relationship between excitation and re-

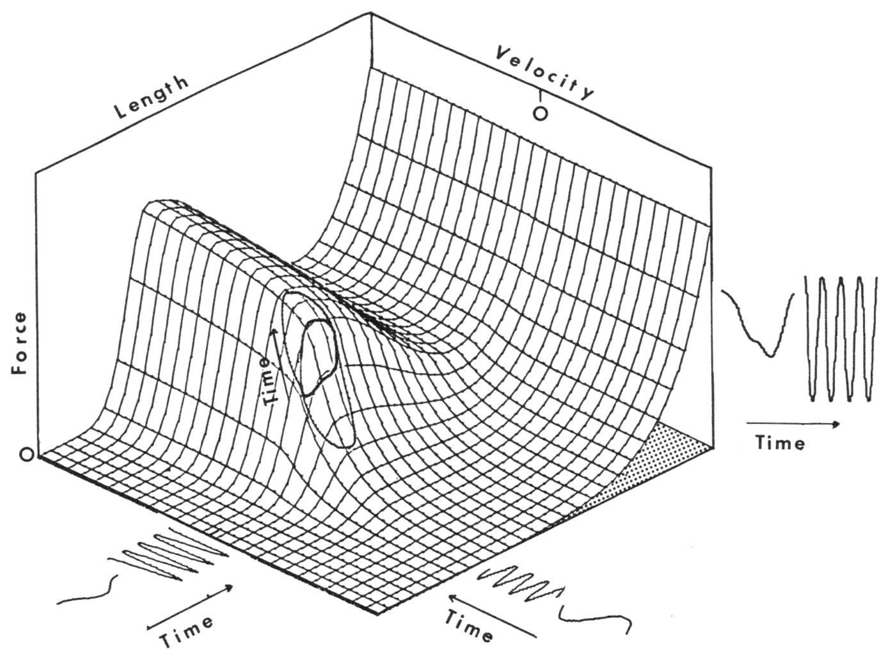

FIG. 8. Sinusoidally driven movement over a maximal force surface. Two different cyclic movements are shown. In both, the length changes are the same. Consequently, the amplitudes of the sinusoidal variation in velocity are proportional to cyclic frequency. Because of the nonlinear relationships involved, the force cycle is not sinusoidal, and its mean value changes with frequency. On the separate time graphs of length, velocity, and force, time progresses uniformly in the direction of the arrows. On the trajectory of the combined functions, time progresses in a clockwise direction, but not uniformly.

sponse, I will define a function of that varying responsiveness of the muscle which depends on operating conditions. "State of activation" is the relative position of a particular force generated by a muscle, and it is measured as the linearly calculated percentage of the span between the maximum and minimum force which that muscle can deliver under neural control in the same operating conditions. Although this definition is related to both Hill's "active state" (28,30) and Pringle's "activation" (47), it differs from both in essential ways for discussion of motor control. Both the force delivered at zero activation and the force delivered at 100% activation are dependent at least on length and velocity of the muscle at the moment, as represented by the force surfaces in Fig. 7. For a particular movement, the trajectory projected onto the maximum force surface for the active muscle and the minimum force surface for the passive muscle would represent a time-changing force for 100% activation and 0% activation, respectively, a pattern unique for that movement. The force span between 0% and 100% activation varies with the length and velocity, being as small as zero where there is no active contraction, a situation found at extreme lengths or with large shortening velocities. That span also reaches values greater than maximal isometric tension with lengthening velocities when near rest length. Several other factors in addition to length and velocity are important parts of the operating conditions of muscle, but these two will serve to represent the group. (Incidentally, the choice of this limited representation was influenced by the fact that a two-dimensional picture of a three-dimensional relationship is much easier to draw than representations of an *n*-dimensional space.)

Figure 8 shows the relationship between time graphs and a trajectory over the maximum force surface, with two different frequencies of sinusoidal movement. In fact, both the maximal and minimal forces available from the muscle vary through the movement cycle as a function of the changing length and velocity. Figure 9 allows comparison, on a time base, of several functions as related to two cycles of a sinusoidal movement of a muscle with active and passive properties like those shown in Fig. 7. Figure 9 shows that the percentage activation needed to produce a constant force through such a movement would vary as the maximum and minimum attainable forces vary through the cycle. The production of a relatively large and constant force equal to 40% of the isometric force at rest length involves a state of activation varying from 26% to 80%, values which depend on the changing lengths and velocities through the cycle. This changing activation is graphed as a function of time at the bottom of Fig. 9. Figure 10 shows how much maximum and minimum force (upon which the state of activation needed for a particular force depends) can vary in one sinusoidal movement.

If the percent activation were to be determined entirely by the neural excitation to the muscle, the excitation cycle, although related, would differ from the activation cycle. Muscle response is nonlinearly related to stimulus and lags behind stimulus, so activation, being defined linearly on the basis of maximal response, also lags and is nonlinearly related to stimulus. To produce the state

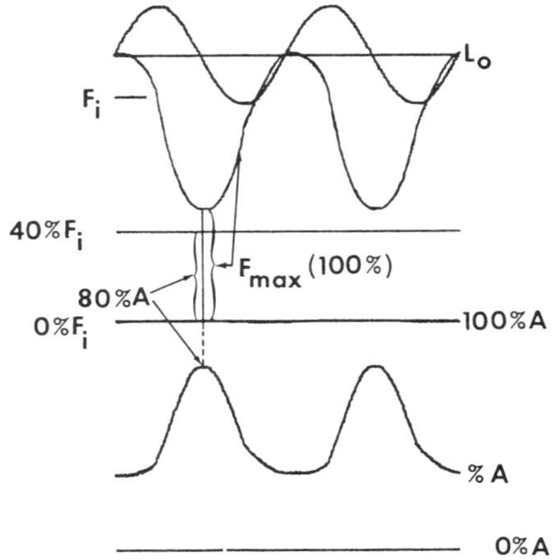

FIG. 9. The maximum and minimum force available through two cycles of a sinusoidal movement, defined in a muscle with properties such as those represented in Fig. 7. The simple sine wave at the top represents the timing of the sinusoidal length change with mid-length marked L_o. In this case, the maximum force available from the muscle (100% activated) as it changes through the movement varies from about 50% of maximum isometric force, F_i, to about 115% F_i. For this particular movement, the minimal (passive) force is almost indistinguishable from zero force (0% F_i). A horizontal line has been drawn to show the force which is equal to 40% of maximal isometric force. Where maximal force is at the lowest point in the cycle, this 40% F_i line is at 80% of the distance between the passive force curve and the maximum force curve. Its relative position varies through the rest of the cycle. The graph line marked % A shows how this constant force varies through the cycle in relative position between the force curves representing zero and maximum activation. This percent activation graph is drawn between the horizontal lines corresponding to zero and maximum state of activation, labeled on the right side as 0% A and 100% A, respectively.

of activation needed to maintain a constant force through the cycle of movement, the neural signal would need to lead the activation curve in a nonlinear way. In the constant-output case, the total variation in both activation and stimulus is related to the changes in the operating state of the muscle, and, of course, does not represent any change of desired output force. However, to accomplish

FIG. 10. Variability of force expected from sinusoidally driven muscle, depending on details of how the sinusoidal length change is defined. For all of the curves, $L = L_o + A \sin \omega t$ when L = length, L_o = mean length, ω = angular frequency = $2\pi f$ when f is frequency in Hz. A is half amplitude of sinusoidal stretch. All distances are measured as percent of rest length, and L_o is measured from rest length. On each family of curves, sinusoidal curve at top is sin ωt and identifies phase of stretch cycles. Zero force is shown by horizontal line at bottom of each family of curves, and maximum isometric force is indicated by upper horizontal line. **A:** Individual curves are produced using stretch at different mean lengths. Cyclic frequency and amplitude were kept constant from one curve to another. Maximal activation was assumed in all of the responses. Labels on left identify order of graphs with respect to mid-length of applied stretch in active muscle. Labels at right identify order for passive graphs. Order labels apply to both ends of graph lines, but since the active graphs cross each other, the order of lines differs in midportion of graphs. **B:** Only the amplitude of the movement was varied to produce this family of curves. Order of ends of graphs as related to amplitude of stretch is tabulated. Order is inverted in midcycle. Passive force curves are in same vertical order as ends of active curves. **C** and **D:** Only the cyclic frequency was varied to produce either of these families of curves. The two families differ in that each was made at a different mean length. For these two families, because frequency was variable, the abscissa represents only phase in the cycle and does not also give a common time scale. Order of lines as related to test frequencies for both graphs is tabulated between graphs. Order of midportion of graphs is reversed. In these cases, passive force is not distinguishable from zero.

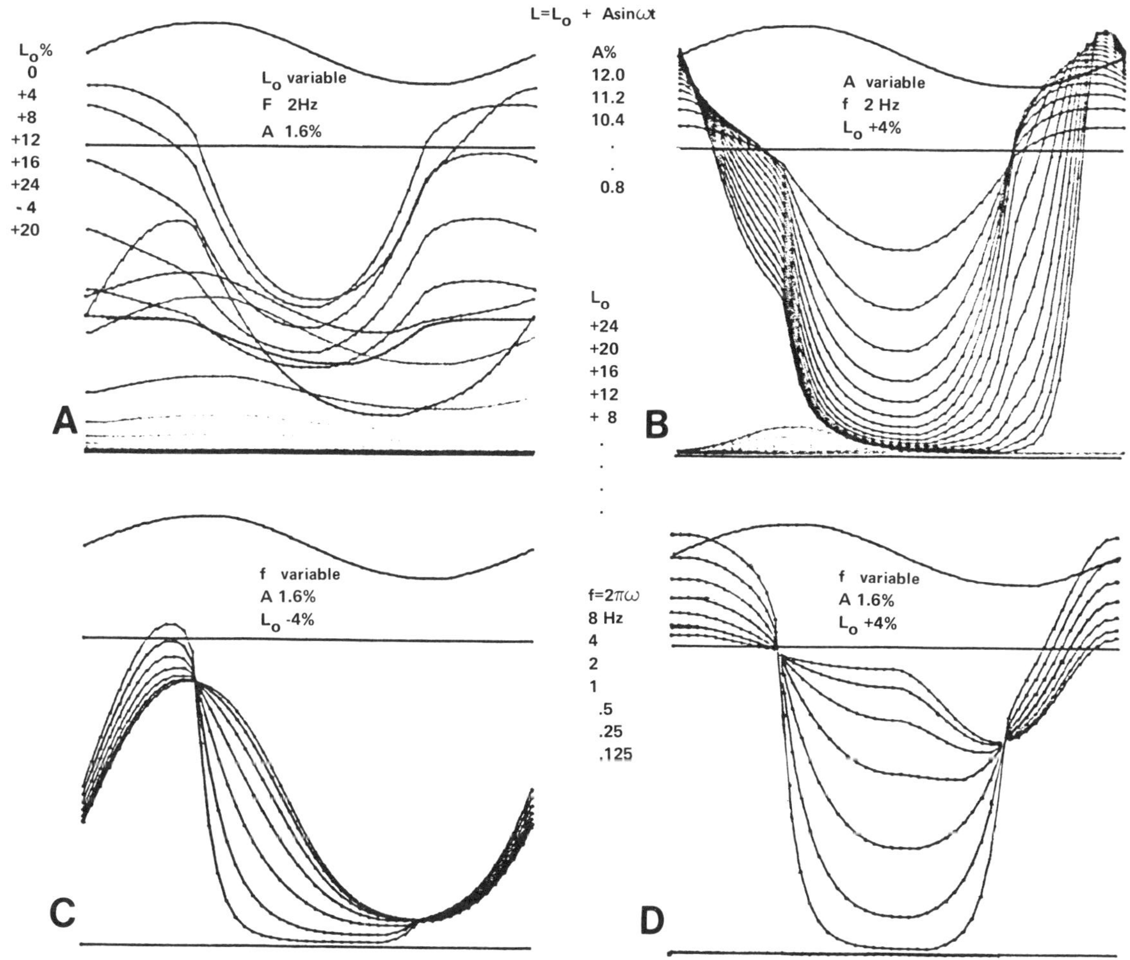

L=L$_o$ + Asinωt
L$_o$%
0
+4
+8
+12
+16
+24
- 4
+20
L$_o$ variable
F 2Hz
A 1.6%
A
A%
12.0
11.2
10.4
0.8
A variable
f 2 Hz
L$_o$ +4%
B
L$_o$
+24
+20
+16
+12
+ 8
f variable
A 1.6%
L$_o$ -4%
C
f=2πω
8 Hz
4
2
1
.5
.25
.125
f variable
A 1.6%
L$_o$ +4%
D

a changing output force, it would be necessary for the activation to vary as a function of: (a) the desired force; (b) the rate of change of the desired force; and (c) the changing rules relating activation to output force, which depend on operating conditions of the muscle. This dependence on operating conditions is of importance in the physiological range. For example, under the conditions in which walking movements occur, the ranges of lengths and velocities make the relationship between neural activity and muscle response quite variable. Consequently, a compensation of the neural signal for the operating conditions of the muscle would be important for walking.

If muscle behaved simply as a group of parallel motor units, each running from one end of the muscle to the other, each independently attached to the load, and each subject to the same conditions in all other ways, the velocity, length, and other terms from the whole muscle would describe the operating conditions of the constituent fibers. However, operating conditions vary through a muscle. For example, temperature gradients often exist (33). Further, mechanical changes can be nonuniform through a muscle. As is well known in muscle physiology, in normal activity a muscle can have considerable internal movement (42)—movement due to stretching of "series elastic elements" (29), movement due to nonparallel forces generated by complex geometric arrangements of motor units, and internal variability of mechanical properties of the load as seen by a particular motor unit, depending on whether other motor units of the same muscle are active or inactive. Motor units may be connected in a series arrangement (25). Even within a single fiber, movement of one sarcomere can be expected to occur by action of forces generated in other sarcomeres in series (29). Consequently, conditions examined with respect to the whole muscle are only crudely representative of the operating conditions of the internal constituent motor units. In any case, the relationship between activation and response of individual motor units is sensitive to length and velocity effects (5,8,49,61,62), so within a single muscle the conditions and therefore the local rules applying to individual units are variable between units. This complicates the job of the investigator, but the motor system seems capable of taking it all in stride.

The response to be expected for a given nerve signal is further dependent on the temperature of the muscle (60), the pH (20), and other factors. In at least the case of temperature, the range that is encountered in normal muscle activity—from the temperature of a limb muscle when swimming in cold water (3,33) to the temperature of the same muscle when subject to fever—is sufficient in experiments on isolated muscles to produce a considerable change in both the amplitude and time rate of change of the response of the muscles (21,27,60). Although it is not customary in motor control discussions to consider inputs to muscles other than mechanical or neural, since these other factors influence the conversion from motor drive to motor response, we need to realize that drive signals will produce faulty motor outputs if nonmechanical effects are not compensated, as well as the mechanical ones.

When we review these properties of muscle in the context of motor control, it becomes obvious that the local rules with which the nervous system must deal in determining motor action vary considerably with the conditions under which the muscles are acting. However, since effective motor control is accomplished, it would appear that the nervous system has the capability of dealing with the problem. Before looking for mechanisms available to compensate for muscle properties, it might be desirable to make some judgment concerning how much error is tolerable in normal operation of muscle without needing compensation. Let us consider a simple, rigid mechanical structure, hinged at one end and having a mass distribution similar to my forearm. If this structure were supported against gravity by a force operating at a distance from the hinge comparable to the attachment point of my biceps tendon, what would a force error do? A straightforward calculation of the Newtonian mechanics reveals that an error of 1% of the needed antigravity force, acting continuously over a 1-sec period, would produce a position error of approximately 10°. (Since only about a 2% activation of the muscle is needed in this case to oppose gravity, this error of 1% of gravity force is only roughly .02% of the maximum muscle force.) The same error sustained over 2 sec would result in a 40° position error. A similar calculation for an error 10 times as large, but acting over only 100 msec leads to an error of 1° in position, and at the end of this 100 msec, an error velocity of 20° per sec, a velocity large enough to have a significant effect on the muscle output at that length. Similar errors in the varying forces during movement would produce a correspondingly large error in the moving positions and velocities. Reproducibility of force output of a muscle from one episode of excitation to another within .02% or even within .2% is a stringent requirement for a muscle operating under constant conditions and even more difficult under changing conditions. The sensitivity of output to operating conditions can be illustrated by the fact that within the range of movement of a cat gastrocnemius muscle as used in walking, a 10% change of maximum force output would result from only a 1% change in muscle length, due simply to the length-tension relationship. Since the excursion of this muscle in walking spans approximately 10% of rest length (57,64), a relatively small difference in position from one step to the next would make important alterations in the response rules. Of course, if successive steps were essentially identical, each step might require the same compensations, and a stylized control pattern could be generated with built-in compensations to produce the necessary precision of output forces.

Let us then look at two recorded steps of a cat measured with respect to the activity of the triceps surae muscle (Fig. 11). These steps were recorded from a cat running smoothly on a treadmill in studies done by John Yager (64). He recorded length of the muscle photographically, EMG with electrodes and preamplifier chips implanted within the cat, and force using a strain-gauge transducer chronically placed on the tendon. The rectified EMG record seen here is digitally filtered without time shift. From his records I have calculated velocity and acceleration using a spline function. Clearly, the force, EMG, length,

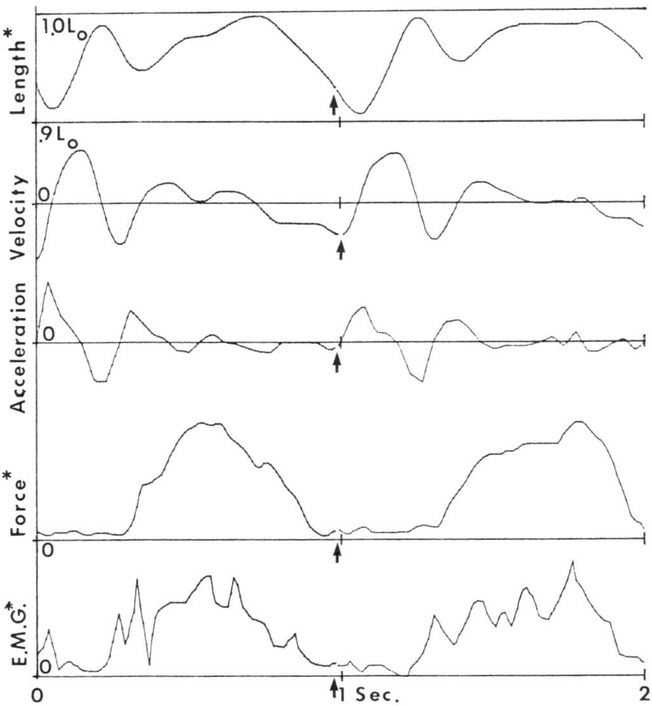

FIG. 11. Time graphs of several measures from two steps of a cat on a treadmill: the rectified EMG measured from triceps surae, length measured from a 32 f.p.s. motion picture, and force measured from a transducer chronically implanted on the tendon (*records redrawn from ref. 64). For this figure, the digitized EMG was smoothed with an optimal filter. Points of measured length were joined by using a second order spline function from which velocity and acceleration data were also derived. The swing phases of the steps can be recognized by the association of low forces with large accelerations.

and velocity all differ appreciably from one step to the next. When force, velocity, and length are plotted (Fig. 12) over the same phase plane as used for the force surface, it can be seen by comparison with Fig. 7 that the activity of this step occurred over an operating region in which considerable variation in muscle responsiveness occurs. Expanding this graph for one step of the force plotted over the phase plane (Fig. 13A), we can better see the low force and large length changes of the swing phase, the rapidly rising and slightly complex force during the stance phase, and the force again falling into the next swing phase. Figure 13B shows the other step in which the swing phase and stance phase both operate through slightly different regions of the phase plane, and the shape of the step is appreciably different. The filtered EMG as a function of length and velocity is shown in Fig. 14. When the generated force is compared with the driving EMG, it is found that at negative velocities and at short lengths

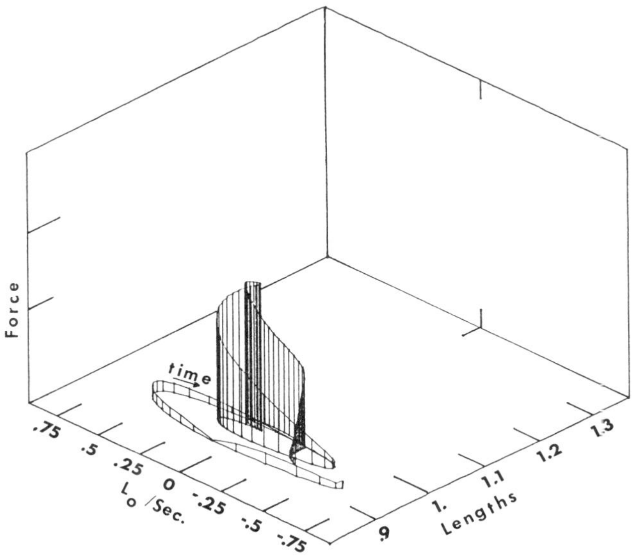

FIG. 12. Length, velocity, and force from the first step of Fig. 11 plotted together in a three-dimensional graph. Length and velocity scales for this graph were adjusted to correspond to scales which were used in Fig. 7A in drawing the maximum force surface of a similar muscle. Data needed to compare relative values of the force scales in these two figures are not available. Vertical "picket" lines between zero force plane and measured force are spaced at 10-msec intervals.

the neural excitation, as represented by the EMG, is rather ineffective in producing force, whereas at lengthening velocities and longer lengths it is appreciably more effective. This comparison is suggestive that at maximum activation the force through these step cycles would follow a pattern rather like the theoretical force surface described previously. From these data, then, it would appear that the theoretical description of the varying relationships between force and activation is pertinent to the actual rules used in the physiological control exercised by nerves over the triceps surae muscles in walking cats.

MECHANICAL FACTORS RELATING MOTOR RESPONSE TO MUSCLE RESPONSE

Muscle is often required to deliver a force against a gravity load while holding an approximately constant position. In fact, this has been one of Brookhart's interests (7). Since (in this case) inertias remain nearly constant, the instantaneous acceleration of the load is almost proportional to the difference between the summed muscle torques and the torque generated by gravity acting on the load. The angular velocity is related to the time integral of this torque difference, and the angular position is proportional to the second time integral. However,

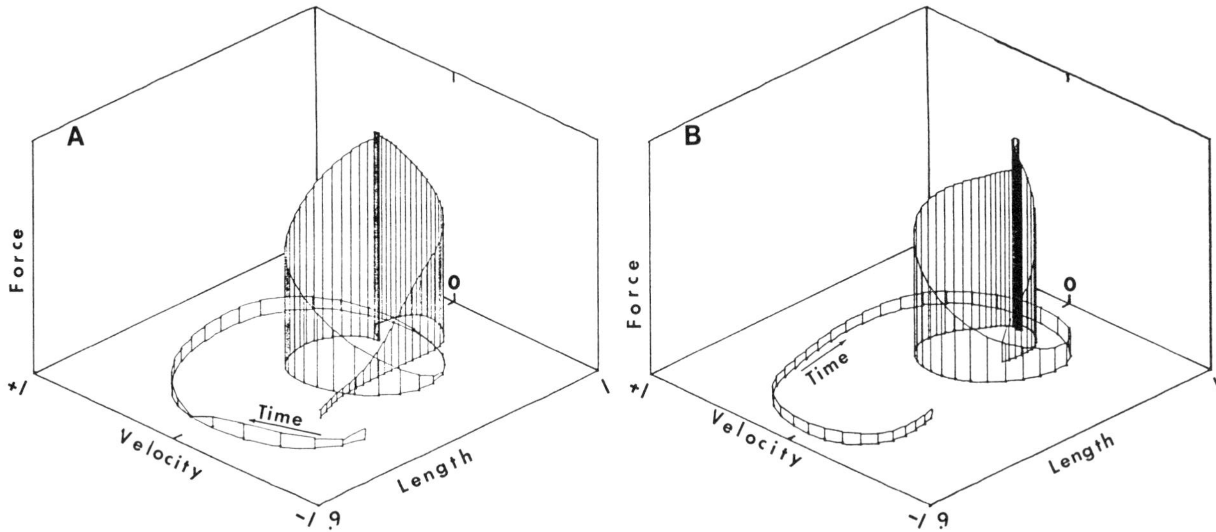

FIG. 13. Data from Figs. 11 and 12 with length scale expanded to allow easier tracing of detail. Vertical pickets are spaced at 10-msec intervals. **A:** First step. Note step does not start and end at the same velocity. **B:** Second step (incomplete if step is defined by return to initial length).

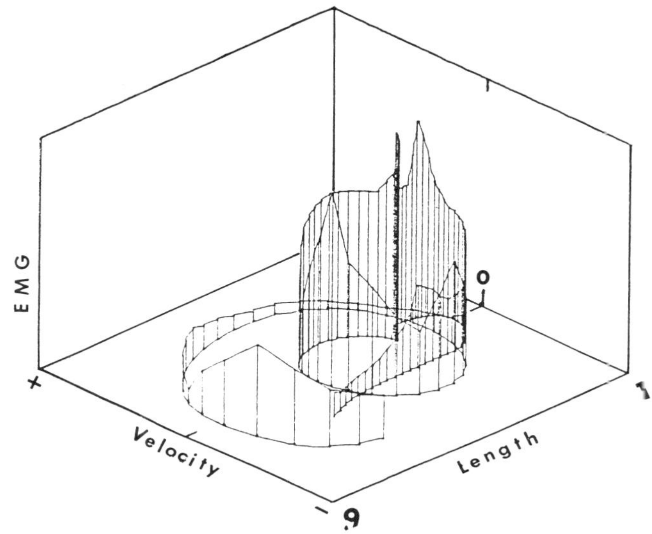

FIG. 14. Combination of length-velocity-EMG plotted on same scales and for the same step as in Fig. 13A. Note high level of EMG activity in early part of step where force was very small.

when a displacement error accumulates, the changing joint angle modifies the effect of gravity on the load, the inertia of the load, the effectiveness of muscle force on the load, and the effectiveness of neural signals to produce muscle force. Thus, if appreciable movement occurs, neither the local mechanical rules nor the local muscle rules remain constant. However, over any long period of time the cumulative effect of mismatched torques must be zero. Often "simple" antigravity action is complicated by the action of an antagonist muscle. If the antagonist muscle which produces movement in the direction of gravity is active (36), the balance is not agonist against gravity, but agonist against the combined torque from gravity plus antagonist. As do changes in load on the limb, changes in activity of the antagonist require corresponding correction by the agonist. Figure 15 illustrates one simple theoretical combination of the force surfaces of an agonist and antagonist. Two identical maximally activated muscles pulling with the same lever ratio could produce the torque surface illustrated here. The figure assumes simple summation of maximal torques of two identical muscles. Clearly, this shows that, from what is known about the force-velocity and length-tension effects, the result of simultaneous action of an antagonist muscle pair is not the simple cancellation of torque. Response in an antagonist pair would seem not adequately described by equating the pair to a single muscle which can both push and pull. On the other hand, it has been shown that at a low excitation rate, extraocular muscles are probably capable of producing near balance of static torque over a few degrees of eye position (51).

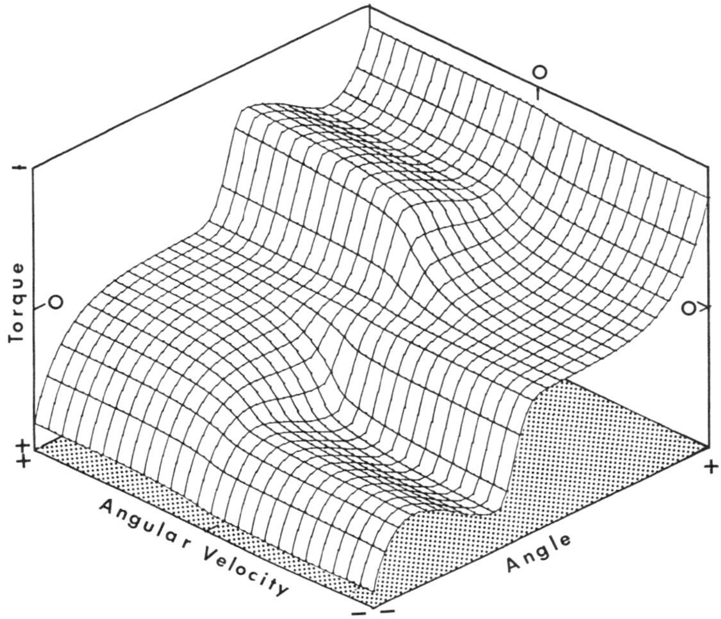

FIG. 15. Torque as a function of joint angle and angular velocity. For this diagram, two identical muscles were presumed, both maximally excited and operating on equal lever arms in exact opposition on a joint with 1 degree of freedom. Note that at zero velocity, the torque effects of the two muscles do not cancel. At finite velocities, the cancellation of opposing muscle effects is minimal. Even if the diagram were redrawn with different assumptions about the relationship between muscle lengths and joint angle, even at zero velocity the muscle actions cannot be made to cancel over the midrange of angles because muscle does not have a symmetrical length-tension curve.

Let us examine a simple specific example of a postural control. To hold the elbow at 90° might be accomplished by pure biceps opposition to gravity for one shoulder position, no muscle torque on the elbow at another position, and a pure triceps action at another position. For this very simple motor function, the requirements for motoneuron output can be varied. The simple task of holding the elbow at right angles requires a precise force balance depending on both orientation with respect to gravity and neural drive to antagonistic muscles, but the particular values involved in different instances do not have either a constant torque direction or constant muscle involvement. If the initial command signal in Fig. 1 represents holding the elbow at a right angle, then the transformation between that signal and the muscle drive must be highly dependent on loading conditions. The problem of variable load is well known, and we have traditionally invoked negative feedback to provide the necessary corrections for the effects of these load variations.

Probably a more practical motor act than holding a fixed joint angle is the holding of an object at a particular position. The position of a rigid object can be described completely in terms of 6 degrees of freedom: the three Cartesian coordinates of its location in space, and the three angular coordinates of its orientation at that location. To place an object, say a cup of coffee, in a particular position, it is necessary to control those 6 degrees of freedom. That control needs considerable precision at the shoulder to grasp the handle, avoid spilling the coffee, and reach the mouth. Temporarily ignoring the flexibility of the hand and the ability to move the shoulder girdle, we find within the arm, including the rotations of the shoulder and the wrist, a total of 7 degrees of freedom, one more than necessary to control the position of our cup. Since each of these joint angles is continuously variable, the extra degree of freedom gives us an infinite number of ways in which we might accomplish this same final cup position. For any muscle in the system, this means that many different lengths might occur at the same end position, and a particular small cup movement might be operated through different ranges for the driving muscles. For example, in two movements starting at the same point but with a change at wrist or elbow compensated by a change in shoulder position, both muscle lengths used and the length changes would differ. Figure 16 illustrates a few options for a limited motor act moving between two points in a two-dimensional, instead of six-dimensional, system. The three angular movements used at wrist, elbow, and shoulder exceed the two dimensions of the space in which the limb movement and the two desired positions are defined. Individual mechanical

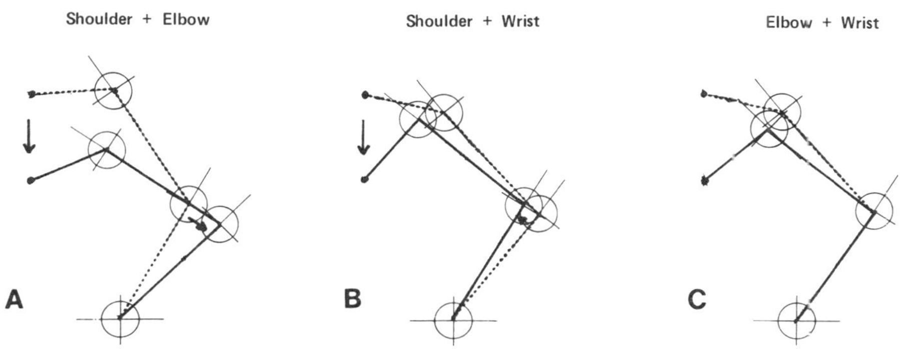

FIG. 16. Location of a point in two dimensions using three joints, but with the whole system restricted to movement in a single plane. This could be visualized as moving a point at the tip of a finger, with the arm moving over a table at shoulder level. Three different ways of moving from a start to an end position are illustrated in **A, B,** and **C.** Each of the illustrated movements uses only two of the three joints. Note that in **A,** the shoulder moved clockwise, whereas in **B** the same end result was accomplished with a counterclockwise shoulder movement. A similar reversal of direction of movement can be shown for each of the other two joints by appropriate selection of pairs of starting and ending positions. Since the angles are continuously variable, the 3 degrees of freedom in the joints allow an infinite number of joint combinations for attaining either position in the two dimensions of the plane. Correspondingly, each muscle involved in the movement could have an infinite number of starting and ending lengths, of which a large number would be at functionally distinguishable different lengths.

results then are not constantly related to the joint positions (24), nor to the nerve signals required for each muscle. Since muscle lengths change with joint angle, even the local rules by which each muscle responds to a nerve signal change. Again, we might invoke proprioceptive feedback to correct the mechanically introduced errors in the task. However, for length feedback to work, all of the needed muscle lengths would have to be determined and represented in fusimotor activity. The feedback could only correct error after the desired positions had been determined by other means.

We have just seen that to bring a cup of coffee to the lips, there are an infinity of ways that the mechanical sequence can be accomplished. We may not use them all, but we do use more than one. For example, when sitting next to a wall we use different positions than when sitting with our elbow over a table, and these are both different from the positions which we might take in the middle of a large space. The extra degree of freedom complicates the programming of the movement, but it gives us the option of adapting our movement to work still when under restrictive conditions. In this connection, the question arises whether the control involves an organization of information in a trigonometric form. The sequences of nerve impulses needed to control the muscles ultimately are distributed in time and over the motor units supplying the whole arm and much of the trunk. Is it necessary that in generating this distributed output for the muscles the information be separated at any neural level into information that discretely represents the needed lengths or forces in the individual muscles?

In controlling movement, the factors in static position are further complicated by the need to balance the time course of energy exchanges for the moving structures. An error in energy transfer will result in a corresponding error of movement. The kinetic and potential energies of the load change through a movement of that load with position, velocity, and configuration; and from one occurrence to another the energy exchanges must be adjusted to the mechanical properties of the load, including the limbs, as well as anything held by the limbs (1). The obvious source of the energy is the agonist muscle. However, it must be realized that passive elastic elements associated with that muscle also will deliver energy when the muscle shortens (1,5,12,22). Further, it is easy to forget that a very important aspect of motor control involves the absorption by the muscle of either kinetic or potential energy which was originally in the load. At any instant, the effectiveness of either energy output or absorption by a muscle is dependent on the level of excitation and also on the point over the phase plane at which the muscle is operating. If the movement originates at a different starting point, even though it ends at the same terminal point, the energy-exchange sequence, activation, and the neural control must be altered accordingly. When both an agonist and its antagonist are active together, energy output from one muscle will be absorbed by the other with only the difference exchanging with the load. Even though two movements may have the same starting and ending points, any variation of antagonist activity or route taken would alter the essential details of energy exchange. These energy exchanges are related to the products of velocity and force. Thus, proprioceptive types of information can (after the fact) still theoretically provide the nervous system with the identification of appropriate details of the operating states which define the local rules applicable.

The types of information available from proprioceptors, although requiring a complex analysis, would be sufficient to define the mechanical results of a movement in a system composed of multiple well-defined joints. On the other hand, all movements are not so simply evaluated in terms of trigonometric operation on information from muscles, tendons, and joints. In many species, jaw movements are not restricted to a rotation around a fixed hinge. Sliding movements change joint position, particularly in ungulates, where chewing involves considerable lateral, as well as vertical, movement with a diversity of pattern (38). The temporomandibular "joint" is not a true hinge (38,53), and the simple trigonometric type of analysis that might control a limb movement is no longer so plausible for control of the force couples of the muscles acting on the mandible. Is the neural strategy for jaw control different from that for the leg? At least the snake seems not to use a basic knee control pattern for some of its jaw actions. Likewise, those sliding movements of the shoulder girdle that are so beautifully displayed in a slow-motion picture of locomotion in members of the cat family are in no sense angular movements, although they do contribute appreciably to limb position. The relationship between the activity of muscles producing such sliding movements and the position of an extremity is mathematically somewhat different from that of angular movements. How is the translation of a command signal for a particular end result organized to smoothly trade off between rotational and sliding movements when both can share variably in producing a chosen motor effect? For some reason, most of us still tend to diagram a prototype of the peripheral system of motor control as a simple joint, without any gross sliding elements.

Other movements differ from the simple trigonometric problems of hinge joints. The eye, with its small and relatively constant load, is such a straightforward pivot joint that it has been possible to develop an elegant trigonometric description of static control by muscle (51,52). However, in that case, the important result usually controlled is probably not mechanical position of the eye, but position of an image within the eye. The original form and the meaning of the visual information as it moves from one receptor to another and the transformation of this spatial signal into signals controlling contraction must be processed in some way quite differently from those graded signals of single muscle spindle receptors which are used to alter excitation in other muscles. Then again, the tensor tympani muscle and stapedius muscle would appear to be controlled in relation to their effect on perceived sound intensity, a still different signal, integrated across frequencies. For the motor control physiologist, that complex interwoven structure of muscle fibers that makes up the tongue and which serves to move food in the mouth or to change the acoustic properties of a sound chamber is also not a simple joint to be described or controlled with the same trigonometry as our elbow. Even the two operations of the tongue must be adjusted to different types of sensory references. [Incidentally, is this the reason we so often hear that we "should not talk while our mouths are full"?] A different control relationship would seem to be involved in the control of those skeletal muscle sphincters that serve a social function. Even different control signals are inferred in Brookhart's earliest papers (6) which contributed information about the inputs to the control of the diaphragm and intercostal muscles. One might suspect that on a short-term basis the control of these muscles is related to mechanics of respiratory gas movement, while on a longer-term basis it responds to respiratory gas exchange, and on occasion may deal even with surfactant problems (50). The evaluation of the motor control of the laryngeal muscles and further adjustment of this control is probably based directly on sound production, and only indirectly on mechanical response. In the skeletal muscle of the limb, the control involved in the increased tonus and finally the shivering by which the muscle contributes to temperature regulation seems also not to be based primarily on mechanical movement, but on thermal factors. Even limb muscles seem to control force, velocity, impedence, or position on different occasions. Further, when olfactory input modifies the running of a dog on a trail, the stimulus is not simply related to the response, which in turn is not one of those single-dimensional functions. Thus, there is a great variety of information carried as afferent nerve signals which might eventually be used to adjust motor response. The meaning to muscle control of these sensory signals must be quite different from one case to another, increasing or decreasing response, changing the speed of a response, changing the envelope or periodicity of an alternating response, or modifying the combination of units involved in the response. Consequently, we need to be cautious about our tendency to think of motor control in terms of only one or two dimensions of information at a hinge joint.

DISCUSSION

Ideally, a theory of motor control might be asked to account for all of the variations of motor control that are known to exist. At the least, a general theory of motor control should not be incompatible with any known aspect of motor control. If the principle of parsimony is applicable to the design of the motor system, we might suspect that the method of accomplishing motor control would be based on a few common principles instead of using different rules for each variant of the problem. Unfortunately, at the system level, neither current experimental data nor current theory on motor control is well enough defined for critical comparisons of measurement and prediction. Consequently, a rather loosely defined speculation is a common approach to the theory of motor control, and, I am afraid, will be the approach in the following argument. This discussion deals with motor control only after the desired response has been chosen. Although several different nomenclatures are used, modern concepts of motor control presume some initial event that would correspond with what is labelled in Fig. 1 "command signal." I refer simply to that pattern of activity which first represents the choice of a particular motor response. It may be localized in the nervous system, or diffuse; it probably is usually spread through some period of time, and involves more than one neuron and more than one impulse in most individual neurons. Whether it originates by a combination of sensory inputs or by something called "voluntary action" is unimportant in this context. Its definition is far easier to state than its location is to find, particularly if that initial event should be distributed over a dispersed group of neurons. A wide range of opinion exists today about the closeness of this command signal to the details of the motor unit activation; but, at least to the best of my knowledge, no one has proposed that the command signal first comes into existence in the form of the final output pulses on the motor neurons. As Humphrey indicates in this volume, current theories seem to accept that between the command signal and the impulse pattern distributed over the final common paths, modifications do occur. The nature, extent, and mechanism of these modifications is where we debate.

Strategies for the Control of Muscle

Because of the variability of the response of muscle to any particular pattern of nerve impulses and the complexity of the rules that relate muscle responses to motor actions, it seems inescapable that the particular pattern of nerve impulses which are delivered to the muscle be generated in a system provided with information about the conditions that exist in the periphery. On the other hand, some accommodation to properties of the peripheral system clearly can be provided without current reference to peripheral events. Thus, it could be assumed that an increase of muscle contraction can usually be called up by an increased firing of a motor neuron and a fast increase of muscle response would require transiently more initial excitation than required for a slow response of the same magnitude. Over long periods of time, the muscle driven is likewise consistently related to the central selection and excitation of motor neurons. Further, maximal or minimal activation of particular muscle fibers can be dependably defined in central terms. However, beyond such basic certainties in motor control, the quantitative adjustments of response depend heavily on variable factors in the periphery.

Most of us own some stock in the negative feedback theory of motor control, but I suspect we are facing a falling market. While Wagner made the first offering of the stretch reflex under the title "feedback" (59), the market was poor until Merton (41) reissued the stock with Granit's promotion (23) and started a rapid rise in acceptance. By about a decade ago, the servo theory had become probably more effective in controlling our thinking than the stretch reflexes are in controlling motor activity. We learned that in many cases information fed back about the results of a control can be consistently related to the magnitude and direction of needed correction. In use of error-correcting feedback, the complex causes of the error and operational conditions other than the error are often unimportant. However, the reflexes have been found to be too slow as a negative feedback system to control so-called "ballistic movements." Measurements based on the assumption that the reflex is acting as a servo-system show its gain to be so low that it would be relatively ineffective (39). Reflexes are not necessary for control in other cases (16). Then Houk (31) reminds us that the Golgi tendon reflex operates in opposition and would tend to cancel the effect of the stretch reflex if it were considered to be a servo-system. However, there seems to be no doubt that the stretch reflex and the Golgi tendon organ reflex do exist and do influence the motor neuron activity in physiological conditions.

It makes no difference what the function of the feedback is, whether acting in a servo-system controlling position or in a feedback system controlling mechanical impedence of the muscle. Neither the Golgi nor the spindle reflex can contribute feedback correction in a ballistic movement. For these ballistic movements, the command signal must be transformed into those signals which drive the muscles without benefit of continuing correction from peripheral progress reports. Could the command be modified by a standard transformation to produce the motor neuron outputs? No. The responsiveness of muscle is too dependent on peripheral conditions for the muscle to be dealt with as a constant operator serving the motoneurons. However, it is, as Terzuolo reminded us in this volume, characteristic of ballistic movements such as fast piano-playing or typewriting that owing to preceding activity the conditions existing just before a particular response is required are quite variable. With different initial conditions at the start of a motor response, the neural signals will find different local rules in the muscles. Even though at the final moment of typing a particular letter the same position might be reached, if, as a result of different starting positions the energy put into the inertial elements being moved should differ, the time course of the energy absorption pattern would also have to be appropriately adjusted. It would appear that ballistic movements, without correction for the effect of the operating conditions on the rules, would have poor success in accomplishing a usable accuracy of response.

However, we not only know that successful ballistic movements are regularly accomplished, but I believe that the information is available on which to build some insight into a possible mechanism. As Young reminds us in this volume, it is only necessary to know the present state of a system, the rules of the system, and all future inputs to the system to define all future states through which that system must pass. I would point out that if that part of the rules which apply at any moment are dependent on the state of the system, the expected moment-to-moment changes in local rules will then also be defined. In a practical system, if neither the local rules nor the state are subject to discontinuous changes, an approximate knowledge of any initial state and future inputs will provide an approximate projection of the changing state and local rules which would be good for a short time into the future. The same relationships in some cases can be exploited in an inverse sense to determine the inputs (e.g., motor nerve impulses) necessary to drive one or more dimensions of the state through a desired trajectory. On these considerations, I propose that, because of the variability of the peripheral system in motor control, the neural drive to muscle for ballistic movements not only must, but can be precompensated for the effects of the operating state of the peripheral system and that signals from known sensory receptors together with the command signals appear likely to include all the information needed to identify the precompensations which would be required. Thus, the information needed for precompensation may be available, and a possible algorithm for the computation exists if the machinery does. Further, if precompensation is used for motor control of ballistic movements, the existing machinery could then be used generally in motor control, driving, in effect, a continuing series of brief ballistic movements using continuously updated information for determining the changing rules of precompensation. However, over the longer periods of time, for the types of control subject to cumulative error, a result-based correction would also be needed to prevent error accumulation (i.e., negative feedback).

The precompensation envisioned here should not be contrasted with earlier concepts of feedback control, but looked upon as a more comprehensive class of mechanisms within which simple feedback is but one aspect. In a classical feedback system the controlled variable is sensed and, by fixed additive rules, combined with a reference signal to completely determine that corrected signal which subsequently drives the controlled variable. All of the correction is vested in the error signal, and each system controls that one dimension whose error is sensed. To produce the correction the rules of the effector system need not be known, only the errors resulting from its action and the direction of that action. However, if the signals are subject to large discontinuities or delays, the correction may not be appropriate by the time it is applied. In comparison, the proposed precompensatory process uses information on the command signal and the state of the peripheral system to determine the appropriate signal for muscle drive. For any combination of operating state and command, a defined set of output pulses would form an ideal muscle drive to accomplish the desired result while precompensating those local rules which will apply at the moment of the driving action. The practical system could only compensate for such local rules as are predictable from state and command signals; but with different command signals, the system could even control different output functions as needed, since the system would not be locked to a particular controlled variable by its feedback signal. Within its operating limits an ideal system with perfect information could operate without error. Incidentally, if each sensed variable were used to correct only the effect of that particular factor, each input could operate as if in a feedback loop with a gain of only -1. Barring discontinuities in functions of either time or the operating state, approximate information could produce a good approximation to the ideal outputs. Negative feedback would be required only to avoid development of cumulative errors as found in inertial loads [essentially unneeded in the eye (51)]. However, unlike the simple feedback system, such a precompensatory system would require storage of a representation of the combined effects of each factor in the state and command on the rules for operation.

The operation of a system using state-dependent precompensation would be relatively simple. For a ballistic movement, by definition, only the single set of starting conditions would be available to predict the whole sequence of needed compensations. For a slower movement, the continuing input and command signals would update the basis for signal determination without need for predicting far into the future from any starting data. However, while the difference in speed between a slow and a single-step ballistic movement would probably require different amounts of compensations in the course of the respective movements, the mechanism of accomplishment need not differ. Any function for which the error-producing factors were sensed could be controlled by muscle signals which are defined by those factors and the command signal. (For some functions, previous error contributes to cumulative future error and would therefore be one of the necessary sensory details.)

From what is known about the problem, certain factors for which precompensation is required can be identified. However, the independent conditions that affect the response rules are not independent in their action. As illustrated in Fig. 7, the effects of length and velocity on the force delivered by muscle are not two separate effects superimposed, but a combined effect. Likewise, the speed called for by the command signal determines the requirement placed on the muscle, while the temperature of the muscle affects how fast the muscle would respond (3,27,60). Further, the effect of velocity and length on force depends on stimulus rate (48). Interacting combinations such as these could not be compensated by separate calculations of individual corrections to be added together at the motoneuron, but would require joint calculation of the combined compensations.

For each distinguishable combination of state and command, a specific drive would be ideal; a precompensation should produce a muscle drive that approaches this ideal. At least three distinctively different ways of determining such a precompensation can be imagined. First, with an appropriate mechanism for storage and retrieval, any precompensation pattern could be implemented from a table of point data. However, it should be noted that to represent a relationship between only two inputs and one output with as much detail as shown in Fig. 7 required interpolation between 2,500 points. An equivalent representation of three inputs (e.g., adding temperature or variable stimulus) increases the data points to 125,000, and still requires interpolation. Second, the representation can define the output as a function of the inputs, particularly if the function changes in a continuous way. Such a representation requires the necessary function to be specified and a few coefficients to be established.

(The points in Fig. 7 were, in fact, calculated from a special function of the two inputs which required 10 coefficients.) A critical factor is the initial selection of the appropriate function; but, once defined, the number of stored details is small and interpolation becomes automatic. The third general type of compensation could be based on a piece-wise linear computation in which the whole range of input combinations is subdivided into small enough regions that the output in each case can be well approximated as a specific linear function of all of the inputs through that particular region. For n inputs, at least $(n + 1)$ coefficients are needed for each region, and the region boundaries also must be defined. (In Fig. 7, the interpolations drawn between points were made on this linear basis.) It is clear that the linear regions which are required must be small where curvature is great, but can be large where curvature is small. Correspondingly, in many regions of Fig. 7, it would not have been necessary to use as many fitted pieces as were used. Practically, to implement the piece-wise linear approach, calculation is simple, since all of the functions are simply additive after the coefficients are defined. [In any of these representations, the output dimensions need not be restricted to force, although the particular combination used for illustration here does have an advantage for the purpose of illustration. Incautious choice of presentation can lead, for example, to dealing with the discontinuous type of relationships sometimes described under the label "catastrophe theory," but at least in some muscle problems the catastrophic results have been evaded by appropriate treatment to avoid dealing directly with jump discontinuities (e.g., see 45).]

Because of delays and lags in the neuromuscular system, the command signal and the muscle drive must both represent future events, while the sensory signals can represent only past events. Barring unanticipated transients, however, the ideal trajectory of signals for muscle control is defined from the applicable rules, the recent past state, and the properties of the current command signal. Consequently, the necessary transformations of the command signal to muscle drive can be related directly to currently available information without a separate prediction of future state. In a continuous system, the resulting error need reflect only error in sensory representation of the state and error in stored information about the applicable rules, and it need not reflect any effect of delay and changing local rules.

For precompensation to produce a usable response requires that enough information be available in the centers about the operating state of the system to reduce the total error from the combinations of all factors influencing the peripheral rules to an acceptable residue. The best-known factors influencing muscle force are length, velocity, temperature, and muscle condition. A motor act involving a limb also is affected by current joint angles, length of limb segments, external forces, loading, and kinetic energy of the load. This information might be sensed either directly or indirectly [i.e., inertia of load, of course, is defined as well by combining acceleration and force information, and in some cases one dimension may be adequately represented by entirely different functions that just happen to be highly correlated. Thus, the stretch of a muscle under some conditions might represent the impedence of the load upon which it is acting (34)]. Joint-angle information could be used to solve for compensation in some motor problems; but, for other functions controlled by muscle, this particular information would either not exist or be inadequate. We can recognize a basis which might be used for precompensation of signals to muscles in information as varied as pattern of light, sound, acceleration, contact, temperature, and even bladder or alveolar stretch. For some motor control tasks, the compensation needs to be extended through the characteristics introduced by the tools used. For example, the final control in some cases of rapid and precise movement must be extended through extracorporeal devices as complex as an airplane. It is not obvious what specific information would be used to identify all the conditions which are involved in determination of the local rules of these effector devices, but in most motor actions mechanical information from muscles and joints would not be enough basis to determine most of the precompensations of muscle drive that would be needed to produce dependable response. The information needed for relating different muscle activations to motor acts is characterized by diversity instead of uniformity. It appears that under one or another circumstance almost any information about the body or about its environment could be an important base needed for calculating precompensation.

Perhaps the best evidence of use of sensory information in motor control in a manner different from negative feedback is to be found in the well-known studies of the diverse reflexes which modify motor response. Do not crossed reflexes, vestibular reflexes, plantar reflexes, etc. (15) in effect precompensate the signals controlling muscles? They can hardly be reports of the magnitude of the simple controlled variable. I also read a similar theme of multiple influences on motor response in about 20 of Brookhart's publications. The innovation here is only to suggest that these inputs, in fact, form the specific basis for that reasonably good prediction of how a desired response can be accomplished that is used by the nervous system for producing the selected responses.

Speculations on Possible Mechanisms for Precompensation

A variety of computational systems might produce the type of precompensation action that has been proposed; but if within the range of known neural operations a reasonably simple mechanism can be envisioned, the proposed operation would be more plausible. At the same time, on an evolutionary basis, a system that must be developed in a single complex step instead of by addition of individually useful simple steps would not seem reasonable. The following discussion is then an attempt to examine some known neural processes and compare these to requirements for a precompensation type of action to see if at least one reasonable mechanism capable of producing precompensation can be identified.

Before discussing the processing of control information, however, it is worthwhile to briefly consider the form in which this information is handled. The dimensions in which the nervous system operates to compute appropriate muscle drive need not be the same dimensions which are made familiar in physics. Thus, in experimental studies, each done by varying only one dimension of stimulus, the muscle spindle is found to respond strongly to length, velocity, temperature, or fusimotor action (40,46). In more normal conditions, the response of the spindle receptor seems to represent the magnitude of a vector function combining these dimensions. Differences in the relative responsiveness to different stimuli, not an exclusive sensitivity to one test dimension, define the functionally different classes of receptors. Likewise, at the motor end of the system, the needed compensations are related to combinations of length, velocity, temperature, etc., rather than to just one of these dimensions. On single nerve fibers at either the sensory or motor end, the firing rate can represent only one dimension, but that dimension involves a transformation and combination of the single dimensions usually used to describe experimental data. If neither the input nor output operates with signals separated into the single dimensions of physics, but they do represent the important information in the system with transformed coordinates, there is no obvious reason why the central processing need ever deal with the information isolated in the coordinates of our technology. It is therefore proposed that in motor control the information in individual neurons about the state, command signal, and muscle drive signal exists in a form that, in general, does not correspond to the particular dimensions used for testing purposes (46).

While speculating about possible neural organization that could produce precompensated signals for muscle drive, I suggest that at all stages of organization of a motor signal in vertebrates, the signals are distributed over a group of neurons. Although, on the sensory side, the information needed for compensatory calculation does not appear in the individual single dimensions familiar in physics, the response characteristics of the individual receptors also seem not to carry that weighting of the sensed dimensions that would directly determine the necessary precompensation of signals in motor neurons. Since the firing rate from a receptor is a one-dimensional signal which is a weighted representation of several physics dimensions and for determining the needed precompensation, the various dimensions must be weighted differently than in the individual sensory signals, the sensory input would need to be carried on more than one nerve fiber, not only for redundancy, but simply to provide the basis for this reweighting of factors.

In most motor activities (e.g., see 24,58) (Fig. 16) diverse combinations of muscle actions can accomplish the same end. Different balances of antagonist action represent one way in which the same results can be the product of diverse activities in the same muscles. Further, because in different actions the relative activation of different muscles must differ (2), the activity which is distributed over many motor neurons must provide some degree of separate control of individual motor units in order to coordinate the variety of solutions to any control problem. Under these circumstances, no single motor neuron can or need carry a full representation of motor drive for a particular action. Again, we find the simplest defining activity distributed over many neural units. As the control of a single joint is apparently accomplished by a variable selection and grading of activity from the pools of motor units acting on that joint, so, too, any complex motor act can be looked upon equally as the product of a variable selection and grading of the activity from the total pool of motor units of the body.

With respect to the command signal, the fact that with a continuous process of discrete neuron losses we do not recognize a regular characteristic of abrupt loss of individual motor actions suggests that specific motor commands do not originate in a single neuron. Of course, they could originate in an individual member of some interlocked system of redundant units, but it is also possible that in vertebrates the command ordinarily originates as a distributed activity, meaningful only as the appropriate patterns of activity occur over a number of neurons. Certainly, the control of those special movements that in some teleosts have been found to funnel through only two Mauthner cells (56) is not typical of the usual findings in study of motor control in vertebrates. Thus, in the sensory, the motor, and probably in the command signals, meaningful activity is distributed over many approximately parallel neurons. Further, I know of no specific reason why the organization of any act must be channeled through an individual neuron at any point in its generation. It is thus not obvious, except in some invertebrates, that an activity in any single neuron defines more than a fraction of any motor action.

Now let us turn from the form of the signals available to accomplish a precompensation to the nature of the precompensations needed. To use information about the operating state and the command signal for determining a precompensation of the nerve signal to muscles requires that stored information be available which defines how the compensated muscle drive is related to the various combinations of operating state and command signal. To achieve the effectiveness of control that is observed would require that this information be also accurate. However, the particular transformation rules derived from this information must be fitted to the physical variations among individuals, and within one individual must be changed with growth, muscle hypertrophy, obesity, etc. It is therefore unreasonable to propose a precompensatory control based entirely on either a genetically introduced set of detailed rules or even on a set of rules which, once learned, remain constant. Since the rules all appear to change smoothly with the different dimensions affecting them, the fitting of rules can probably be developed for whole neighborhoods of operating states and used in an interpolating manner. The change of rules occasioned by adding a hammer or a pair of heavy boots to the system controlled might be dealt with as a new operating state rather than as a change of rules to be learned and unlearned each time needed, but that is possible only if the information used to choose compensation includes enough input to identify these as qualitative differences. For precompensation to work in motor control, it seems necessary that it be organized to adapt continually to the changing rules of the peripheral system. Incidentally, the loss of its tail to a developing anuran adult need present no unique problem for adjusting motor control if normal control involves a continuing readjustment which would adapt control to such changes as body weight.

It is therefore proposed that precompensation is based on a learning system that can: (a) improve the resolution of the compensating rules with the repetition of nearly identical situations; (b) use a continuous interpolation to approximate rules when the operation moves into regions of the operating state that fall in relatively large gaps between any that have previously been experienced; and (c) readjust already established rules to correct for changes that have occurred in the peripheral system.

To this point it has been assumed, as is usual in neural theory, that the necessary combinations of signals do occur upon which to base the computation of the responses. For some combinations, the association may have been so invariant as to have established a genetic basis for very specific connections. Even in most of these cases, however, it seems likely that individual fine adjustment of effectiveness would still have to be established by learning. On the other hand, many motor functions require adaptations to conditions so nearly unique to the individual that the necessary connections would not be likely to have been provided on a specific basis. A system with many sets of either relatively random connections or potential connections would seem to be necessary upon which to build a learned adjustment. In such a case, a variety of direct or indirect combinations of individual command and sensory fibers would have to be available to provide contingent combinations. The effectiveness of the combinations then would be adjusted by experience. Again, the individual motor neurons, acting as a final common path, would also seem to require a diversity of firm or potential driving connections representing the various combinations of sensory and command activities. Given a potential for a variety of effective combinations with a possibility of learned adjustment of effectiveness, the necessary set for operation could be chosen by a learning process not necessarily different from that which would adjust the effectiveness of the individual inputs. An input with near-zero effectiveness for a particular action thus could represent a learned disconnect of unrelated or conflicting signals.

The description of a possible system for precompensation of the signals to drive muscle should include a suggestion concerning the type of precompensation visualized. It has already been proposed that individual nerves which carry part of the state and/or command signals converge on other neurons in originally more or less random groupings. It is also known that the grading of output of individual neurons occurs only between the limits of threshold and saturation. Combining these two relationships leads to the probability that, without regard to operating mechanism, different regions in the operating state/command signal space would be characterized by the particular groups of neurons which are within their range of graded response when that state exists. Assuming that learning modifies the effectiveness of neural signals arriving at selected junctions, initially random sets of converging signals could become a basis for a group of different local rules, each acting in a particular region in the operating state/command signal space. This could provide a variety of different regionally applicable rules for transformation of command signals into drive signals for muscles with precompensation for the operating state. It is proposed that the convergence of signals (55) includes the different types of signal processing known to occur in neural systems, namely, delays and lags followed by inhibitory, excitatory, additive (37), and multiplicative interactions (e.g., 34). The multiplicative interactions would be important to: (a) adjust effectiveness of one signal on the basis of magnitude of other signals; (b) combine signals in order to represent interaction between different dimensions; and (c) develop a power type of nonlinear function of one or more signals. This proposed mechanism for generating precompensated signals falls between the use of a defined compensatory function and a piece-wise linear approximation. With a nonlinear but piece-wise fitting, the fitted regions could be relatively large; but the complexity of the fitting equations need not be great (e.g., no more than second degree). For example, the surface of Fig. 7 was actually generated by what amounted to a piece-wise, nonlinear function. Only four fitted regions were used, and multiplying and additive combinations were used with coefficients adjusted in each region in order to approximate the experimental data. The nonlinear curves were simple, curves having no more than one inflection point. The intersection of regions was, however, carefully established to avoid discontinuities, in spite of the sharp boundaries used between regions. No overall function was identified which was complex enough to fit the total data range.

Summarizing the details provides an outline of one possible way that precompensated sets of nerve impulses could be delivered to muscle for producing a "desired" response in spite of variability of the operating conditions of the effector system. Only plausibility, not necessity, is claimed for the parts of this particular mechanism. The scheme is: Various parts of a spatially and temporally distributed command signal are mixed with a continually changing and distributed set of nerve impulses originating in activities of most, if not all, of the body sensory receptors. The receptors have diverse enough properties and locations that their total activity at any moment is quite characteristic of the particular combinations of conditions which are likely to influence the relationship at that moment between motor act and the ideal neural output code. Combinations of nerve signals which occurred in initially rather nonspecific groupings with either existing or potential connections are capable of combining on different neurons various groups of neural signals in either a multiplying or adding manner. The adjustments (with learning) of the effectiveness of the individual connections or potential connections give the system the specific ability to relate past state and present command to the muscle drive that will probably be required in the near future. The fitting involves an adjustment of fit of simple functions to merging local regions of the operating state, rather than either a point-by-point fitting or an overall fitting by a simple complex function. The initially distributed sensory and command functions combined in a distributed system finally produce a distributed pattern of motor nerve output to the motor units which will approximately accomplish the commanded motor response, barring unpredicted transient interference. For any command signal, a complex muscle drive pattern represents an ideal response, depending on operating conditions. In different motor acts, the control of a particular muscle can determine either position, velocity, force, mechanical impedence, or even a temporally changing sequence or combination of these functions, as long as a distinctive pattern is used in the command signal to call up each of the specific functions, and provided that these functions fall within the capabilities of the peripheral system. In fact, a single command would often control different functions in different muscles at a single moment. The total body response to a command-state combination would be distributed over the whole motor output without reference to the fact that different fibers act in different muscles.

Let us briefly look at this system as it might be described over drinks. An innovation of hard-wired circuitry may have been given evolutionary approval for effecting a successful escape by relating a swinging body position to moving branches of a tree. The same circuit, when installed in a new model and trained to a new program, could contribute to an equally important escape from destruction by controlling the intersection of two other moving objects in the forms of a pitching carrier deck and the wheels of a jet plane. The apparent mathematical differences in the two motor problems and the sensory augmentation of radar might be treated as variants of the same general problem. Thus, both might be reduced simply to the general plan that when some part of an array of neural activity which falls between previously experienced activities occurs, residual effects of previously "rewarded" performances will increase the probability of generation of a pattern of impulses for muscle control with relationships between those successfully tested on the previous occasions. Patterns associated with previous penalties would have low priority. If the input represents enough information, if the outputs are sufficiently versatile, if the experiences are closely enough related, and if the transformation functionals are resolved and stored well enough, then the outcome again probably will be reasonably successful and add a mite to the stored rules. Muscle length, joint angle, wave direction, jet dynamics, branch elasticity, muscle compliances, and cosine of angles may not need to be isolated to solve either of the motor control problems; and the differences between the computations involved could all be learned and stored as different parameter adjustments in the two networks built from the same plan. That original plan, laid out by chance, would require neither built-in information about the dynamics of both systems nor redesign when the new problem appeared. It would, however, be required to have the ability to sort pulse patterns and also be able to adjust its transfer properties on the basis of some judgment of success.

CONCLUSIONS

Seriously, the properties of muscles and the motor linkage provide a basis for defining some limits on the control of the motor in motor control:

1. To accomplish an accurate motor act requires the signals to the muscle to contain details that compensate for the variability of the rules of muscle and motor linkages. Further, in many cases, the muscles must be controlled to adjust nonpositional results.
2. If the movements occur under variable peripheral conditions, they cannot be accomplished by a fixed pattern of neural drive to the muscle, but they require compensation based on the operating state of the muscle and its linkages at the time the neural drive acts.
3. For the first two criteria to be met in a fast motor act, a prediction of need and from this a precompensation of the muscle driving signals is necessary. Time would not be available to use correction based on error sensed in the not-yet-started response.
4. If the sensory signals and central processing required for a precompensated fast and accurate motor response in variable conditions do continue to operate in slow responses or constant conditions, in effect a succession of precompensation should occur.
5. The organization of a system for precompensation does not obviously require anything of the nervous system that is incompatible with known properties of neural systems, nor does it require a system design based on needs in specific tasks.
6. Precompensation of control requires information about the operating state and how that state is likely to change the rules of the peripheral system, in addition to that information about outcome which by itself must suffice in simple feedback control.
7. Precompensation based on multidimensional information would not be bound necessarily to serve a single controlled dimension.

ACKNOWLEDGMENTS

Research in the author's laboratory has been supported by grants from NIH and from the Easter Seal Research Foundation.

The author wishes to express his appreciation to L. A. Benton for reviewing the manuscript and making many helpful criticisms and to J. Emerson for deciphering the text into readable typescript.

REFERENCES

1. Alexander, R. McN. (1973): Muscle performance in locomotion and other strenuous activities. In: *Comparative Physiology,* edited by I. Bolis, K. Schmidt-Nielsen, and S. H. P. Maddrell, pp. 1–21. North-Holland Publishing Co., Amsterdam.
2. Alexander, R. McN. (1975): Evolution of integrated design. *Am. Zool.,* 15:419–425.
3. Ausmussen, E., Bonde-Peterson, F., and Jørgensen, K. (1976): Mechano-elastic properties of human muscles at different temperatures. *Acta Physiol. Scand.,* 96:83–93.
4. Bawa, P., and Stein, R. B. (1976): Frequency response of human soleus muscle. *J. Neurophysiol.,* 39:788–793.
5. Blix, M. (1895): Die länge und spannung des muskels. *Skand. Arch. f. Physiol.,* 3:295–318, 1892; 4:249–409, 1893; 5:150–172, 173–206, 1895.
6. Brookhart, J. M. (1936): Combined effects of vagal and other afferent impulses on respiratory act. *Univ. Hosp. Bull. Ann Arbor, Mich.,* 2:28–29.
7. Brookhart, J. M., and Talbott, R. E. (1974): The postural response of normal dogs to sinusoidal displacement. *J. Physiol. (Lond.),* 243:287–307.
8. Buchthal, F. (1942): The mechanical properties of the single striated muscle fibre at rest and during contraction and their structural interpretation. *Det. Kgl. Danske. Videnska bernes Selskab, Biologiske Meddelelser,* 17(2):1–140.
9. Burke, R. E., Rudomin, P., and Zajac, III, F. E. (1976): The effect of activation history on tension production by individual muscle units. *Brain Res.,* 109:515–529.
10. Burke, R. E., Rymer, W. Z., and Walsh, J. V., Jr. (1976): Relative strength of synaptic input from short-latency pathways to motor units of defined type in cat medial gastrocnemius. *J. Neurophysiol.,* 39:447–458.
11. Carlson, F. D., and Wilkie, D. R. (1974): *Muscle Physiology.* Prentice Hall, Englewood Cliffs, New Jersey.
12. Cavenaugh, G. A. (1978): Storage and utilization of elastic energy in skeletal muscle. In: *Exercise and Sports Science Reviews,* edited by R. S. Hutton *(in press).*
13. Cogshall, J. C., and Bekey, G. A. (1970): Emg-force dynamics in human skeletal muscle. *Med. Biol. Eng.,* 8:265–270.
14. Cooper, S., and Eccles, J. C. (1930): The isometric responses of mammalian muscles. *J. Physiol. (Lond.),* 69:377–385.
15. Creed, R. S., Denny-Brown, D., Eccles, J. C., Liddell, E. G. T., and Sherrington, C. S. (1932): *Reflex Activity of the Spinal Cord.* Oxford University Press, London.
16. Denier van der Gon, J. J., and Thuring, J. Ph. (1965): The guiding of human writing movements. *Kybernetik,* 2:145–148.
17. Desmedt, J. E. (1977): Fast motor units are not preferentially activated in rapid voluntary contractions in man. *Nature,* 267:717–719.
18. Evanich, M. J., and Lourenço, R. V. (1976): Frequency response analysis of the diaphragm *in vivo. J. Appl. Physiol.,* 40:729–735.
19. Fenn, W. O., and Marsh, B. S. (1935): Muscular force at different speeds of shortening. *J. Physiol. (Lond.),* 85:277–297.
20. Finerty, J. C., and Gesell, R. (1945): The effect of pH on humoral stimulation of striated muscle and its application to the chemical control of breathing. *Am. J. Physiol.,* 145:1–15.
21. Fowler, W. S., and Crowe, A. (1976): Effect of temperature on resistance to stretch in tortoise muscle. *Am. J. Physiol.,* 231:1349–1355.
22. Goubel, F. (1972): Evaluation de l'énergie élastique emmagasinée par le muscle lors d'une contraction isométrique. *J. Physiol. (Paris),* 65:238A.
23. Granit, R. (1955): *Receptors and Sensory Perception.* Yale University Press, New Haven.
24. Gray, J. (1944): Studies in the mechanics of the tetrapod skeleton. *J. Exp. Biol.,* 20:88–116.
25. Grillner, S., and Kashin, S. (1976): On the generation and performance of swimming in fish. In: *Neural Control of Locomotion,* edited by R. M. Herman, S. Grillner, P. S. G. Stein, and D. G. Stuart. Plenum, New York.
26. Guttman, S. A., Horton, R. G., and Wilber, D. T. (1937): Enhancement of muscle contraction after tetanus. *Am. J. Physiol.,* 119:463–473.
27. Hartree, W., and Hill, A. V. (1921): The nature of the isometric twitch. *J. Physiol. (Lond.),* 55:389–411.

28. Hill, A. V. (1944): The abrupt transition from rest to activity in muscle. *Proc. R. Soc.*, 136B:399–420.
29. Hill, A. V. (1953): The mechanics of active muscle. *Proc. R. Soc.*, 141B:104–117.
30. Hill, A. V. (1964): The effect of tension in prolonging the active state in a twitch. *Proc. R. Soc.*, 159B:584–595.
31. Houk, J. C., Singer, J. J., and Goldman, M. R. (1970): An evaluation of length and force feedback to soleus muscles of decerebrate cats. *J. Neurophysiol.*, 33:784–811.
32. Joyce, G. C., Rack, P. M. H., and Westbury, D. R. (1969): The mechanical properties of cat soleus muscle during controlled lengthening and shortening movements. *J. Physiol. (Lond.)*, 204:461–474.
33. Keatinge, W. R. (1969): *Survival in Cold Water.* Blackwell Sci. Public., Oxford.
34. Kim, J. H., and Partridge, L. D. (1969): Observations on types of response to combinations of neck, vestibular, and muscle stretch signals. *J. Neurophysiol.*, 32:239–250.
35. Kumazawa, T., and Mizumura, K. (1977): Thin-fiber receptors responding to mechanical, chemical and thermal stimulation in the skeletal muscle of the dog. *J. Physiol. (Lond.)*, 273:179–194.
36. Lestienne, F., and Bouisset, S. (1968): Pattern temporel de la mise en jeu d'un agoniste et d'un antagoniste en fonction de la tension de l'agoniste. *Rev. Neurol. (Paris)*, 118:550–554.
37. Liddell, E. G. T., and Sherrington, C. S. (1924): A comparison between certain features of the spinal flexor reflex and of the decerebrate extensor reflex respectively. *Proc. R. Soc.*, 95B:299–339.
38. Luschei, E. S., and Goodwin, G. M. (1974): Pattern of mandibular movement and jaw muscle activity during mastication in the monkey. *J. Neurophysiol.*, 37:954–966.
39. Matthews, P. B. C. (1967): The reflex response to muscle vibration in the decerebrate cat. *J. Physiol. (Lond.)*, 192:18P.
40. Matthews, P. B. C. (1972): *Mammalian Muscle Receptors and Their Central Actions.* Williams & Wilkins Co., Baltimore.
41. Merton, P. A. (1953): Speculations on the servo-control of movement. In: *The Spinal Cord,* edited by G. E. W. Wolstenholme, pp. 247–255. Churchill, London.
42. Meyer-Lohmann, J., Riebold, W., and Robrecht, D. (1974): Mechanical influence of the extrafusal muscle on the static behavior of deefferented primary muscle spindle endings in cat. *Pflüg. Arch.*, 352:267–278.
43. Moss, R. L., and Halpern, W. (1977): Elastic and viscous properties of resting frog skeletal muscle. *Biophysics J.*, 17:213–228.
44. Partridge, L. D. (1966): Signal-handling characteristics of load-moving skeletal muscle. *Am. J. Physiol.*, 210:1178–1191.
45. Partridge, L. D. (1972): Interrelationships studied in a semibiological "reflex." *Am. J. Physiol.*, 223:144–158.
46. Partridge, L. D. (1974): Methods in the study of proprioceptors. In: *Handbook of Bioengineering, Vol. 2,* edited by D. G. Fleming and B. N. Feinberg. Chemical Rubber Co. *(in press).*
47. Pringle, J. W. S. (1960): Models of muscle. *Symp. Soc. Exp. Biol.*, 14:41–68.
48. Rack, P. M. H., and Westbury, D. R. (1969): The effects of length and stimulus rate on tension in the isometric cat soleus muscle. *J. Physiol. (Lond.)*, 204:443–460.
49. Ramsey, R. W., and Street, S. F. (1940): The isometric length-tension diagram of isolated skeletal muscle fibers of the frog. *J. Cell. Comp. Physiol.*, 15:11–34.
50. Reynolds, L. B., and Hibgeson, M. D. (1965): Increase in breathing frequency following the reflex deep breath in anesthetized cats. *J. Appl. Physiol.*, 26:491–495.
51. Robinson, D. A. (1964): The mechanics of human saccadic eye movement. *J. Physiol. (Lond.)*, 174:245–264.
52. Robinson, D. A. (1975): A quantitative analysis of extraocular muscle cooperation and squint. *Invest. Ophthalmol.*, 14:801–825.
53. Robinson, M. (1946): The temporo-mandibular joint: Theory of reflex controlled nonlever action of the mandible. *J. Am. Dent. Assoc.*, 33:1260–1271.
54. Rosenthal, N. P., McKean, T. A., Roberts, W. J., and Terzuolo, C. A. (1970): Frequency analysis of stretch reflex and its main subsystems in triceps surae muscles of the cat. *J. Neurophysiol.*, 33:713–749.
55. Sherrington, C. S. (1930): Some functional problems attaching to convergence. *Proc. R. Soc.*, 105B:332–362.

56. Stefanelli, A. (1951): The Mauthnerian apparatus in the icthycpsida; its nature and function and correlated problems of neurohistogenesis. *Quart. Rev. Biol.*, 26:17–34.
57. Stephen, J. A., Renking, R. M., and Stuart, D. G. (1975): The motor unit of cat medial gastrocnemius: Electrical and mechanical properties as a function of muscle length. *J. Morphol.*, 146:495–512.
58. Townsend, M. A., Shiavi, R., Lainhart, S. P., and Caylor, J. (1978): Variability and biomechanics of synergy patterns of some lower limb muscles during ascending and descending stairs and level walking. *Med. Biol. Eng. (in press).*
59. Wagner, R. (1927): Arbeitsdiagramme bei willkürbewegung, II. Mittelung. *Z. Biol.*, 86:397–426.
60. Walker, S. M. (1960): The relation of stretch and of temperature to contraction of skeletal muscle, Part I and Part II. *Am. J. Phys. Med.*, 39:191–215 and 234–258.
61. Weber, E. (1846): Muskelbewegung. In: *Handwörterbuch d. Physiol., Vol. 3,* edited by R. Wagner. Braunschweig.
62. Wilkie, D. R. (1950): The relation between force and velocity in human muscle. *J. Physiol. (Lond.),* 110:249–280.
63. Wilson, D. M., and Larimer, J. L. (1968): The catch property of ordinary muscle. *Proc. Natl. Acad. Sci.,* 61:909–916.
64. Yager, J. G. (1972): The electromyogram as a predictor of muscle mechanical response in locomotion. Ph.D. Thesis, University of Tennessee Center for the Health Sciences. *Dissert. Abstr. Int., University Microfilms* 33:3909B–3910B, Ann Arbor, Mich.

Posture and Movement, edited by
R. E. Talbott and D. R. Humphrey.
Raven Press, New York © 1979.

Motor Control Processes: New Data Concerning Motoservo Mechanisms and a Tentative Model for Stimulus-Response Processing

James C. Houk

Department of Physiology, Northwestern University Medical and Dental Schools, Chicago, Illinois 60611

INTRODUCTION

In a recent article (14) I reviewed evidence concerning the specific operations performed by skeletomotor control processes and suggested an organizational scheme involving two rather different types of neural processes, one analogous to a servomechanism and another based on logical operations. The purposes of this paper are firstly to summarize recent work from my laboratory that helps to define the neural mechanisms underlying the servo process and secondly to outline a theory that attempts to pinpoint the neural mechanisms that may be involved in the logical process.

Figure 1 illustrates my current working hypothesis concerning the organization of the skeletomotor control system. This model consists of two CNS processes each of which performs quite different operations on sensory signals, a summation point (representing motoneurons and interneurons in segmental reflex pathways) where the actions of the two parallel processes are combined to form motor output, and the controlled system to which motor output is sent (the muscles and the mechanical loads upon which the muscles act).

The continuous processor represents the segmental (and perhaps suprasegmental) servoregulatory pathways that amplify and combine length and force signals from muscle stretch receptors to produce a continuous modulation of motor output. This process is the one mediating the classical stretch reflex and the reflex response to unloading; it provides continuous negative-feedback regulation of muscle performance in a closed-loop subsystem that I refer to as the *motor servo* (boundaries defined by the dashed enclosure in Fig. 1). Experimental evidence and theoretical considerations suggest that the motor servo regulates stiffness, which is a combined property constituted by the ratio of force change to length change, rather than controlling either muscle length or force individually (cf. ref. 13). The best present evidence for stiffness-regulation has been

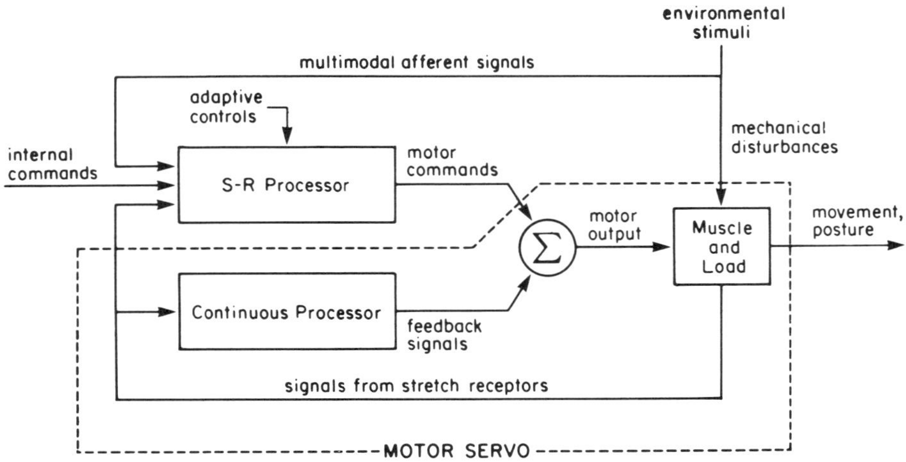

FIG. 1. Two-stage model of sensorimotor processing. See Introduction for details.

obtained in the decerebrate cat (25), but supporting evidence has also been obtained in experiments with normal human subjects (4).

Human and intact animal subjects are capable of adaptive modification of their responsiveness to disturbances in load, i.e., rather different responses can result from identical mechanical stimuli when the postural goals are altered by instructions or by past experience. In a landmark paper on this topic Hammond (10) suggested that the adaptive features are attributable to two mechanisms: (a) preset changes in gain of the servoregulatory pathways and (b) longer-latency voluntary reactions. Subsequent to this report a tremendous amount of attention has been given to the former possibility, even though the evidence for the proposed adaptation in servo pathways has never been conclusive (cf. ref. 14). Our recent studies of human stretch responses (4) have failed to confirm an adaptive control of motoservo gain and have led to the suggestion that all of the observed modifications may be the result of reaction-time movements, many of which are too prompt and too automatic to be properly called "voluntary" responses. The Stimulus-Response (S-R) Processor in Fig. 1 represents a proposed subsystem for this type of sensorimotor integration, a neural process that produces the central motor commands which initiate triggered movements.

The S-R Processor is a classic reaction-time mechanism in the sense that it is characterized by a response latency that may become longer and more variable when choice is introduced in the task (cf. ref. 14). Furthermore, some of the observed errors in the operation of this process are ones in which completely inappropriate responses are produced. Neither of these properties can be explained on the basis of servoregulatory actions—they require the intervention of mechanisms that perform logical operations. The automaticity and short minimal latency at which outputs are produced, particularly when proprioceptive stimuli are used to trigger responses, suggest the involvement of a simple decision-making process that selects one from a group of available preprogrammed responses. Unlike the operations of the motor servo, those of the S-R Processor do not employ continuous feedback; instead, afferent signals appear to be analyzed only to detect the occurrences of environmental stimuli. Once detection is achieved, the process is somehow triggered to release open-loop motor commands that provide inputs to the motor servos regulating the activities of individual muscles. The resultant actions upon the motor servos are well characterized by an alteration of the equilibrium length of the regulated stiffness, i.e., a shift in the threshold of the stretch reflex (8,13,21).

The adaptive controls applied to the S-R Processor appear to be capable of altering either the particular sensory cues required to trigger a given response or the quantity or quality of the response released by any given sensory cue. One imagines a set of neural detectors that can be tuned to respond to different sensory cues or to internal commands; the ouput signals from these detectors might then be used to trigger transitions in state of a neural network that controls the levels of descending motor commands. Some substance will be given to this speculation later in the paper. For the moment, however, I wish only to emphasize that the moment-to-moment operations of the S-R Processor are probably simple ones (e.g., detection-readout), since the latencies at which reaction-time movements can be initiated may be remarkably short. Elsewhere I estimated that central latencies for human subjects may be as brief as 50 to 60 msec when proprioceptive stimuli are used as cues (14).

AFFERENT MECHANISMS CONTRIBUTING TO STIFFNESS-REGULATION

Nonlinear operator theory can be used to demonstrate that length feedback from muscle spindle receptors combined with force feedback from Golgi tendon organs should give rise to a system in which stiffness is regulated (15,25). However, Nichols and Houk (25) calculated the degree of regulation expected, based on the loop gain of the system, and found it to be less than the degree actually observed. One of the mechanisms proposed to explain this discrepancy required the assumption that muscle stretch receptors respond in a particular nonlinear manner to changes in muscle length or force. Unfortunately, previous studies of spindle receptors and tendon organs did not furnish the information required to evaluate the validity of this assumption.

Most of the available literature on muscle receptors describes their properties under conditions that are unphysiological, since a major purpose has been to evaluate receptor mechanisms and this requires precise control of individual inputs. However, information about functional properties in the operative system is required for the assessment of theories of motor control, and techniques are now available to support such studies. Microelectrodes have been used to record from stretch receptor afferents in human peripheral nerves (35) and in dorsal roots in intact cats (27). Unfortunately, the techniques for recording and identifying receptor type are difficult to use, and, at present, only very limited observations have been reported. An alternative technique is to use hook electrodes to record muscle afferents isolated from small filaments that are teased from otherwise intact dorsal roots (32). Since these recordings can be carried out in decerebrate animals without any appreciable interference with reflex function, one can isolate clearly identified muscle afferents and then study their responses for several hours as force is graded physiologically with crossed extensor and stretch reflexes. Recently Drs. Crago, Rymer, and I have applied this technique to the study of several hundred muscle afferents identified as primary or secondary spindle receptors or Golgi tendon organs in the soleus muscle of the decerebrate cat. We were able to study the responses of as many as four muscle receptors simultaneously, over periods of several hours, while also recording the length, force, and electromyographic activity of the functionally isolated soleus muscle.

Golgi Tendon Organs

Previous studies had shown that tendon organs are selectively sensitive to the forces produced by a small group of muscle fibers that connect directly in series with the receptor (17). Since these in-series fibers constitute less than 0.1% of the total number of fibers in a muscle, an individual receptor receives as input a very restricted spatial sample of total muscle force. A variety of anatomical and electrophysiological evidence suggests that this sample, though restricted, is a random one, but the quality of the sample was not known. Our direct observations of tendon organ discharge as contractile force that was graded physiologically demonstrated that the individual receptor provides a surprisingly reliable and linear measure of total muscle force throughout most of the physiological range of force gradation (5,16). In the past most authors have neglected these sensitive force signals from tendon organs when considering theories of motor control, an omission that should be rectified in the future. These observations add important support to the theory that stiffness is regulated by a balanced interplay between length and force feedback; but, since the response properties of tendon organs are approximately linear, they do not explain the extra degree of regulation mentioned earlier.

Muscle Spindle Receptors

It is usually assumed that the dynamic response of primary endings in muscle spindles conveys velocity information used by the CNS in movement control. However, velocity-sensitivity has never been adequately evaluated. Most studies have been with de-efferented spindles, or with spindles in which single gamma fibers have been stimulated, often at abnormally high frequencies, whereas in the normal state a spindle receives steady spontaneous input from several gamma fibers. Furthermore, the range of velocities over which the responses have been evaluated is relatively small. Finally, there is evidence that the dynamic response may be an expression of nonlinear behavior rather than a specific sensitivity to velocity (11,12). In our recent studies the normal spontaneous activity of gamma fibers was preserved while responses were evaluated over a 200-fold range of stretch velocity.

Our results indicate that the dynamic response of primary endings is only very weakly dependent upon the velocity of stretch (31). A 100-fold increase in velocity produces only a 3- to 4-fold increase in response. The results we obtained at different constant stretch velocities, and in response to abrupt transitions in velocity, indicate that the nonlinear expression (length) $\times$ (velocity)$^{0.3}$ provides a quite general description of the salient features of dynamic responsiveness, and this holds for secondary as well as for primary endings. The dynamic response of the secondary ending is simply an attenuated version of the response of primary endings; thus, there seems to be little possibility that the CNS can extract a pure velocity signal from either individually or by combining the two. It appears that the large dynamic response of primary endings may simply signal to the CNS that the muscle is being stretched, rather independently of the velocity of that stretch.

Correlations with Motor Output

We also studied the velocity dependence of the reflex changes in motor output by analyzing rectified and filtered electromyographic responses. The EMG was found to depend on stretch velocity and muscle length in the same nonlinear manner as did the responses of muscle spindles (18). The mechanistic implication of this finding is that motoservo pathways do not greatly alter the dynamics of the afferent input, in agreement with the sinusoidal results of Poppele and Terzuolo (26). The functional implication is that feedback from primary endings does not provide viscous damping of the motor servo, except in a limited range of small-amplitude stretch. Instead this feedback would be expected to produce reflexly a type of damping analogous to friction, and, in fact, such behavior has been observed previously by Roberts (28).

One of the peculiar features of stiffness-regulation that was not well understood prior to our studies of muscle receptors is the remarkable compensation for the abrupt yielding of muscle mechanical stiffness during stretch. As illustrated in Fig. 2, the reflex in the cat soleus muscle appears to act precisely at the time when the yielding occurs, rather than being subject to the expected loop delay of a more conventional negative feedback system (25). The result is a complete masking of this failure in muscle stiffness at stretch velocities in the range encountered during walking or slow running. [At higher velocities, yielding occurs a few milliseconds before the onset of reflex action (25).] A variety of results taken together disqualified signals from tendon organs and secondary endings as being responsible for this particular compensatory action, whereas

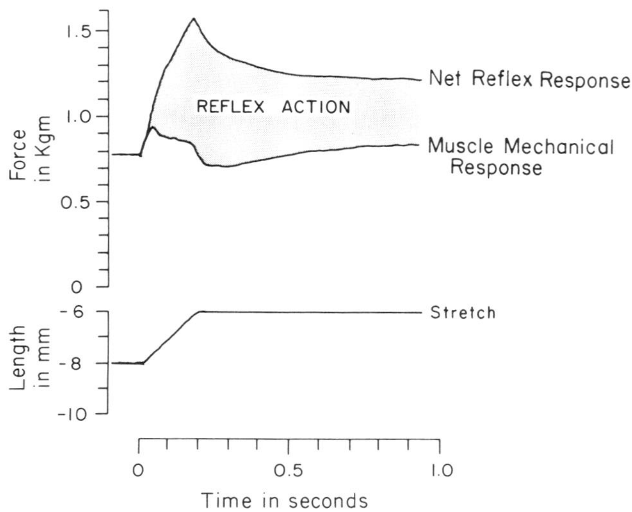

FIG. 2. Compensation for yielding (abrupt failure in muscular stiffness) in the functionally isolated soleus muscle of the decerebrate cat.

the signals from primary endings seemed ideally suited. Any external disturbance that causes muscle stretch would be expected to produce a large, dynamic receptor response before muscle yielding actually begins. At the time of yielding this signal would have already propagated around the stretch reflex arc, and its actions might therefore mask the yielding. This hypothesis treats the nonlinear feedback from primary endings as a type of anticipatory control mechanism. If this effect were sufficiently large, it would also explain the extra degree of regulation mentioned earlier.

We tested these possibilities by using longitudinal vibration (22) to clamp primary-ending discharge at the frequency of the vibration, thus occluding the stretch responses of primary endings. The efficacy of the clamp on primary-ending discharge was checked with dorsal root recordings (often simultaneous). We found (30) that the early transient phase of compensation for muscle yielding was greatly attenuated during periods of vibration, providing strong support for the stated hypothesis. A later phase of compensatory action that often was not occluded by vibration must be attributed to stretch receptors other than primary endings, and we agree with Matthews (22) that the stretch responses of secondary endings are most likely to be responsible for it.

NEURAL MECHANISMS FOR STIMULUS-RESPONSE PROCESSING

There is little hard information at present concerning the neural mechanisms that control reaction-time movements, although there is a vast body of literature relevant to this question. I will not attempt to summarize this literature here, but I would like to emphasize some of the characteristic features of reaction-time movements that appear to correlate particularly well with the properties of a unique category of brainstem neurons, those of the inferior olive. Figure 3 shows a tentative model that incorporates olivary neurons as detectors for S-R processing.

The *psychological refractory period* is a term used to describe the longer reaction time of a subject to a second stimulus when it is delivered shortly after the first (36,38). The existence of this refractory state suggests that stimuli may be analyzed in a serial manner, and that new stimuli ordinarily do not have access to the analytic process until it has completed its current computation (39). The psychological refractory period appears also to be the basis of an intermittency in the correction of errors during tracking of continuous targets (6,24).

Earlier I suggested that the automaticity and short latency of many reaction-time movements might best be explained if the controlling operations consisted in a simple detection stage that triggers state transitions in a motor network. Consideration of this idea in combination with the psychological refractory period leads one to search for CNS neurons that might serve the function of detection, and once having fired, would be refractory for a few hundred milliseconds, the approximate duration of the psychological refractory period.

Candidate neurons having these properties are to be found in the inferior olivary nucleus. These cells are somewhat unique in showing a refractory period of 100 to 200 msec following each response (1,7). Furthermore, the single spike or brief burst characteristic of the responses of these cells is appropriate as a trigger signal generated by event detectors (29,33), and the projections to the cerebellum as climbing fibers have been shown to control long-lasting changes

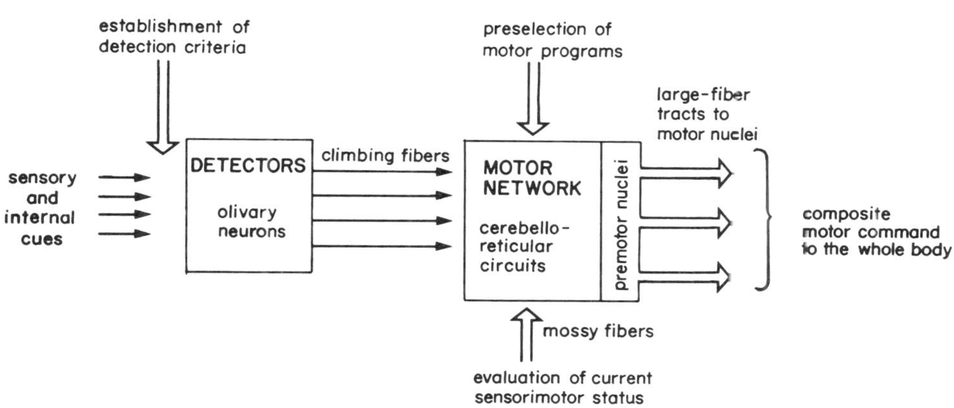

FIG. 3. Tentative model of the neuronal mechanisms that may underlie the subsystem identified in Fig. 1 as an S-R processor.

in the state of cerebellar output (3) conveyed in turn to motoneurons via the large-fiber descending pathways originating from premotor nuclei (motor cortex, red nucleus, reticular and vestibular nuclei). The electrotonic coupling between clusters of olivary neurons (20) might be an important mechanism for the synchronization of motor commands required to produce ballistic movements (cf. ref. 19). The known differences in latency associated with different probabilities of stimuli in choice reaction tasks could result from more stringent criteria (and hence higher thresholds) for detection when there are a greater number of alternative stimuli. The model in Fig. 3 attributes the establishment of detection criteria to inputs (perhaps cortico-olivary) that gate alternative sources of input to olivary neurons. In summary, there are a number of unique features of the olivary system that correspond very well with the characteristic properties of reaction-time movements, although it must be pointed out that previous attempts to correlate the times of occurrence of climbing fiber discharge with the onsets of movements have often been unsuccessful.

A general theory of S-R processing should also account for the tracking of continuous targets, which involves the production of slowly changing modulations of motor commands. The suggestion offered here is that these continuous movements may be the result of a large number of very small discrete movements spread out over time. Each of the proposed micromovements might be triggered by the discharge of individual olivary neurons that in this case are not synchronized with discharges of neighboring cells. The feasibility of this notion is supported by Barmack's (ref. 2 and *this volume*) studies of units in the dorsal cap of the olive in relation to slowly moving visual stimuli and optokinetic nystagmus. The olivary neuron responses are distributed over time in the manner just proposed, and there is reasonable evidence that this neural system does participate in the control of the resultant slow eye movements. Other investigators have pointed out that the subsequent cerebellar discharge has features of function generation (33).

The introduction of the previous example raises the interesting possibility that eye-movement control may employ processes that are quite similar to those suggested here for the control of the skeletal musculature. I would further propose that the set points for homeostatic systems are established in a similar fashion. For example, the fastigial pressor response (23) may represent one component of the exercise program. Thus, the composite motor command to the whole body can be viewed as a single, multidimensional signal established by the overall discharge of all of the large-fiber tracts from premotor to motor nuclei.

The proposed theory has the further advantage that it accounts for predictive mode tracking and the tracking of randomly varying targets with the same neural mechanisms. In effect, it treats all skeletomotor control as being predictive, as previously suggested on other grounds (9), and this may apply to oculomotor control as well (37).

I would like also to indicate the potential relation between this model of

S-R processing and basic learning paradigms, namely classical and operant conditioning. Classical conditioning represents a fundamental mechanism for sensory learning—through association the subject learns to use new and highly abstract sensory cues to produce the same motor response. This is readily accounted for by the adaptive controls in Fig. 3 that provide for the establishment of detection criteria, perhaps by gating olivary inputs and/or by modifying the thresholds for the generation of climbing-fiber spikes in olivary neurons.

At the other end of the spectrum, operant conditioning represents a fundamental mechanism for motor learning—new motor programs are developed independently (and often in the absence) of specific sensory cues. In Fig. 3 this is accounted for by the adaptive controls that preselect motor programs. These preselection inputs are postulated to produce subtle changes in state internal to the cerebelloreticular network that then become manifested as changes in motor output when major state transitions are triggered by olivary discharge. During the acquisition of an operant behavior internal cues to olivary detectors might be used, but the same state transitions could later be triggered by a variety of sensory cues learned through the mechanisms of classical conditioning.

Finally, I would like to emphasize that the model proposed accounts only for the automatic mode of S-R processing. It is well known that patients with major cerebellar lesions are nevertheless capable of producing voluntary movements. However, anyone who has observed these patients in action is struck by the intense voluntary effort that is required to make even rather simple movements. Presumably this is due to the loss, or malfunction, of their automatic mechanisms for stimulus-response processing.

SUMMARY

A brief summary was given of an organizational model of skeletomotor control in which motor output to muscles is controlled by the concurrent operations of two neural processes. One, called the *motor servo,* is thought to be a stereotypic mechanism that operates on the basis of continuous afferent feedback from muscle spindles and tendon organs. The motor servo appears to regulate stiffness, the ratio of force change to length change (in analogy with a spring). Recent evidence suggests that nonlinear feedback from primary spindle receptors is the basis for an anticipatory mechanism that compensates for the failure in muscle stiffness that may occur during stretch. Feedback from secondary endings and tendon organs is more nearly linear and probably contributes more to stiffness-regulation in the steady state.

The second process, here referred to as a Stimulus-Response Processor, appears to operate on the basis of a detection-readout mechanism. The speculation is advanced that olivary neurons may serve as the detectors, and the readout may result from state transitions in the cerebelloreticular circuitry. Other inputs are required to account for the remarkable adaptive features of this system, and a primary source of these signals may be conveyed in cortico-olivary and corticopontine pathways. It is important to emphasize that present evidence for this relation between reaction-time movements and neuronal mechanisms is essentially circumstantial and more definitive studies must be designed.

REFERENCES

1. Armstrong, D. M. (1974): Functional significance of connections of the inferior olive. *Physiol. Rev.*, 51:358–417.
2. Barmack, N. H. (1977): Visually evoked activity of neurons in the dorsal cap of the inferior olive and its relationship to the control of eye movements. In: *Developments in Neuroscience, Vol. I: Control of Gaze by Brain Stem Neurons,* edited by Baker and Berthoz, pp. 361–370. Elsevier/North-Holland Biomedical Press, Amsterdam/New York.
3. Boylls, C. C. (1978): Prolonged alterations of muscle activity induced in locomoting, premammillary cats by microstimulation of the inferior olive. *Brain Res.*, 159:445–450.
4. Crago, P. E., Houk, J. C., and Hasan, Z. (1976): Regulatory actions of the human stretch reflex. *J. Neurophysiol.*, 30:925–935.
5. Crago, P. E., Houk, J. C., and Rymer, W. Z. (1976): Tendon organ responses during crossed extensor and stretch reflexes. *The Physiologist,* 19:160.
6. Craik, K. J. W. (1947): Theory of the human operator in control systems. I. The operator as an engineering system. *Br. J. Psychol.*, 38(P. 2):56–61.
7. Eccles, J. C., Provini, L., Strata, P., and Taborikova, H. (1966): Analysis of electrical potentials evoked in the cerebellar anterior lobe by stimulation of hindlimb and forelimb nerves. *Exp. Brain Res.*, 6:171–194.
8. Fel'dman, A. G. (1966): Functional tuning of the nervous system with control of movement or maintenance of a steady posture. II. Controllable parameters of the muscle. *Biophysics,* 11:565–578.
9. Gottsdanker, R. M. (1952): The accuracy of prediction motion. *J. Exp. Psychol.*, 43:26–36.
10. Hammond, P. H. (1960): An experimental study of servo action in human muscular control. In: *Proceedings of the IIIrd International Congress of Medical Electronics,* pp. 190–199. Institution of Electrical Engineers, London.
11. Hasan, Z., and Houk, J. C. (1975): Analysis of response properties of deefferented mammalian spindle receptors based on frequency response. *J. Neurophysiol.*, 38:663–672.
12. Hasan, Z., and Houk, J. C. (1975): The transition in the sensitivity of spindle receptors that occurs when the muscle is stretched more than a fraction of a millimeter. *J. Neurophysiol.*, 38:673–689.
13. Houk, J. C. (1979): Regulation of stiffness by skeletomotor reflexes. *Annu. Rev. Physiol.*, 41:99–114.
14. Houk, J. C. (1978): Participation of reflex mechanisms and reaction-time processes in the compensatory adjustments to mechanical disturbances. In: *Progress in Clinical Neurophysiology, Vol. 4: Cerebral Motor Control in Man: Long Loop Mechanisms,* edited by J. E. Desmedt, pp. 193–215. Karger, Basel.
15. Houk, J. C. (1978): A two-stage model of neural processes controlling motor output. In: *Cybernetics 1977,* edited by G. Hauske and E. Butenandt, pp. 35–46. R. Oldenbourg Verlag, München/Wien.
16. Houk, J. C., Crago, P. E., and Rymer, W. Z. (1978): Functional properties of Golgi tendon organs. In: *Progress in Clinical Neurophysiology: Segmental Motor Control in Man,* edited by J. E. Desmedt. Karger, Basel *(in press).*
17. Houk, J., and Henneman, E. (1967): Responses of Golgi tendon organs to active contractions of the soleus muscle of the cat. *J. Neurophysiol.*, 30:466–481.
18. Houk, J. C., Rymer, W. Z., and Crago, P. E. (1977): Complex velocity dependence of the electromyographic component of the stretch reflex. *Proc. XXVIII Int. Cong. Physiol. Sci.*, Paris.
19. Llinas, R. (1970): Neuronal operations in cerebellar transactions. In: *The Neurosciences, Second Study Program,* edited by F. O. Schmitt, pp. 409–426. Rockefeller Press, New York.
20. Llinas, R., Baker, R., and Sotelo, C. (1974): Electrotonic coupling between neurons in cat inferior olive. *J. Neurophysiol.*, 37:560–571.
21. Matthews, P. B. C. (1959): A study of certain factors influencing the stretch reflex of the decerebrate cat. *J. Physiol.*, 147:547–564.

22. Matthews, P. B. C. (1969): Evidence that the secondary as well as the primary endings of the muscle spindles may be responsible for the tonic stretch reflex of the decerebrate cat. *J. Physiol.,* 204:365–393.
23. Miura, M., and Reis, D. J. (1970): A blood pressure response from fastigial nucleus and its relay pathway in the brainstem. *Am. J. Physiol.,* 219:1330–1336.
24. Navas, F., and Stark, L. (1968): Sampling or intermittency in hand control system dynamics. *Biophys. J.,* 8:252–302.
25. Nichols, T. R., and Houk, J. C. (1976): The improvement in linearity and the regulation of stiffness that results from the actions of the stretch reflex. *J. Neurophysiol.,* 39:119–142.
26. Poppele, R. E., and Terzuolo, C. A. (1968): Myotatic reflex: Its input-output relation. *Science,* 159:743–745.
27. Prochazka, A., Westerman, R. Z., and Ziccone, S. P. (1976): Discharges of single hindlimb afferents in the freely moving cat. *J. Neurophysiol.,* 39:1090–1104.
28. Roberts, T. D. M. (1963): Rhythmic excitation of a stretch reflex, revealing (a) hysteresis and (b) a difference between responses to pulling and to stretching. *Quart. J. Exp. Physiol.,* 48:328–345.
29. Rushmer, D. S., Roberts, W. J., and Augter, G. K. (1976): Climbing fiber responses of cerebellar Purkinje cells to passive movement of the cat forepaw. *Brain Res.,* 106:1–20.
30. Rymer, W. Z., Houk, J. C., and Crago, P. E. (1976): Vibratory occlusion of compensation for muscle yielding. *Soc. Neurosci., Abs.* #758.
31. Rymer, W. Z., Houk, J. C., and Crago, P. E. (1977): The relation between dynamic response and velocity sensitivity for muscle spindle receptors. *Proc. XXVIII Int. Cong. Physiol. Sci.,* Paris.
32. Severin, F. V., Orlovskii, G. N., and Shik, M. L. (1967): Work of muscle receptors during controlled locomotion. *Biophysics,* 12:575–586.
33. Simpson, J. I., and Alley, K. E. (1974): Visual climbing fiber input to rabbit vestibulo-cerebellum: A source of direction-specific information. *Brain Res.,* 82:302–308.
34. Simpson, J. I., and Hess, R. (1977): Complex and simple visual messages in the flocculus. In: *Developments in Neuroscience, Vol. I: Control of Gaze by Brain Stem Neurons,* edited by Baker and Berthoz, pp. 351–360. Elsevier/North-Holland Biomedical Press, Amsterdam/New York.
35. Valbo, A. B. (1970): Slowly adapting muscle receptors in man. *Acta Physiol. Scand.,* 78:315–333.
36. Vince, M. A. (1948): The intermittency of control movements and the psychological refractory period. *Br. J. Psychol.,* 38(P. 3):149–157.
37. Young, L. R. (1977): Pursuit eye movement—What is being pursued? In: *Developments in Neuroscience, Vol. I: Control of Gaze by Brain Stem Neurons,* edited by Baker and Berthoz, pp. 29–36. Elsevier/North-Holland Biomedical Press, Amsterdam/New York.
38. Welford, A. T. (1973): Attention, strategy, and reaction time: A tentative metric. In: *Attention and Performance IV,* edited by S. Kornblum, pp. 37–54. Academic Press, New York.
39. Welford, A. T. (1974): On the sequencing of action. *Brain Res.,* 71:381–392.

Posture and Movement, edited by
R. E. Talbott and D. R. Humphrey.
Raven Press, New York © 1979.

The Organization of Rapid Postural Adjustments of Standing Humans: An Experimental-Conceptual Model

Lewis M. Nashner and Marjorie Woollacott

Neurological Sciences Institute, Portland, Oregon 97209

During the generation of purposeful movements, the sensorimotor system must simultaneously regulate and control three important activities: (a) the generation and coordination of the basic movement patterns among the limbs, (b) the adaptation of these muscular controls to the external loading and sensory conditions of the task, and (c) the maintenance of the equilibrium of the body throughout the movement. By analyzing the experimental results from a number of posture-control studies with human subjects, this chapter will examine some organizational concepts which describe the way in which the sensorimotor control system performs these activities.

Because human subjects are easily incorporated into experimental paradigms which do not place severe constraints upon the CNS functions or upon the external conditions of the task, the human results are a valuable complement to animal studies, which have revealed many details about control of the basic movement patterns but little about the adaptive controls and the maintenance of postural equilibrium. By focusing upon what happens to sensorimotor controls during well-rehearsed tasks and immediately following unexpected changes in the external conditions, the human posture-control studies have made some quantitative statements about the adaptive sensorimotor controls and about how somatosensory, vestibular, and visual inputs are collectively used in the maintenance of balance during movements.

In recent years, the predominant concept has been that the sensorimotor system expresses purposeful movement, movement adaptation, and balance control utilizing a hierarchically organized group of specialized subsystems. Bernstein (2) argued on purely theoretical grounds that hierarchical simplification of these activities was essential, because the brain could not independently regulate the many muscles and mechanical linkages which are collectively necessary to compose movements. Based upon these theoretical arguments, Gelfand et al. (6), proposed that complex locomotor movements are broken down into much simpler, stereotyped patterns, each of which is executed by a spinal "function generator" upon the issuance of a single command from the brain. Greene (7) described how hierarchically organized function generators can simplify movement-control tasks by drawing the analogy of a graphic display system which utilizes a family of preprogrammed, curvilinear function generators to produce any desired complex figure. Rather than having to prescribe each figure on a point-by-point basis, the central system specifies the initial conditions and gains for the appropriate sequence of the curvilinear functions (executed entirely by the peripherally preprogrammed generators), thereby significantly reducing the number of centrally prescribed points. According to the hierarchical theory, the brain generates movements in a similar way by prescribing a sequence of stereotyped patterns which are then executed by peripheral generators and fine-tuned by the somatosensory inputs to suit the particular conditions of the task. Thus, the central movement-control centers of the brain do not regulate the activity of muscles directly and are therefore freer to assess the overall performance of the movement task and to adapt the strategies for movement control and balance following any unexpected perturbations and/or changes in the external conditions. Also, because the brain specifies sequences of stereotyped patterns rather than individual muscular activities, the task of composing new movements is presumably much simpler. The major disadvantage of hierarchical control is that the range of movement and balance behaviors is severely limited to combinations of the available stereotyped patterns, and this number must be kept manageably small. However, investigators are of the opinion that the range of actual movement behaviors exhibited by animals and humans is in fact much narrower than that of which they are theoretically capable (4,10).

The study of spinalized, mesencephalic, and decorticate cats walking upon treadmills has proven to be one of the most useful approaches for studying the neural mechanisms which generate and coordinate locomotor movements. Two detailed reviews of this literature have been written recently by Grillner (8) and by Wetzel and Stuart (21). Both reviews cite the extensive evidence that spinal programs alone can produce the basic rhythmic stepping movements in individual limbs and coordinate the limbs during the various gaits. During treadmill walking a major role of the supraspinal descending commands is to provide the nonspecific excitation which is needed to activate and sustain the spinal locomotor programs. Because this emerging view of locomotor control is consistent with the theoretical concepts proposed by Bernstein (2) and elaborated by Gelfand et al. (6), many investigators believe that similar hierarchically organized neural structures are also responsible for almost all of the perceptual, motor, and learning functions associated with the sensorimotor system (20). In keeping with these convincing arguments, the results presented here are also interpreted utilizing the concept of hierarchically organized structures.

Several investigators have shown that postural motions of human subjects are organized into fixed patterns of interaction or "synergies." For example, the absence of stabilographic oscillations correlated with respiration (which rhythmically changes the position of the center of mass of the thorax during upright stance) was ascribed to a synergic compensation for this internal, postural disturbance (9). Angular measures revealed antiphase movements of the torso and hips which were in precise phase with the respiratory rhythm and in fixed proportion to the tidal volume. The lack of respiratory sway under normal conditions was shown to be related to these antiphasic movements by experimentally constraining the hips: This interrupted the "respiratory synergy" and produced sway correlated with the respiratory rhythm. A second example involves the motor skills used in marksmanship. Because aiming the arm precisely is not an inherent capability of man, Arutyunyan et al. (1), examined the way in which it was learned in order to show that subjects could program a new fixed pattern of interaction among the links of the body. During aiming, they observed that the relative motions among the body links of skilled marksmen were tightly organized into a fixed pattern, whereas interactions among the links were much less organized in the unpracticed subjects. Although neither of these studies was able to establish directly a neural basis for the fixed patterns of motion, the authors used kinematic models to argue that the compensatory, patterned movements were produced by the organization of muscular activities into fixed patterns, a conclusion which is entirely consistent with concepts of hierarchical organization.

The research discussed in the following section involves analysis of the organization of the sensorimotor system as revealed through the observation of postural adjustments elicited in the leg muscles of subjects standing upon a movable platform. The commonly shared property of all the experimental observations reported is that they are made upon the earliest functionally useful postural adjustments, which normally occur in the leg muscles of standing subjects at latencies of 100 to 110 msec following the onset of postural perturbations. In analyzing the spinal and central organization of postural adjustments, these functionally useful responses have the decided advantage of being too rapid for subjects to have exercised any significant voluntary influences over them (3). However, because their latencies are long compared with the myotatic stretch reflexes, it is likely that these responses are mediated by the intersegmental and central structures of the sensorimotor system (5,19), a property which is necessary if they are to reveal the organization within these structures. These responses have been given various names in the literature such as "functional stretch responses," "long-loop," "long-latency," and "transcortical" stretch reflexes (11,12), probably because they were elicited by stretching muscles and were too delayed to be mediated by the segmental reflex pathways alone. However, results presented here suggest that the response of a given muscle is much too complexly interwoven into an organizational pattern to have been elicited only by its own stretch input. For lack of a better term, we will refer to perturbationally elicited responses of 100- to 110-msec latency as "triggered responses" (TRs) and avoid until later in the discussion any speculation about organization of sensory inputs.

EXPERIMENTAL METHODS

The experimental methods used to elicit triggered responses have already been described in detail in previous publications (15,16), and need only be briefly outlined here. The principal experimental paradigm has been to measure the triggered postural responses of subjects while they stood upon a platform programmed to move unexpectedly. Six independent degrees-of-motion (horizontal, vertical, and rotational displacements of each foot independently) were produced by proportionately controlled hydraulic servomotors. Figure 1 depicts the four modes of stimulation used to elicit postural reactions and shows the general features of the postural responses to each mode: in A the platform bases both translate to induce anteroposterior (AP) sway rotation which occurs principally about the ankle joints; in B the bases are both rotated so that the ankle joints rotate independent of AP sway motion (for the first few hundred millliseconds); in C the bases are both vertically displaced to extend or flex the legs and vertically perturb the subject; whereas in D the vertical displacements of the bases are reciprocal so that one leg extends, the other flexes, and the body sways laterally.

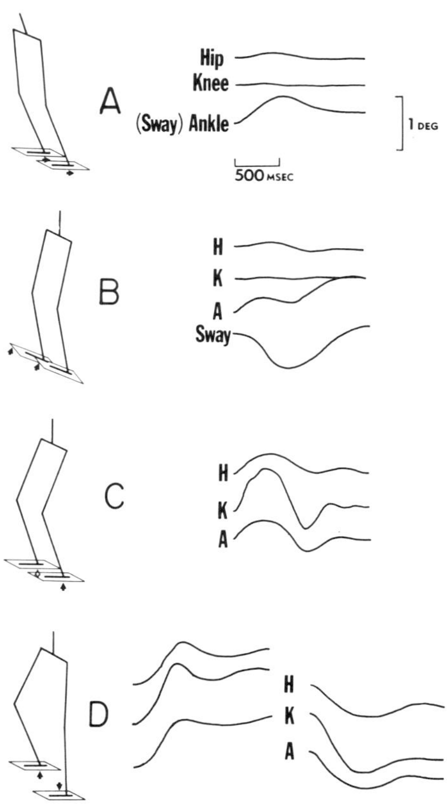

FIG. 1. Four modes of perturbation and associated angular rotations about the leg joints. **A:** AP sway induced by horizontal translation of the two platforms causes sway, principally about the ankles. **B:** Directly rotating the ankle joints causes sway in the opposite direction. **C:** Elevating both platforms rotates the ankles, knees, and hips together. **D:** Reciprocally elevating and lowering the platforms flexes the three joints of one leg and extends those of the other leg.

Each of these platform motions was designed to perturb the subject's posture in a way that required the contraction of different combinations of proximal and distal leg muscles. Also, in some instances, contraction of the stretching muscles was advantageous for stabilization, whereas in other instances contraction of the shortening muscles improved the subject's stability. Specifically, during translationally induced AP sway (Fig. 1A) contraction of the stretching ankle muscles stabilized the sway motion, whereas similar contraction of the stretched ankle muscles when the ankle joints were rotated directly (Fig. 1B) had a destabilizing effect. During both of these motions, the ankles and hips rotated antiphasically, and the simultaneous contraction of ankle and hip extensor or flexor muscle pairs helped stabilize these coupled motions. The contraction of the stretched ankle muscles during the simultaneous vertical displacements of the body (Fig. 1C) opposed any changes in the vertical height of the body. However, the shortening ankle muscles had to contract during the reciprocal, vertical displacements of the two platform bases (Fig. 1D) in order to prevent the body from falling sideways. Because the ankles and knees flexed or extended simultaneously during vertical perturbation, the contraction of ankle and knee extensor or flexor muscle pairs would help stabilize these motions.

The presence of fixed patterns of contraction among proximal and distal leg muscles and the adaptation of these patterns following unexpected changes in the mode of perturbation were analyzed by recording the EMG activity of four leg muscles which control ankle, knee, and hip motions. EMG activities of the gastrocnemius (G), tibialis anterior (T), quadriceps (Q), and hamstring (H) muscles (using surface electrodes with 2.5-sq-cm area, spaced 3 to 5 cm apart) were each processed by full-wave rectification and filtered to give a DC signal whose level was approximately in proportion to contractile strength of the muscle. The vertical and torsional forces exerted by each foot were measured by strain gauges mounted within each platform base. The ankle, knee, and hip joint angles were measured using potentiometers which were attached to the leg with rigid, light rods. During each test, the subject was asked to stand upon the platform, relaxed, one foot (base) upon each platform base, and with her arms folded at the waist. Also, the subject was asked to flex her legs slightly (approximately a 10° rotation of the knees) so that the knees would not be locked. During 1-hr sessions, perturbations were imposed at random intervals and with rest periods between groups of approximately 20 trials (about 50 to 60 trials per session). To observe the effect of unexpected changes in the conditions of the task, the mode of perturbation was changed unexpectedly every 3 to 10 trials during a sequence.

THE SYNERGIC ORGANIZATION OF TRIGGERED POSTURAL RESPONSES

Before introducing the experimental results, it will be useful to examine the functional relations between the forces generated by the leg muscles and the resulting postural motions of the body. During AP sway motions, the knee joints remain approximately stationary and the ankle and hip joints tend to rotate in an antiphasic relation. When the knees are stationary, the G and H muscles function as a complementary pair to extend the ankles and hips, while the T and Q muscles flex these two joints. In marked contrast, ankle, knee, and hip joints are all either extended or flexed (with the largest angular changes occurring about the knees) whenever the body moves vertically up or down. Under these conditions the G and Q muscles function as a complementary pair to extend the ankles and knees (lifting the body), while the complementary pairing of the T and Q muscles flexes these two joints and lowers the body. Figure 2 illustrates these functional properties by showing the EMG activities produced by voluntary AP sway and vertical motions.

During TR adjustments elicited by translationally induced AP sway, the ankle muscle stretched and the complementary hip muscle (H coupled with G or Q coupled with T) contracted in fixed proportion as shown in Fig. 3, independently of the magnitude of the sway adjustment (16). To show that this fixed linkage between functionally complementary proximal and distal leg muscles was neurally programmed by the system and was not the result of the antiphase, mechanical coupling between ankle and hip motions, TR sway adjustments were elicited by directly rotating the ankle joints. Following the imposition of single, unexpected direct rotations during sequences of sway trials, TR adjustments occurred among the same complementary pairs of proximal and distal muscles and in the same fixed proportions as during the AP sway perturbations. Then, to show that rotation of the ankle joints was a necessary input to elicit responses within this fixed linkage, the platforms were rotated during AP sway so that the ankle joint angles remained stationary. Whenever AP sway perturbations were not accompanied by ankle rotations, the earliest adjustments were always delayed

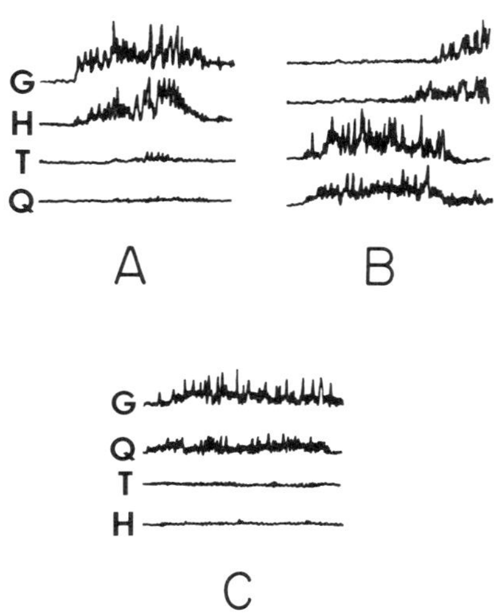

FIG. 2. The function of the four leg muscles during voluntarily executed postural motions. **A:** Forward sway. **B:** Backward sway. **C:** Stooping down by flexing the legs.

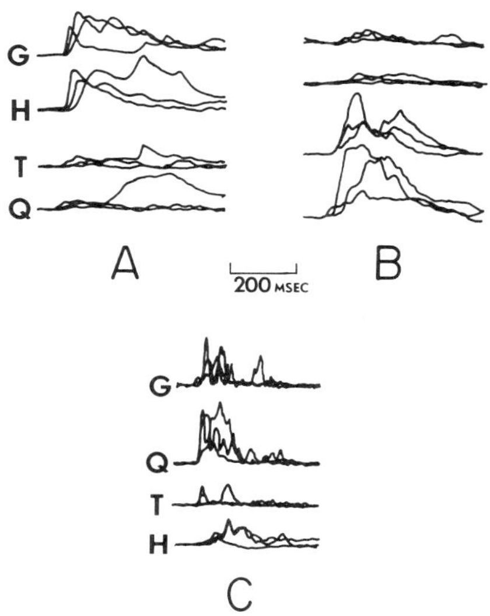

FIG. 3. TR contractions of the four leg muscles elicited by perturbations. **A:** Forward induced sway. **B:** Backward induced sway. **C:** Elevating the platforms.

until 180 to 250 msec following the onset of sway. These longer-latency responses were believed to be initiated by inputs primarily from the vestibular system (13,14). A single sensory input which activates in fixed proportion the complementary ankle and hip muscles is functionally useful to produce the torques necessary to arrest the AP sway motions about the ankles and, at the same time, to counteract the resulting antiphase rotations of the hip joints. Because the fixed linkage of ankle and hip extensor muscles occurs during sway adjustments and appears to perform this useful stabilizing function, the organizational structure has been termed the "sway synergy."

During TR adjustments elicited by vertically induced changes in body height, the opposite relations between proximal and distal leg muscles were observed. The muscles which extend and flex the ankle and knee joints together (G coupled with Q and T with H) contracted in fixed proportion as illustrated in Fig. 3 (18). To illustrate that this fixed pattern of EMG activity among proximal and distal leg muscles was also neurally programmed by the system and was not caused independently by the stretching of proximal and distal muscles, one of the platform bases was rotated during a vertical change in height so that the ankle motion was unexpectedly and completely disassociated from that of the knees and hips. Because the response pattern was unaltered by this ankle rotational input (in marked contrast to sway-synergic responses which were dependent upon ankle rotations), a plausible conclusion is that the suspensory synergy is programmed by the system to occur whenever the predominant somatosensory input is the rotation of the knees. Functionally, a single command to activate in fixed proportion the ankle and knee muscles which extend and flex the entire leg helps stabilize the vertical height of the body. For this reason, this combination of muscular activity has been termed the "suspensory" synergy.

Figure 4 shows the way in which EMG activity among the leg muscles is reorganized from the sway into the suspensory synergy following unexpected changes from sway to vertical perturbations. Each 3-trial sequence illustrates the first three responses following an unexpected change from another mode of stimulation. Sequences A, B, and C (simultaneous and reciprocal vertical perturbations) were all imposed following AP sway tests. Sequence D (direct rotational perturbations) was also imposed after AP sway tests, whereas the three E trials (AP sway tests) were imposed after direct rotational perturbations. Surprisingly, the reorganization from the sway into the suspensory synergy was clearly evident during the first trials, although some "fine tuning" of the antagonist muscle activities continued during subsequent trials. During transitions

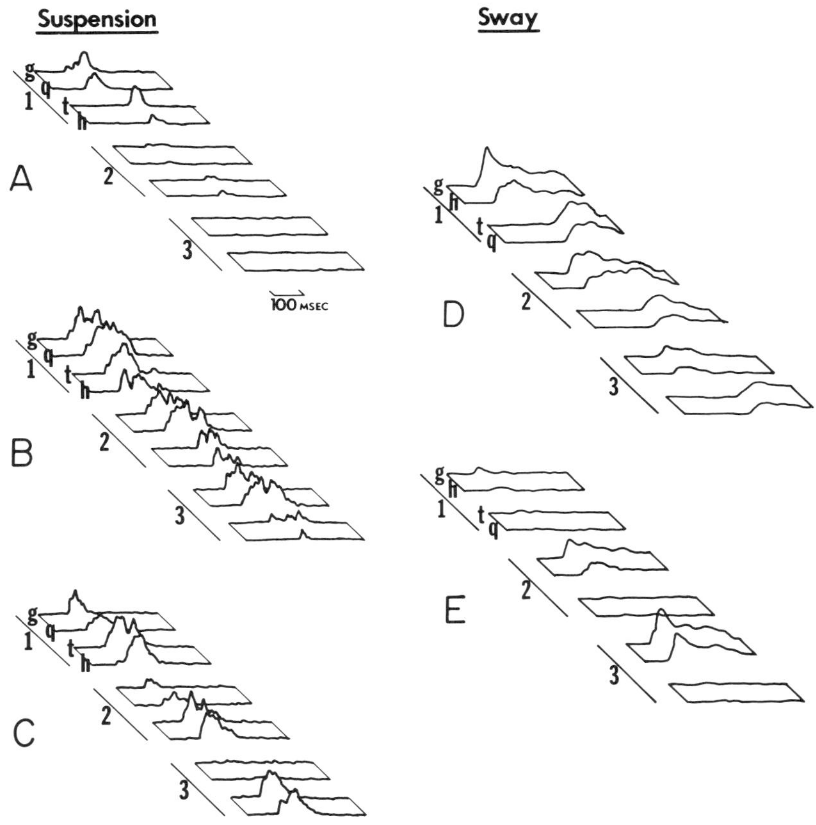

FIG. 4. Sequences of the first three TR adjustments elicited following unexpected changes in the mode of perturbation. **A:** The platforms are both elevated (following swaying). **B** and **C:** The platforms are reciprocally elevated and lowered with the measured leg lowering in **B** and elevating in **C** (both following simultaneous vertical perturbations). **D:** Direct ankle rotation (following sway). **E:** Induced swaying (following direct ankle rotations).

from suspensory to sway perturbations, the major synergic reorganization was also evident during the first trial, but again with some subsequent "fine tuning." Observations of TR organization suggest that during stance the muscles of the leg were always organized by one or the other of the two synergies, although these results do not rule out the possibility that other synergic arrangements among leg muscles may be elicited under different task conditions This conclusion is consistent with the concept of a hierarchical organization in which lower-level programs organize muscles into functionally useful patterns with a single input regulating the level of activity within the program and another selecting the appropriate pattern. Because reorganization of the synergic pattern occurred largely during the first trials after changes in the task, the process of selection is probably an integral part of the lower-level program which generates these synergies. The simplest hypothesis which is consistent with these experimental observations is that the sway synergy is elicited during stance whenever movements occur largely about the ankle joints, whereas predominance of knee movements elicits activity organized into the suspensory synergy. The next section explores the ways in which sensory inputs and conditions of the task regulate the level of TR activity, independent of the particular synergic arrangement among the muscles.

REGULATING THE LEVEL OF ACTIVITY OF TR ADJUSTMENT

As already mentioned, the contraction of stretching leg muscles was not always useful to help stabilize posture during perturbations. During some modes of stimulation either no contraction or contraction of muscles subjected to shortening was more useful. This section explores some of the ways in which the relations between somatosensory stretch inputs and TR adjustments adapted to the particular conditions of the task.

The experimental observations presented in the previous section already established that changes in muscle length were a necessary input to elicit TR adjustments. However, experimental observations have shown that TR adjustments within a functionally related group of muscles do not necessarily occur in those subjected to stretching, as the idea of a stretch response would imply. For example, when ankle rotational disturbances were imposed, the consequent stretching of ankle muscles was not related to AP sway, and contraction of the stretching ankle (and the coupled proximal) muscles during initial unexpected trials was destabilizing. However, following a sequence of such perturbations, TR adjustments (always organized into the sway synergy) were successively reduced in amplitude until they were completely absent after 3 to 5 trials as shown in Fig. 4D and 4E (15). Then, when a sequence of translationally induced AP sway perturbations were unexpectedly reintroduced, no TR adjustments were evident during the first trials, but these adjustments returned to full amplitude (again organized into the sway synergy) within 3 to 5 trials. This progressive attenuation and facilitation of stretch-elicited TR sway adjustments has shown that the relations between stretch inputs and muscular contractions are highly adaptable. Furthermore, the adaptive process which alters the TR sway activity elicited by ankle rotation is functionally independent of the motor program which maintains synergic arrangement among the leg muscles.

The activation of leg muscles organized into the "suspensory synergy" will normally compensate for any changes in the vertical height of the body. Indeed, when both platforms were unexpectedly raised (flexing both legs), contraction of G and Q muscles resisted this motion by extending the legs. The activation of T and H muscles whenever both platforms were lowered tended to resist the extension of the legs. However, active resistance to such changes in platform height were difficult to produce in many subjects and were attenuated rapidly in those who did evidence them (Fig. 4). The progressive adaptation of height-induced suspensory responses was functionally useful, since resisting changes in platform height only increased the vertical perturbation of the body (in much the same way as resisting direct ankle rotations increased the sway and were progressively attenuated). Presumably, an experimental paradigm which unexpectedly changes only the body height (rather than platform height) would facilitate the resisting suspensory adjustments. We tested this by measuring suspensory adjustments just after downward motions of both platforms (at constant velocity for 300 msec) were suddenly halted. Although the downward motion could not be sustained long enough for the vestibular system to accommodate to this moving inertial reference frame, halting the motion (causing the body to be thrust downward relative to this frame of reference) elicited TR suspensory adjustments which were not adaptively attenuated after repeated trials.

Figure 4 shows the EMG activity (organized into the suspensory synergy) elicited by reciprocal, vertical displacements of the two platform bases. Surprisingly, the muscles in each leg subjected to shortening contracted even during the first unexpected trials of reciprocal motion. In other words the contractile activity of muscles was now directed so as to follow rather than to resist the motion of each leg. Adjustments which followed rather than resisted the direction of platform motions were functionally stabilizing in this task, since any resistance of reciprocal leg flexions and extensions would increase the lateral swaying of the subject. However, the significant aspect of this observation was that adaptive changes in the direction of suspensory adjustments (from resisting simultaneous vertical perturbations to following reciprocal ones) were evident within the latent period of single TR adjustments. Therefore, the peripheral motor program probably includes another selection process which reverses the relationship between the direction of motion and the direction of response.

The testing of selected patients with cerebellar gait ataxia and patients with spasticity in their legs has been a complementary part of our experimental program.[1] Although these studies are in progress, some of the preliminary find-

[1] These studies have been conducted in collaboration with R. J. Grimm and R. D. Rafal.

ings have helped to reveal more about the organization of postural adjustments. Although the tendon reflexes of 4 patients with multiple sclerosis (with neuropathies confined principally to peripheral systems) were clinically much more active than normal, their TR adjustments occurred at normal latency and grossly normal amplitude. Usually, however, all their leg muscles contracted synchronously in response to perturbations rather than being synergically organized. Spasticity was confined to one side in a patient who suffered a unilateral stroke. This patient evidenced largely normal synergic organization on the unaffected side and a pattern of synchronous contraction of all leg muscles on the affected side.

The observations made upon the postural adjustments of cerebellar patients have revealed quite different organizational changes (17, and L. M. Nashner and R. J. Grimm, *unpublished studies*). These patients were tested for "vestibular" posture control function (standing with ankles stabilized to eliminate somatosensory inputs from sway), for adaptive changes in TR gain during direct ankle rotations, and for the synergic organization among leg muscles. One 29-year-old patient with a relatively "pure" anterior lobe lesion (caused by hemorrhages) with relatively mild cerebellar signs evidenced normal synergic organization of responses among his leg muscles during all modes of stimulation. However, his TR adjustments did not adapt during direct ankle rotations. Also, this patient could not stand with his eyes closed during the vestibular tests. Loss of TR adaptation and vestibular deficits were common observations among all the cerebellar patients tested, whereas only the more severe cases evidenced synergic deficiencies. One generalization which thus far appears to be valid is that cerebellar lesions affect adaptation and visual-vestibular controls, whereas lesions within the more peripheral spinal and brain stem systems have a greater effect upon the synergic programs which coordinate muscular activities.

A CONCEPTUAL MODEL FOR THE ORGANIZATION OF POSTURAL ADJUSTMENTS

The experimental observations have revealed a number of particularly prominent features of organization which are the primary focus of the conceptual model. Somatosensory inputs (from rotation of leg joints) were necessary to elicit TR adjustments within the interval of 100 to 200 msec following a perturbation (although vestibular and visual inputs probably have a modulating influence). A second prominent feature was that the selection of the synergic organization was always appropriate, even following unexpected changes in the mode of perturbation. Thirdly, the direction of suspensory adjustments made the appropriate change from resisting simultaneous vertical motions to following reciprocal ones immediately. In marked contrast, the magnitude of sway adjustments elicited by direct rotations and of suspensory adjustments elicited by direct vertical motions were inappropriate during the initial unexpected trials and adapted appropriately only after 3 to 5 trials. Our reasoning in synthesizing an organizational model from these observations is that the processes which are appropriately organized during the latent period of single TR adjustments are programmed within the peripheral elements of the sensorimotor system and are activated primarily by intersegmental somatosensory inputs. In contrast, organizational and adaptive changes which require several trials involve the integrative functions of the central nervous system and utilize more complex combinations of somatosensory, vestibular, and visual inputs.

Perhaps the neurally programmed sway and suspensory synergies operating in standing human subjects are organizationally parts of the same spinal motor programs which also produce coordinated locomotor movements in spinalized cats. If this hypothesis is correct, we can then begin to distinguish functionally among the activities which generate the basic patterns of movement and those which select these patterns and regulate the levels of activity within them. In isolation, the peripheral motor programs must generate the basic patterns of movement using only somatosensory inputs, which alone can only distinguish movements of one part of the body *relative* to other body parts. Thus, for example, these peripheral motor programs can distinguish among AP sway motions (ankles rotate primarily), simultaneous vertical motions (primarily knees rotate in phase, but also ankles and hips), and reciprocal vertical motions (knees rotate in antiphase). However, they alone cannot resolve any perturbation which requires a distinction between relative body motions and motions of external objects, because the resolution of body and environmental motions can be performed only with the complex combinations of visual, vestibular, and somatosensory inputs. For example, the peripheral motor program alone cannot distinguish between AP sway motion and rotations of the platform, and the initial TR adjustments were erroneous. The central motor programs resolve these sensory conflicts utilizing vestibular and/or visual information and then modify the peripheral program to attenuate the erroneous ankle rotational inputs. In other words, the peripheral motor programs rapidly organize a pattern of movement using the intersegmental somatosensory inputs, and then the central programs reorganize parts of the peripheral programs to suit the external conditions, utilizing complex combinations of information from all sensory modalities and requiring considerably more time to be executed.

The conceptual posture control model is shown in Fig. 5. The biomechanical description of body motions and the effect of muscle forces is simplified into a body with three major segments and three joint angles. The direction of angular motion produced by the force of each muscle is indicated by the plus signs (for extensors) and the minus signs (for flexors). Note that all but the T muscle affects the motion about two adjacent joints.

The activity of the four leg muscles is organized either by the sway or the suspensory function generator, with plus and minus signs indicating the patterns of EMG activity produced by each generator. Each function generator has a single input which regulates the level of activity. The simplest organizational scheme which is consistent with the experimental observations is used to express

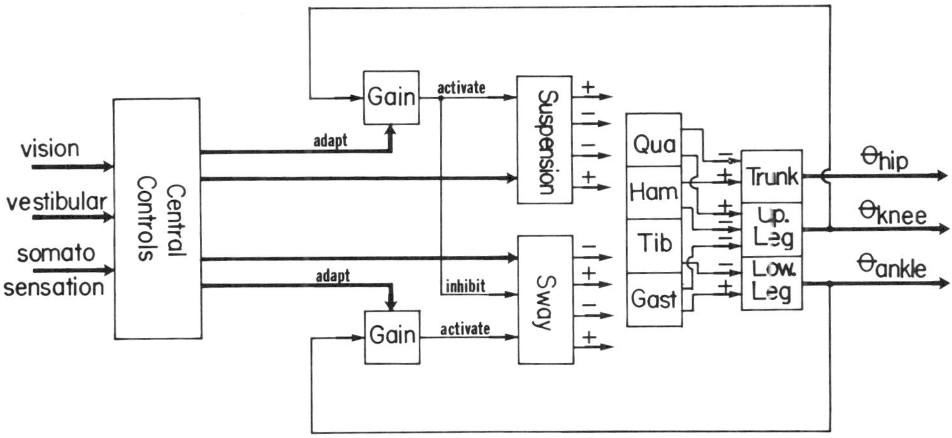

FIG. 5. Organizational model for control of TR adjustments during quiet stance. Plus signs indicate extensor muscles, and minus signs, flexors.

the way in which somatosensory inputs regulate the synergic levels of activity. In this scheme the ankle rotational input (presumably including muscle proprioceptors and joint afferents) regulates the sway activity through a simple "gain" element, whereas the knee rotational input (also including muscle proprioceptors and joint afferents) regulates the level of activity of the suspensory synergy through a second "gain" element. To prevent both synergies from being activated at once, sway activity is inhibited by activity in the suspensory synergy. At the very simplest level, this scheme will immediately produce the appropriate pattern of muscle activity during either sway or suspensory perturbations. Also, direct ankle rotation alone will produce sway organized activity, whereas direct ankle rotational inputs into the sway synergy are inhibited whenever there is a significant rotational input from the knee joint.

The central control centers receive somatosensory, vestibular, and visual inputs relating the relative motions of the body with those of the external environment. The central control centers then adapt the peripheral programs to suit the external conditions of the task and to maintain the balance of the body by adaptively changing the "gain" elements which peripherally regulate motor activity and/or by providing direct regulating inputs into the synergic generators. Adaptation of TR adjustments to direct ankle rotation is one example of a central adaptive process which changes the gain of the ankle rotational input. The regulation of erect stance when a subject's eyes are closed and his ankles stabilized is an example of direct vestibular input into the sway synergy when somatosensory inputs from the ankle joints have been disrupted.

Of course, as it now stands, this model gives only a representative sampling among the many complex functions of the postural motor system. Clearly, the system is probably capable of more than just two synergic patterns of muscular activity, although the sway and suspensory patterns did characterize the majority of activities seen during our erect-stance tasks. When a standing subject assumes other stance postures, carries loads, or performs voluntary acts, the function of the individual muscles changes and other forms of organization can be seen. Also, other patterns of activity can be programmed centrally rather than peripherally during the performance of voluntarily learned tasks or of complex combinations of postural and voluntary movements.

In its simplicity, however, this conceptual model can be useful in attempts to interpret some of the complex physiological observations of central and peripheral motor controls. It suggests the great importance of relative intersegmental inputs in producing the basic patterns of coordination and of multisensory inputs in producing the more complex adaptive and balance functions of the motor system. Also, the model provides a format with which to interpret observations of patients with disorders of posture and locomotion. The results of platform tests and conceptual models have already suggested some ways to differentiate among deficits within the basic peripheral motor programs, the adaptive central motor programs, and the sensory inputs to these systems.

ACKNOWLEDGMENTS

The authors wish to thank Gloria Tuma for her assistance during some of these experiments and Joyce Maeda for her assistance in preparation of the manuscript.

L. M. Nashner is supported in full by Grant NS00148. M. Woollacott and the research project are supported by Grant NS12661. Both grants are received from the NINCDS.

REFERENCES

1. Arutyunyan, G. A., Gurfinkel, V. S., and Mirskii, M. L. (1969): Organization of movements on execution by man of an exact postural task. *Biophysics,* 14:1162–1167.
2. Bernstein, N. (1967): *Coordination and Regulation of Movements.* Pergamon Press, New York.
3. El'ner, A. N. (1973): Possibilities of correcting the urgent voluntary movements and the associated postural activity of human muscles. *Biophysics,* 18:966–971.
4. Engberg, I., and Lundberg, A. (1969): An electromyographic analysis of muscular activity in the hindlimb of the cat during unrestrained locomotion. *Acta Physiol. Scand.,* 75:614–630.
5. Evarts, E. V. (1973): Motor cortex reflexes associated with learned movement. *Science,* 179:501–503.
6. Gelfand, I. M., Gurfinkel, V. S., Fomin, S. V., and Tsetlin, M. L. (1971): *Models of the Structural-Functional Organization of Certain Biological Systems.* MIT Press, Cambridge, Mass.
7. Greene, P. H. (1972): Problems of organization of motor systems. In: *Progress in Theoretical Biology, Vol. 2.* Academic Press, New York.
8. Grillner, S. (1975): Locomotion in vertebrates: central mechanisms and reflex interaction. *Physiol. Rev.,* 55(2):247–304.
9. Gurfinkel, V. S., Kots, Y. M., Paltsev, Y. I., and Feldman, A. G. (1971): The compensation of respiratory disturbances of the erect posture of man as an example of the organization of interarticular interaction. In: *Models of the Structural-Functional Organization of Certain Biological Systems,* edited by I. M. Gelfand, V. S. Gurfinkel, S. V. Fomin, and M. L. Tsetlin, pp. 382–395. MIT Press, Cambridge, Mass.

10. Kots, Y. M., Krinskiy, V. I., Naydin, V. L., and Shik, M. L. (1971): The control of movements of the joints and kinesthetic afferentation. In: *Models of the Structural-Functional Organization of Certain Biological Systems,* edited by I. M. Gelfand, V. S. Gurfinkel, S V. Fomin, and M. L. Tsetlin, pp. 371–381. MIT Press, Cambridge, Mass.
11. Marsden, C. D., Merton, P. A., and Morton, H. B. (1973): Is the human stretch reflex cortical rather than spinal? *Lancet,* 7806:759–762.
12. Melvill Jones, G., and Watt, D. G. D. (1971): Observations on the control of stepping and hopping movements in man. *J. Physiol. (Lond.),* 219:709–727.
13. Nashner, L. M. (1971): A model describing vestibular detection of body sway motion. *Acta Oto-Laryngol.,* 72:429–436.
14. Nashner, L. M. (1972): Vestibular posture control model. *Kybernetik,* 10:106–110.
15. Nashner, L. M. (1976): Adapting reflexes controlling the human posture. *Exp. Brain Res.,* 26:59–72.
16. Nashner, L. M. (1977): Fixed patterns of rapid postural responses among leg muscles during stance. *Exp. Brain Res.,* 30:13–24.
17. Nashner, L. M., and Grimm, R. J. (1977): Analysis of multiloop dyscontrols in standing cerebellar patients. In: *Progress in Clinical Neurophysiology, Vol. 4,* edited by J. E. Desmedt. S. Karger, Basel.
18. Nashner, L. M., and Woollacott, M. The organization of rapid postural adjustments to vertical perturbations during stance. *(Submitted for publication.)*
19. Phillips, C. G. (1969): Motor apparatus of the baboon's hand. *Proc. R. Soc. Lond. [Biol.],* 173:141–174.
20. Szentágathai, J., and Arbib, M. A. (1972): Conceptual model of neural organization. *Neurol. Res. Prog. Bull.,* 12:313–510.
21. Wetzel, M. C., and Stuart, D. G. (1976): Ensemble characteristics of cat locomotion and its neural control. In: *Progress in Neurobiology, Vol. 7,* edited by G. A. Kerkut and J. W. Phillis, pp. 1–98. Pergamon Press, London.

Posture and Movement, edited by
R. E. Talbott and D. R. Humphrey.
Raven Press, New York © 1979.

Novel Experimental and Theoretical Approaches To Study the Neural Control of Locomotion and Jumping

Felix E. Zajac and William S. Levine

Electrical Engineering Department, University of Maryland, College Park, Maryland 20742

INTRODUCTION

During most of this century neuroscientists have studied a variety of animals in "motionless" states. As a result, their research has emphasized the structural and functional connections of the neural and muscular systems which were presumed to participate in the control of movement. The emphasis has more recently been on the study of signals emanating from the nervous system and, to a lesser extent, from the muscular system during different movements in a variety of species, including man, to define mechanisms of neural control. The properties of neural circuitry based on measurements from animals in motionless states have been helpful in the interpretation of data obtained from moving animals. It has become clear, however, that the functional mechanisms of nervous system control during movement are not necessarily similar to and in some instances are in opposition to those mechanisms expected from previous studies in inert animals [e.g., phasic control of reflexes during locomotion (13,14,15)].

Movement of the body and its components has also been investigated using biomechanics. Scientists in this field, in contrast to neuroscientists, have almost always studied the human body while it is in motion. Their techniques have utilized cinematography, ground reaction force measurements, dynamics, and electrical signals from muscles to estimate stresses in bones, to understand the coordination of body parts during movement, to find the etiology of orthopedic injuries, and to discover fundamental differences between athletes and nonathletes performing sporting events. These objectives are different from that of the neuroscientist, which is to comprehend the mechanisms for nervous system control of movement. The different objectives of the fields of biomechanics and neuroscience have thus led scientists in the former to use devices external to the body and scientists in the latter to use internal ones during the study of movement in man and animals.

The study of locomotion has the potential of effecting a symbiosis of many disciplines, including biomechanics and neuroscience, since propulsion by legged animals is an important element of each animal's movement repertoire. Kinesiologists have traditionally investigated human locomotion, whereas it has been only recently that neuroscientists have renewed studies into the neural mechanisms subserving mammalian locomotion. The principal reason for this rejuvenation is that locomotion can be induced in decerebrate and in spinal animals and signals originating in identified neural structures can be recorded and correlated with specific events in the locomotor cycle. It is thus possible to associate the biomechanical events occurring in intact animals during locomotion with the neurological events obtained from "unconscious" walking animals. Comparative studies have shown that similarities exist between human and animal locomotion (9,24) and suggest that through animal studies one might gain insight into the neural control of human locomotion.

Nervous System Strategy During Control of Locomotion

It is not evident from all these studies, however, what task the central nervous system (CNS) is attempting to accomplish during locomotion. For example, it is not difficult to envision that command signals in the CNS are generated, at one extreme, to steadily propel an animal while minimizing energy consumption and, at the other extreme, to propel an animal as fast as possible while assuming that only a finite energy supply exists. These two commands are associated with different performance criteria, and thus their CNS strategies for inter- and intramuscular coordination might not be congruent. The dependence of CNS control strategy on motor task is emphasized in various chapters of this book. Our comprehension of CNS mechanisms effecting movement and our analysis of data should be enhanced by knowledge of the specific task which is under control or, to state it differently, by knowledge of a quantity with which to measure performance (the performance criterion).

The central nervous system in designing a strategy to maximize performance of a given motor task must also be cognizant both of the properties of the skeletomuscular system and of the limitations imposed by the CNS on itself. With respect to the latter, for instance, it has been suggested that only a few kinematic or kinetic variables are independently controlled during movement for simplicity of control (4,7,11,12,31,36). Such hypotheses imply intermuscular synergies rather than independent control of each muscle. The orderly recruitment of motoneurons which innervate a given muscle according to the size of either the neural or muscle unit portion of the motor unit (22,23) would also impose constraints on admissible CNS control strategies. The degradation, if any, in performance of various motor tasks, or of different locomotor gaits, that would result from such constraints on CNS control signals has not been investigated.

Rationale for Studying Jumping Animals and Humans

We have chosen jumping of animals (and eventually, man) to their maximum achievable heights as a subject for interdisciplinary study for a variety of reasons.

First, the quantitative and theoretical techniques developed during a study of jumping cats (or human subjects) will provide the foundation for similar research on galloping of cats (or running of humans) via studies of horizontal jumping. Horizontal jumping in quadrupeds, a logical extension of our present studies of vertical jumps, probably has characteristics similar to galloping. Goslow et al. (19) showed that the hindlimb trajectory of a cat jumping onto a table is quite similar to that developed during galloping, hinting that knowledge of the former might give a better understanding of locomotion. The running long jump of the dog shows a clear demarcation between forces exerted by the fore- and hindpaws on the platform (2). The hindpaw forces provide the horizontal movement essential in the dog's achievement of the long jump (2), just as the propulsion phase of the gallop is probably a result of hindlimb muscular contraction, based on the similarities of body trajectories during these two movements (2,16). We expect a similar connection between human jumping and running. Later work will continue the study of locomotion with emphasis on slower gaits.

Secondly, the performance criterion is trivially specified for animals jumping to their extreme. For example, maximization of the height of the animal's center of mass is the performance measure for animals jumping to their maximum achievable heights. Usually, however, rigid specification of the specific motor task which is under CNS control is difficult or impossible. It is not hard to imagine that energy expenditure, performance time, and distance traversed are all significant variables to locomotion, with knowledge of the relative importance of each one necessary for the specification of the particular locomotor task. Even if energy expenditure is not included in the performance measure, it must be considered as a constraint (just as it is in the study of animals executing maximal jumps) around which the CNS must develop its strategy. The comprehension of mechanisms, both muscular and neural, which control fast gaits should provide insight into the criterion utilized during slow locomotion.

Thirdly, jumping maximally to vertical heights or to horizontal distances provides an excellent subject for multidisciplinary, experimental research. Comparative studies of cat and human jumping are possible since trained cats, like humans, use only their hindlimbs for propulsion. The cat's hindlimbs (or human legs) are bilaterally synchronized and appear to involve no complex interlimb relationships, such as are found in normal locomotion or swimming (29,30). The maximal (vertical) jump is highly stereotyped and reproducible in the cat (42) and presumably in the human too. The principles of biomechanics and neuromuscular systems can be used to study the kinematics and kinetics of either human or animal jumping via analysis of data from cinematography, force plates, and surface or implanted electromyographic electrodes. Recordings from, or stimulation of, neural structures during jumping might be possible in decerebrate animals, since elicitation of jumping has been shown in such preparations (20,35). Unfortunately, jumping behavior in decerebrate cats has hardly been studied, and comparison of such behavior with that of intact cats is thus not yet possible. The decerebrate jumping cat should be an excellent experimental model for use in the study of neural mechanisms controlling jumping and an adjunct to the investigations on intact animals, just as the decerebrate cat has been helpful in the elucidation of CNS control of walking (17,33,36,38).

Fourthly, maximal jumping is also amenable to analytical study, mathematical modeling, and computer simulation. The jumping cat has been formulated as an optimal control problem [(41); see below]. That is, the dynamics of the hindlimbs are modeled using Newtonian mechanics; the inputs to the model are the controls and represent activation patterns of muscles; the outputs are any measurable variables (e.g., joint trajectories and torques); and the performance criterion is to maximize the peak height of the center of body mass. The computer solution to this problem exists since the performance criterion is specified, which in the past has been a major difficulty in the application of optimal control to physiological systems [however, see (21)]. It should also be noted that the emphasis has previously been on the development of linear mathematical models to describe system components. The choice of linear models has been based primarily on the ease with which linear systems can be analyzed and not necessarily on the physiological realities. The development of modern systems theory, including optimal control theory, during the past 20 years has been extremely helpful in the analysis and understanding of highly complex systems, such as satellite control (27), aircraft control (5), and socioeconomic systems (3). There are two primary reasons that permit the mathematical treatment of such complex systems: i) Models for the system components do not need to be linear; that is, they can be nonlinear, time-varying, and constrained to take on only limited values; and ii) computers allow for the solution of these problems even though no analytical solution is readily apparent. Optimal control is an extension of modern systems theory and enables one to precisely model a control problem as a combination of system dynamics, constraints, and performance measure. Digital computers are then utilized to solve this precise mathematical problem to find the response(s) resulting from the optimal control(s).

A multidisciplinary approach to the study of jumping should provide for a quantitative understanding of neuromuscular control mechanisms, including intermuscular coordination, that function during this task. The computer solution to the jumping cat is under investigation and will give the inter- and intramuscular activation patterns that produce maximal jumps. These patterns will be compared with experimental ones to gain confidence in the efficacy of the theoretical model of the jumping cat. The mathematical problem will then be trivially solved by repeated computer trials to study, for example, the effect of intermuscular synergies, the orderly recruitment of motor units, and the sensitivity of parameters (e.g., limb lengths; muscular forces and dynamics) on the animal's performance. These computer results will identify the important variables for future study and will suggest control strategies used by the nervous system for muscular coordination of this task. It would be at least difficult, and probably impossible, to perform physiological and biomechanical experiments to satisfy similar objectives. It is through this kind of interdisciplinary approach that future studies may solve the reverse and more difficult problem of specifying the performance criterion during slow gaits of locomotion.

JUMPING CAT AS AN OPTIMAL CONTROL PROBLEM

This section first reviews the elements of an optimal control problem and its solution. The formulation of the jumping cat into such a problem is then presented. Finally, evidence is given to support the optimal control formulation.

Structure of an Optimal Control Problem

Components

An optimal control problem consists of four precisely defined components (6):

(i) Plant. The "plant" maps whatever "input" or control is applied into a unique "state" and "output." For example, in missile control problems the "plant" can be regarded as the missile itself. The "input" is the thrust at each instant of time. The "state" and "output" coincide and correspond to the position and velocity of the missile.

Generally, the plant is described by a vector differential equation

$$\dot{x}(t) = f\Big(x(t),\ u(t)\Big) \qquad [1]$$

where $x(t)$ is an n-vector, the system state,
$u(t)$ is an m-vector, the system controls,

and a vector equation

$$y(t) = g\Big(x(t)\Big) \qquad [2]$$

where $y(t)$ is an r-vector, the system outputs,

which maps states into outputs. A good intuitive idea of the system state is the information which, if known at time t and combined with knowledge of the control from time t on, will let you compute the state and output from time t on.

(ii) Control set. In defining the plant one actually defines the type of controls. However, one also needs to define precisely the set of possible controls. Usually, this will be

$$u_{\min} \leq u(t) \leq u_{\max} \qquad [3]$$

where $u(\cdot)$ is a piecewise continuous function of time.

Specification of the control set limits the choice of the optimal control to this physically reasonable set.

(iii) Performance measure. It is necessary to specify a criterion for evaluating the performance of a given control. In order to facilitate comparison of one control with another the performance criterion must map each control that lies in the control set into a unique real number. Generally, this is done by defining

$$J(u) = \int_{t_0}^{T} \ell\Big(x(\tau), u(\tau)\Big)d\tau + K\Big(x(T), T\Big) \qquad [4]$$

where $\ell\Big(x(\tau), u(\tau)\Big)$ and $K\Big(x(T), T\Big)$ are scalars, and t_0 and T are defined below.

(iv) Terminal conditions. The time at which the problem starts, t_0, and the time at which it ends, T, must be specified. In addition, the initial state of the system and the final state must be defined. Often, the initial state and time are given, and the final state and time are defined implicitly by the rest of the problem.

Once the above four components are defined, the optimal control problem becomes

maximize $J(u)$,

subject to the constraints imposed by Eqs. [1], [2], [3] and the terminal conditions.

Intellectually, the above problem has the advantage of being mathematically precise. Thus, the problem can, in principle, be mathematically resolved. If a solution exists, it should be possible to find it. If no solution exists, it should be possible to prove this. From the viewpoint of the physiologist, the appealing aspect is that it is not necessary to specify the control or the state as a specific time function in order to model the physiological problem. It is necessary only that the task be precisely defined, and then the mathematical machinery gives the input and state time functions.

Solutions

The solution may still depend on relatively fine details of the model. For example, if the performance criterion takes the form

$$\ell\Big(x(\tau), u(\tau)\Big) = a_1\ell_1\Big(x(\tau), u(\tau)\Big) + a_2\ell_2\Big(x(\tau), u(\tau)\Big)$$

then the ratio a_1/a_2 may drastically affect the optimal control. An example of interest in locomotion might have

$$a_1 \int_{t_0}^{T} \ell_1\Big(x(\tau), u(\tau)\Big)d\tau$$

reflect energy consumption and

$$a_2 \int_{t_0}^{T} dt$$

reflect the time to walk the desired distance. One would expect a slow gait as the optimum to correspond with $a_1/a_2 >> 1$ and, conversely, a fast gait when $a_2/a_1 >> 1$.

Often, however, the gross structure of the optimal control problem determines the gross structure of the optimal control. Two notable examples of this are (6,8):

(a) If the plant is linear

$$\dot{x}(t) = Ax(t) + Bu(t)$$

and the performance criterion is quadratic

$$J = \int_{t_0}^{T} [x^T(\tau)Qx(\tau) + u^T(\tau)\, Ru(\tau)]\, d\tau + x^T(T)Kx(T)$$

and the control is unconstrained

$$-\infty < u(t) < \infty$$

then a linear feedback control

$$u^*(t) = F(t)x^*(t)$$

is optimal.

(b) If the plant is linear in the control

$$\dot{x}(t) = f_1\left(x(t)\right) + f_2\left(x(t)\right) u(t)$$

and the performance criterion is linear in the control

$$\ell\left(x(\tau), u(\tau)\right) = \ell_1\left(x(\tau)\right) + \ell_2\left(x(\tau)\right) u(\tau)$$

and the control is constrained as in Eq. [3], then the optimal control is "bang-bang." That is, $u(t)$ switches back and forth between u_{max} and u_{min}.

One more gross structural result is important. As the optimal control problem is formulated here there is no advantage to a feedback control. In order for feedback to be advantageous there must be some uncertainty or "noise" in the problem. Furthermore, if the feedback observations are not equivalent to knowledge of the state, then feedback control can be worse than open-loop control.

Jumping Cat Formulation

The major interest in this article is the application of such theory to neural control of jumping. A discussion of a general formulation as it applies to the cat is presented next. Afterwards, the details of our model are given to specify mathematically the optimal control problem.

General Structure

The plant is the skeletomuscular system of the hindlimb and torso. The inputs to the plant are the motoneuronal outputs to hindlimb and torso muscles. The state of the plant is regarded as the output.

If the controls are the output of motoneurons, then the control set is a set of pulse sequences in which only the pulse times are significant. This is different from the set of bounded, piecewise continuous signals described earlier. However, analytical and computational techniques exist for pulsed controls (28,32), so motoneuronal controls cause no major problem.

The performance criterion is the height of the center of mass (CM) at the peak of the jump. This is to be maximized subject to the constraints imposed by the rest of the problem.

Finally, the terminal conditions are twofold. The terminal time is the instant the toes leave the ground. The terminal state is free. It can be anything that is convenient. The initial time and initial state are mildly ambiguous, since the cat can start to jump and then change its mind. Furthermore, there will be a best initial state for the jump. Thus, the optimization problem needs to be solved for several different initial conditions.

It should be noted that, even before the maximal jump has been mathematically modeled, several simplifying assumptions have been made. In particular, the peak height of the CM is not exactly the height of the jump so far as the experiment is concerned. The assumption is that the cat has so much control of its body that the need to hit a target and land on its feet does not significantly affect its actions during take-off. This assumption is supported by data which suggest that work developed during propulsion and used to move the body after lift-off in other than a vertical direction is very small.

A much more serious approximation is involved in assuming that the outputs are equivalent to the state. Physiologically, this ignores neuronal delays that could be important. The reason for making the approximation is that, in the initial stages of this research, open-loop and feedback controls (if the state is fed back with no delay) are equivalent. There seems to be no point in complicating matters by including delays until the question of how the outputs are utilized becomes important.

Four-Segment Approximation

The model can be made mathematically precise by reference to Fig. 1, where the skeletomuscular system of the cat is approximated by a four-segment inverted pendulum. The inputs are ideal torque generators operating on the segments. The state, which coincides with the observations, is the angular position and

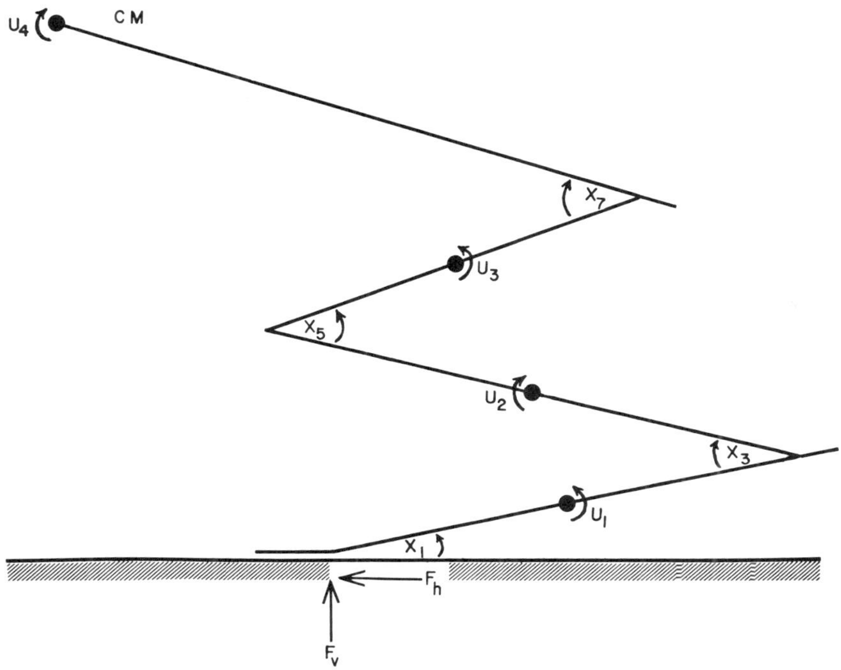

FIG 1. Four-segment inverted pendulum as a model for jumping. Torque generators, u_1, u_2, u_3, and u_4, operate on the foot, lower leg, thigh, and trunk of the animal, respectively, and are conveniently defined in the extensor direction about the center of mass of each segment *(large dots)*. The center of mass of the trunk can be considered about equal to the CM of the whole animal (at least for a cat). Toe, ankle, knee, and hip joint angles are x_1, x_3, x_5, x_7. Velocities of these angles are defined by $x_i = \dot{x}_{i-1}$ for $i = 2,4,6$, and 8. The state of the animal is represented by the vector x where $x^T = [x_1, x_2, \ldots, x_8]$. Motion and weights of these segments produce ground reaction forces in the vertical (F_v) and horizontal (F_h) directions (lateral motion in the other horizontal direction is neglected). F_v and F_h act on the toe joint when the heel is off the ground (shown here) and on a point between the toes and heel otherwise. The animal remains on the ground as long as $F_v \geqslant 0$ and leaves the ground (lift-off) at the instant $F_v < 0$.

angular velocity of each segment. Obviously, there are additional simplifications that are inherent in modeling the cat's skeletomuscular system this way.

One key assumption in this approximation is that a four-segment model suffices; and supportive evidence is provided for this in the next section, though additional segments can be added if necessary. Another assumption is that muscle action can be replaced by ideal torque generators. The assumption of ideal torque generators is only for convenience. These torque generators will be replaced by ideal force generators to model muscles with appropriate origins and insertions. The ideal force generators will later be replaced as necessary by more accurate muscle models which include such properties as force-length, force-velocity, and elasticity (1,25,34). The use of ideal torque generators as a first approximation might still provide qualitative insight into mechanisms controlling the jump (see next section).

For the present model with ideal torque generators it is straightforward to write the plant equations. These are linear in the control and of the form (41)

$$\dot{x} = A(x) + B(x)u$$

where x is an 8-vector.

This linearity in the control will still hold when one uses ideal force generators or other muscle models based on the properties stated above, though x may have higher dimension since it will include internal muscular variables.

The control set is taken to be

$$u(t) \text{ piecewise continuous}$$

where $$u_{\min} \leq u(t) \leq u_{\max}$$

This also extends to the better models of muscle until one has motoneuronal output as the control.

The performance criterion has the form

$$J = K_1\Big(x(T)\Big) + K_2\Big(x(T)\Big)$$

where $K_1\Big(x(T)\Big)$ is the height of the cat's CM at the instant the cat leaves the ground and $K_2\Big(x(T)\Big)$ is proportional to the square of the vertical component of the cat's velocity at the same instant. This choice of J gives the maximum height reached by the cat's CM.

Finally, the terminal time is given by the earliest time, T, at which the vertical force $F_v\Big(x(t), u(t)\Big)$ exerted by the ground on the cat would become negative. The initial conditions are arbitrary as was mentioned earlier.

Note that the optimization problem is completely defined once one agrees to maximize J subject to the constraints imposed by the plant, control set, and terminal conditions. Unfortunately, one cannot solve this problem analytically. The best analytic result available is that the control is "bang-bang" (though this statement must be hedged by the possibility of some exceptions).

Thus, the problem must be solved on the computer. There exist many algorithms that one might use to do this. Two of these are being used. The first is basically a hill-climbing or gradient algorithm. Intuitively, it seeks the direction of change in u that maximizes the rate of increase in J. The algorithm moves to a maximum in that direction and then finds the new direction of steepest ascent. A version of this algorithm has already been used successfully on the above problem. The algorithm is being revised to make it faster and more efficient.

A second algorithm, differential dynamic programming (26) is also being tested. It is harder to give an intuitive description of the operation of this algorithm. Let it suffice to say that Hatze (21) has used it successfully to study a time optimal control problem involving motion of the human leg.

In any case, it is clear that the problem can be solved computationally. At the moment emphasis is on finding an algorithm that will be efficient for any problem of this type.

Evidence for Optimal Control Model

On–Off Control

It was shown in the previous section that the solution to the control of the maximal jump will most likely require that muscles be fully activated, or not be activated at all. Our experimental results are consistent with this on–off type of muscular activation. Electromyographic recordings, during propulsion of the cat performing a maximal vertical jump, from semitendinosus (ST), vastus lateralis (VL), and medial gastrocnemius (MG) muscles (40,42) suggest that these muscles are initially in a state of no to very little activity. They are then abruptly activated about 50 msec prior to the vertical dashed lines in Fig. 2, and finally cease firing approximately at lift-off. The electrical activities from these muscles during the jump were so great that only minimal, if any, activities were observed during walking with the amplifier gain settings unaltered. The quick and strong muscular contractions in ST, VL, and MG (and presumably other synergistic extensor muscles) resulted in rapid rise in torques about the hip, knee, and ankle joints (Fig. 2). The rise times of these torques are in agreement with those times obtained from a terminal experiment on the same animal during *in situ,* maximum isometric contractions of hip, knee, and ankle extensor muscles to high frequency tetani (see Fig. 2 for more details). The fast changes in joint torques imply that our model of the optimal control problem with ideal torque generators as the controls might indeed be a good approximation for initial studies.

Other evidence also suggests that the hip, knee, and ankle extensor muscles are fully activated during the maximal jump. Fig. 2 shows the torques about each of the three joints based on calculations from external force-plate and high-speed (380 fps) cinematographic measurements [e.g., (2)]. The hip and ankle torques occurring during the jump are in agreement with those torques which result from maximum isometric contractions of the appropriate extensor muscle group (torque comparisons should be made only for times corresponding to those after the dashed lines to account for speed of muscular contractions), though a stretch-shorten contraction in ankle extensor muscles (42) complicates the interpretation. Recent force recordings from the MG tendon during intact cat jumping (37) also suggest that MG muscle is fully activated. The disparity of the two torque curves about the knee joint, which results from isometric torque measurements greater than the calculated ones, might not necessarily imply muscular activation patterns contrary to the on–off type. First, all single-

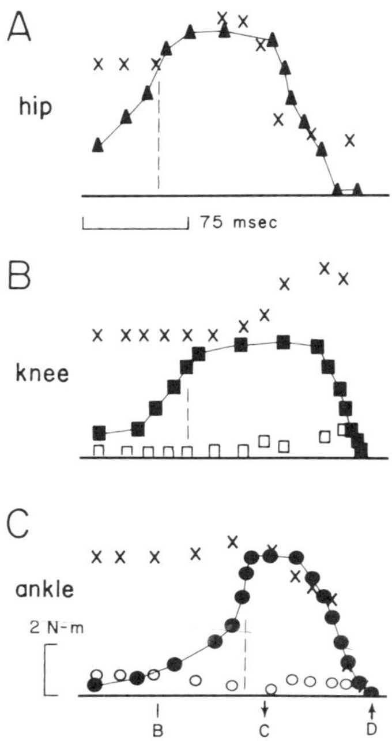

FIG. 2. Comparison of torques about the hip, knee, and ankle joints during jumping with isometric torques during maximum excitation of extensor (or flexor) muscles. Joint torques during propulsion were calculated for this 4.5-kg cat from measurements of the resultant force vector (composed of vertical and anteroposterior horizontal ground reaction forces) and of the corresponding hindlimb position. The force vector was assumed to act through the toes after the heel leaves the platform (*point C;* indicated by ↓) and before lift-off (*point D;* indicated by ↑) and through the animal's center of mass when the foot was on the force-plate (time prior to *point C*). Most propulsion occurs during the stereotypic launching phase [time *B* to *D;* (40,42)] rather than during the variable preparatory phase (time before *B*). *Dashed lines* show the times at which 80% of peak torque about the hip (▲), knee (■), and ankle (●) are reached during the jump. Electromyograms (40,42) indicate that muscles extending these joints begin their activity about 50 msec sooner. Isometric torques about these three joints were measured later in the same cat with its hindlimb set to those positions which occurred during propulsion. Supramaximal trains of stimuli (100 pps for about 0.5–1.0 sec.) were applied to muscle nerves with long intervals between trains to prevent fatigue. **A,B,C:** Isometric extensor torques (x) about the hips, knee, and ankle joints from stimulation of the hamstring, femoral, and tibial nerves, respectively. (Tibial nerve was cut after innervation of plantaris muscle.) Flexor torques about the knee (□) and ankle (○) joints from stimulation of the hamstring and peroneal nerves, respectively. (*Unpublished data* from M. R. Zomlefer, F. E. Zajac, and W. S. Levine.)

joint quadricep muscles undergo a rapid shortening during the latter part of the propulsion phase (time C to D in Fig. 2) without previously undergoing lengthening. A significant reduction in force from its maximum isometric value is thus expected owing to the force-velocity relationship of muscle (25). Secondly, rectus femoris and sartorius muscles might be inactivated owing to their hip-flexion action. Thirdly, the hamstring muscles are presumably maximally activated to produce extension torque about the hip joint (Fig. 2), and such activation will also contribute to flexion torque about the knee (Fig. 2B, black squares). It seems reasonable, therefore, to conclude tentatively that muscles providing contributions to propulsion are maximally activated at times to ensure a maximum-height jump.

Four-Segment Structure

The optimal control problem formulated in the previous section assumes that the skeletomuscular system can be approximated by a four-segment inverted pendulum (Fig. 1). This approximation implies a rigid torso, which is certainly not true based on cinematographic analysis; since the back is flexed initially and then accelerates during the preparatory phase of the jump ($t < 0$; Fig. 2) to cause back extension. The following analysis reveals, however, that back motion probably does not contribute much to propulsion.

The height that the animal can attain during the jump can be predicted based on the four-segment model and on the assumption that work produced by such a model during propulsion is converted to a change in potential energy which results from the animal's raising his CM from its initial position corresponding with the crouch to a position corresponding with the peak height reached by the cat's CM. The total work is given by

$$\sum_i \int_{t_0}^{T} T_i dx_i$$

where $i = 1, 3, 5, 7,$
$x_i =$ joint angles (see Fig. 1),
and $T_i =$ torque about each of the joint angles x_i.

The work was computed using the calculated torques given in Fig. 2 (solid symbols) and cinematographic measurements of the joint angles. We found (neglecting work produced about the toe joint) that the predicted peak height was 10% greater than the actual height reached by the cat. If back extension were important to propulsion, a larger disparity between heights would occur (work about the toe joint will also increase the difference). Back extension would also contribute to hip extension torque, which can be attributed to full activation of the hamstrings (see above) without activation of back musculature. Finally, torques producing back extension must occur during the early part of the jump ($t \leq 50$ msec) when the back is accelerating; however, the work developed by the animal during this time is only 14% of the total work produced by lift-off.

There exist factors which would lower the predicted height. These factors include work used to move the cat horizontally in the air, to rotate the cat in the air, to rotate the cat (or any of its limb segments) during propulsion, and to overcome frictional losses at the joints. Our estimates show that the work to account for all these factors, except for the last one for which information is scarce, is less than 5% of the work developed during propulsion. It does not appear that the height predicted from work output will be less (or at least significantly less) than the experimental observation. We have thus tentatively concluded that back motion can be neglected as a first approximation and, as a result, that a four-segment model of the cat is valid.

There exist other comparisons which support the validity of the four-segment model. We find experimentally that a wave of activation proceeds from proximal to distal hindlimb musculature to cause torque generation about the joints to follow a similar pattern (Fig. 2). This sequence of torque production is compatible with kinematics. That is, detectable motion of the various parts of the animal proceeds in time from vertebral column, to femur, to tibia, to metatarsals, and finally to tarsals. Preliminary results from a computer solution to the four-segment model show that the sequence of activation of the ideal torque generators is also proximal to distal (41). This sequence of muscular activation appears consistent with proposed spinal mechanisms for controlling locomotion in the cat (10,38) and with the rostral to caudal pattern of muscular activation in the dogfish (18).

Analytical Solution to Simplified Model

Although the major emphasis of this research is the understanding of the control of complex movement, there is considerable value in analyzing a very simplified model of maximal-height jumping. Such a model and its solution are illustrated in Fig. 3. The mathematical description of the problem and its analytical solution are given below.

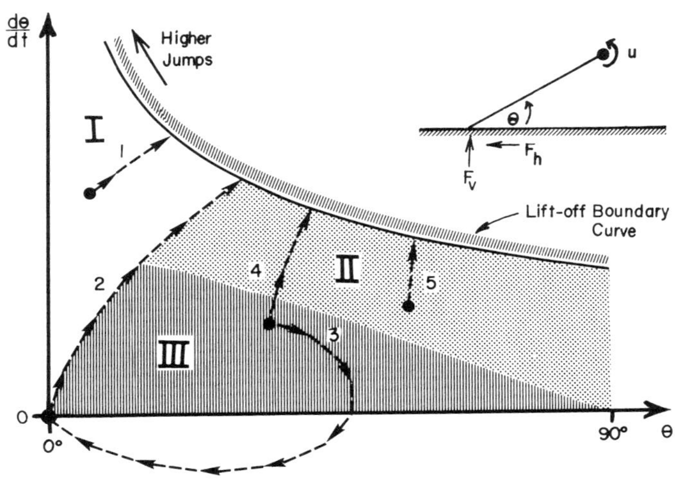

FIG. 3. Trajectories and solution to optimal control problem for a simple model of jumping. **Inset:** Diagram of the model, which is a massless rod with point mass (●) at one end. Torque generator *(u)* causes angular rotation (θ) of the rod and ground reaction forces $(F_v;F_h)$. **Graph:** A point in the $d\theta/dt-\theta$ plane corresponds to one state of the rod and, together with subsequent values of *u,* determines the ensuing motion. Trajectories 1 to 5 specify motion of the rod for maximal jumps which begin from different initial positions (●). The rod leaves the ground when a trajectory intersects the lift-off boundary curve. Higher jumps result for intersections which have a higher $d\theta/dt$, or a lower θ. See text for more details. (*Unpublished data* from W. M. Roberts, W. S. Levine, and F. E. Zajac.)

Model Formulation and Rationale

The plant, under the assumption of a massless rod is described by

$$\ddot{\theta}(t) = \begin{cases} \dfrac{u(t)}{M\ell^2} - \dfrac{g}{\ell}\cos\theta(t) & \pi > \theta > 0 \\ \dfrac{u(t)}{M\ell^2} & \theta = 0 \end{cases} \qquad [5]$$

where M = mass
ℓ = length of massless rod
g = acceleration due to gravity
$u(t)$ = control
and $\begin{bmatrix}\theta(t)\\ \dot{\theta}(t)\end{bmatrix}$ = the observations and the state space.

The control set is defined by

$$0 \le u(t) \le u_{\max} \qquad [6]$$

where $u(t)$ is piecewise continuous.

The performance criterion to be maximized is the maximum height achieved during the jump, or

$$J = \ell\sin\theta(T) + \frac{\ell^2}{2g}\dot{\theta}^2(T)\cos^2\theta(T) \qquad [7]$$

Note that the term $\ell \sin \theta(T)$ is the height of the CM at lift-off and $\ell\dot{\theta}(T) \cos\theta(T)$ is the vertical component of the velocity of the CM at lift-off. Finally, lift-off is defined by the terminal conditions. Assume that:

(i) The rod rests on the ground but is not fastened to the ground.
(ii) The rod does not slip, regardless of the horizontal force exerted. (This is a reasonable assumption since, for example, cats have claws.)

Then, T is determined by the instant at which F_v first becomes negative (since the ground cannot exert negative force), or

$$Mg\sin^2\theta(T) - M\ell\dot{\theta}^2(T)\sin\theta(T) + \frac{u(T)}{\ell}\cos\theta(T) < 0 \qquad [8]$$

The problem is solved for all possible initial conditions.

It is worthwhile to emphasize the advantages of studying such a simplified model. First, the simple problem with a known analytical solution provides a very good check on the computer algorithms and programs that are used to solve the complex, realistic problem. Second, the two-dimensional, one-segment problem clarifies the formulation of the complex (at least 8-dimensional) four-segment model problem. For instance, it is much easier to understand the lift-off condition as well as the performance criterion in the simple problem from which extrapolation to more complicated models becomes obvious. It should also be noted that inertias and limb masses can be neglected in the simple problem but not in the real problem in order to uniquely specify hindlimb position. Finally, it will be seen below that a great deal of qualitative insight into optimal strategies can be obtained from the solution to the simple problem.

Results

Figure 3 shows the solution to the optimization problem as a function of initial conditions. That is, if $\theta(0)$ and $\dot{\theta}(0)$, the initial conditions, lie in region I, the unique optimal control is to set $u(t) = u_{\max}$ until lift-off occurs. Trajectories 1 and 2, for example, specify motion of the rod for two initial conditions in this region. An optimal control in region II is also to set $u(t) = u_{\max}$ until either lift-off occurs (e.g., trajectory 5) or $\theta = \pi/2$. It should be noted that other controls, which give identical performance to $u(t) = u_{\max}$ are also optimal in parts of region II that are close to $\pi/2$ and for which the rod does not jump.

Region III is the most interesting, because at every point in region III there are two distinct, locally optimal controls. That is $u(t) = u_{\max}$ until lift-off occurs (e.g., trajectory 4) is a better control than any other control that is, in a sense that can be made precise, "near" it. However, much better performance can be achieved by a control that is $u(t) < u_{\max}$ [one can give precise values for $u(t)$] until $\theta(t) = \dot{\theta}(t) = 0$. Trajectory 3 is an example of resulting rod motion for such a control; from then on, $u(t) = u_{\max}$ (see trajectory 2) until lift-off [for more details and a proof, see (34a)]. The basic idea of the proof, valid for all regions, is to show that it is always best to have lift-off occur with $u(t) = u_{\max}$ and that moving to the left on the lift-off boundary curve (Fig. 3) is always advantageous. This implies that, if it is possible to choose the initial condition, it is always advantageous to have $\dot{\theta}(0)$ as large as possible and $\theta(0)$ as small as possible. The height of sucessive jumps would thus be progressively raised with resetting of initial conditions which lie in regions I, II, and III, respectively.

Comparison with Experiments

These analytical results give considerable insight into a number of experimental observations of jumping cats. In particular, cats exhibit a number of false starts in attempting to jump maximally high. The above results suggest that cats are attempting to achieve an optimal initial state, and if the preparatory phase does not achieve this state, they reiterate this phase. In fact, all of region III can be viewed as a set of wrong initial states which are "corrected" by returning to the right initial state, which is $\theta(0) = \dot{\theta}(0) = 0$. Similarly, region II can be viewed as a set of wrong initial states which cannot be corrected, so one just makes the best of it and jumps to a lower height than that achieved by starting at $\theta(0) = \dot{\theta}(0) = 0$.

A more dramatic analytical result corresponds to the response of cats to unexpected auditory stimuli. It has been found that if a stimulus occurs early enough, the cat aborts the jump (39). However, if the stimulus occurs after a well-defined point in the propulsion phase (about point B in Fig. 2), the cat jumps almost maximally despite the stimulus. This behavior corresponds quite closely to what one would expect from the analytic solution to the simple problem. It can be shown that a continuous boundary curve exists in regions I and II and that an aborted jump is only possible for states below this curve. This model suggests, therefore, that a cat should be able to abort a jump if a stimulus is given early in the jump before the state of the cat reaches the analogous boundary surface, and that the cat should not be able to refrain from jumping otherwise.

It should also be noted that the analytical results agree with the conventional wisdom in the Olympic high jump (21a) where competitors strive for (i) high-speed run-up; (ii) CM as low as possible at the start of the final step; and (iii) follow through. This corresponds to: (i) maximizing $\dot{\theta}(0)$, (ii) minimizing $\theta(0)$, and (iii) keeping $u(t) = u_{\max}$ until lift-off.

In summary, it is believed that considerable insight has been obtained by the study of a very simple model of maximum-height jumping. It must be emphasized that these are insights into both the solution of the more complex and meaningful optimization problem and the experiments. The solution to the simple problem is not regarded as, in any sense, a solution to the "jumping problem."

PLANNED RESEARCH

The highest priority item of future research is the solution to the optimal control problem with muscle force generators as the control. It is believed that the solution to this problem will define the intermuscular coordination pattern that would exist in the real cat in the absence of CNS constraints on intermuscular coordination. Experimentally, these results will be compared with the intermuscular coordination patterns observed in maximally jumping cats by means of EMG recordings. It is hoped that these comparisons will provide evidence for, or against, CNS constraints.

Other work that is planned for the immediate future has already been described. In particular, the development of efficient algorithms for solving this type of optimization problem has been mentioned. This is motivated by planned research that involves solving optimization problems over and over. For example, it would be very desirable to know how sensitive the optimal controls are to changes in the parameters of the cat. Based on these sensitivity studies, critical parameters will be identified, and experiments to evaluate them more accurately will be designed and performed.

Another planned series of sensitivity studies is to remove specific muscles from the computer cat and recalculate the optimal jump. This is designed primarily to discover whether significantly different, intermuscular coordination patterns occur. If possible this question will also be explored experimentally.

Still another extension of the present effort, which can be viewed as a form of sensitivity study, will be a study of maximum-height jumps of humans. From the viewpoint of the computer, this involves only changing the parameters. It will be possible, for instance, to find the initial position of the human (e.g., crouch vs. standing) which, together with the optimal intermuscular activation pattern, produces the highest jump. Experimentally, such a study will also be possible in humans, but it would be difficult, if not impossible, in cats. This type of interdisciplinary study will determine the different neural and skeletomuscular mechanisms involved in humans who initiate jumping from various positions. It is possible that hitherto unknown modes of jumping might emerge that would produce the highest jump by using the globally optimal muscular coordination and start pattern.

The research described above does not test the sensitivity of the results to changes in the performance criterion. In fact, the effect of changes in performance criterion on the control is absolutely fundamental. The *only* difference between walking, galloping, and jumping is the performance criterion. The techniques described above apply immediately to any other locomotion task for which the performance criterion can be defined. In the long run, it is expected that this research can be extended to gaits of locomotion. One possibility is to impose on the animal a performance criterion such as, for example, maximum-speed run. Before this, research is planned on maximal horizontal jumps for comparison with maximal vertical jumps. In all these studies of locomotion it is hoped that the presence or absence of CNS constraints can be determined and comparisons made to determine if there exist universal laws governing neuromuscular control of locomotion.

All of the above research is either experimental or computational. In addition, analytical work is being carried out on a two-segment model of a jumping cat. Since this problem has essentially only three state variables, there is a real hope of an analytical or partly analytical solution. Solution of this problem would greatly enhance the qualitative understanding of the many-segment problem. For example, it should be possible, in three dimensions, to find the equivalents of regions I, II, and III for the "baton" problem. This would give a much better understanding of the role of initial conditions.

In conclusion, it is believed that useful information about a wide range of locomotory activities can be obtained by the interdisciplinary approach outlined in this chapter.

SUMMARY

Research is described which provides insight into the neural control of locomotion via the interdisciplinary study of cats and humans as they jump to their maximum achievable heights. Definition of the performance criterion, a maximum-height jump, results in highly stereotyped behavior of the experimental animal and simultaneously facilitates construction of an equivalent optimal control problem. The optimal control problem is based on a four-segment inverted pendulum to model the animal's dynamics. The preliminary computer solution of this optimal control problem is compared with experimental results, and the comparison implies that the optimal control problem is a good model for the cat as it performs a maximal-height jump. Theoretical and experimental results suggest that muscles are either fully activated or relaxed during the propulsion phase of the maximal jump. In addition, analytical results for a simplified, one-segment, model of the jumping cat are used to provide qualitative insight into the experimental results, including some that, otherwise, seem difficult to explain.

Planned research will emphasize studies of the sensitivity of the above results to changes in animal anatomy, dynamic properties of muscles, and task. It is suggested that this combined analytic, computational, and experimental approach can be applied to a number of related problems such as other forms of locomotion.

ACKNOWLEDGMENTS

We thank Dr. Michael R. Zomlefer and Mr. William M. Roberts for their permission to use experimental data obtained from jumping cats and analytical results from the simplified jumping model, respectively. Mr. David Dungan assisted in the preparation of the figures. This work was supported by NIH grant NS11971 and the University of Maryland Computer Science Center.

REFERENCES

1. Abbott, B. C., and Wilkie, D. R. (1953): The relation between velocity of shortening and the tension-length curve of skeletal muscle. *J. Physiol. (Lond.)*, 117:77–86.
2. Alexander, R. McN. (1974): The mechanics of jumping by a dog *(Canis familiaris)*. *J. Zool.*, 173:549–573.
3. Aoki, M. (1976): Stochastic control in economic theory and economic systems. *IEEE Transactions on Automatic Control,* Vol. AC-21, No. 2, pp. 213–220.
4. Asatryan, D. G., and Fel'dman, A. G. (1965): Functional tuning of the nervous system with control of movement or maintenance of a steady posture. I. Mechanographic analysis of the work of a joint on execution of a postural task. *Biophysics,* 10:925–935.
5. Athans, M. (1977): Adaptive control of NASA's F-8C aircraft using multiple model adaptive control method. *IEEE Transactions on Automatic Control,* Vol. AC-22, No. 5, pp. 768–780.
6. Athans, M., and Falb, P. L. (1966): *Optimal Control, An Introduction to the Theory and Its Applications.* McGraw-Hill, New York.
7. Bernstein, N. (1967): *The Coordination and Regulation of Movements.* Pergamon Press, Oxford.
8. Bryson, A. E., and Ho, Y. C. (1969): *Applied Optimal Control.* Blaisdell Publishing Company, Waltham, Mass.
9. Craik, R., Herman, R., and Finley, F. R. (1976): Human solutions for locomotion. II. Interlimb coordination. In: *Neural Control of Locomotion,* edited by R. M. Herman, S. Grillner, P. S. G. Stein, and D. G. Stuart, pp. 51–64. Plenum Press, New York.
10. Edgerton, V. R., Grillner, S., Sjöstrom, A., and Zangger, P. (1976): Central generation of locomotion in vertebrates. In: *Neural Control of Locomotion,* edited by R. M. Herman, S. Grillner, P. S. G. Stein, and D. G. Stuart, pp. 439–464. Plenum Press, New York.

11. Fel'dman, A. G. (1966): Functional tuning of the nervous system with control of movement or maintenance of a steady posture. II. Controllable parameters of the muscles. *Biophysics,* 11:565–578.
12. Fel'dman, A. G. (1966): Functional tuning of the nervous system with control of movement or maintenance of a steady posture. III. Mechanographic analysis of the execution by man of the simplest motor tasks. *Biophysics,* 11:766–775.
13. Forssberg, H., Grillner, S., and Rossignol, S. (1975): Phase dependent reflex reversal during walking in chronic spinal cats. *Brain Res.,* 85:103–107.
14. Forssberg, H., Grillner, S., Rossignol, S., and Wallen, P. (1976): Phasic control of reflexes during locomotion in vertebrates. In: *Neural Control of Locomotion,* edited by R. M. Herman, S. Grillner, P. S. G. Stein, and D. G. Stuart, pp. 647–674. Plenum Press, New York.
15. Forssberg, H., Grillner, S., and Rossignol, S. (1977): Phasic gain control of reflexes from the dorsum of the paw during spinal locomotion. *Brain Res.,* 132:121–139.
16. Gambarjan, P. P., Orlovskii, G. N., Protopopova, T. Y., Severin, F. V., and Shik, M. L. (1971): The activity of muscles during different forms of locomotion of cats and the adaptive function of the (hindlimb) musculature in the Family Felidae. *Proc. Inst. Zool. Acad. Sci. USSR,* 48:220–239 *(in Russian).*
17. Grillner, S. (1975): Locomotion in vertebrates: Central mechanisms and reflex interaction. *Physiol. Rev.,* 55:247–304.
18. Grillner, S., Perret, C., and Zangger, P. (1976): Central generation of locomotion in the spinal dogfish. *Brain Res.,* 109:255–269.
19. Goslow, G. E., Jr., Reinking, R. M., and Stuart, D. G. (1973): The cat step cycle: Hind limb joint angles and muscle lengths during unrestrained locomotion. *J. Morphol.,* 141:1–42.
20. Guzman, F., and Del Pozo, E. C. (1953): Jump reflex in hypothalamic cat. *J. Neurophysiol.,* 16:376–380.
21. Hatze, H. (1976): The complete optimization of a human motion. *Math. Biosci.,* 28:99–135.
21a. Hay, J. G. (1975): Biomechanical aspects of jumping. In: *Exercise and Sport Sciences Reviews, Vol. 3,* edited by J. H. Wilmore and J. F. Keogh, pp. 135–161. Academic Press, New York.
22. Henneman, E., and Olson, C. B. (1965): Relations between structure and function in the design of skeletal muscles. *J. Neurophysiol.,* 28:581–598.
23. Henneman, E., Somjen, G., and Carpenter, D. O. (1965): Functional significance of cell size in spinal motoneurons. *J. Neurophysiol.,* 28:560–580.
24. Herman, R., Wirta, R., Bampton, S., and Finley, F. R. (1976): Human solutions for locomotion. I. Single limb analysis. In: *Neural Control of Locomotion,* edited by R. M. Herman, S. Grillner, P. S. G. Stein, and D. G. Stuart, pp. 13–49. Plenum Press, New York.
25. Hill, A. V. (1938): The heat of shortening and the dynamic constants of muscle. *Proc. R. Soc. B,* 137:237–280.
26. Jacobson, D. H., and Mayne, D. Q. (1970): *Differential Dynamic Programming.* Elsevier Press, New York.
27. Lemay, J. L., Brogan, W. L., Seal, C. E., and Hendrickson, H. T. (1975): Performance predictions for high altitude self-contained satellite navigation systems. *IEEE Transactions on Automatic Control,* Vol. AC-20, No. 6, pp. 739–747.
28. Lermontov, M., (1973): A gradient method for the optimization of PFM systems. *IEEE Transactions on Automatic Control,* Vol. AC-18, No. 2, pp. 177–179.
29. Miller, S., Van der Burg, J., and Van der Meché, F. G. A. (1975): Locomotion in the cat: Basic programmes of movement. *Brain Res.,* 91:239–253.
30. Miller, S., Van der Burg, J., and Van der Meché, F. G. A. (1975): Coordination of movements of the hindlimbs and forelimbs in different forms of locomotion in normal and decerebrate cats. *Brain Res.,* 91:217–237.
31. Nichols, T. R., and Houk, J. C. (1976): Improvement in linearity and regulation of stiffness that results from actions of the stretch reflex. *J. Neurophysiol.,* 39:119–142.
32. Onyshko, S., and Noges, E. (1968): Optimization of pulse frequency modulated control systems via modified maximum principle. *IEEE Transactions on Automatic Control,* Vol. AC-13, No. 2, pp. 144–150.
33. Orlovsky, G. N., and Shik, M. L. (1976): Control of locomotion: A neurophysiological analysis of the cat locomotor system. In: *International Review of Physiology, Vol. 10: Neurophysiology II,* edited by R. Porter, pp. 281–317. University Park Press, Baltimore.
34. Ritchie, J. M., and Wilkie, D. R. (1958): The dynamics of muscular contraction. *J. Physiol. (Lond.),* 143:104–113.

34a. Roberts, W. M. (1978): Detailed differential dynamic programming algorithm for bang-bang control problems. M.S. thesis. Electrical Engineering Department, University of Maryland, College Park, 88 pp.
35. Schaltenbrand, G., and Girndt, O. (1925): Physiological observations on thalamic cats. I. General behavior in acute experiments. *Pflügers Arch.*, 209:333–361 *(in German).*
36. Shik, M. L., and Orlovsky, G. N. (1976): Neurophysiology of locomotor automatism. *Physiol. Rev.*, 56:465–501.
37. Walmsley, B., Hodgson, J. A., and Burke, R. E. (1977): The forces produced by hindlimb muscles in freely moving cats. *Neurosci. Abstr.*, 3:280.
38. Wetzel, M. C., and Stuart, D. G. (1976): Ensemble characteristics of cat locomotion and its neural control. *Prog. Neurobiol.*, 7:1–98.
39. Zajac, F. E., Zomlefer, M. R., and Levine, W. S. (1977): Effects of auditory stimuli on the control of maximum height jumping. *Proc. Int. Union Physiol. Sci.*, 13:832.
40. Zomlefer, M. R. (1976): Biomechanics of the jumping cat. Ph.D. dissertation. Electrical Engineering Department, University of Maryland, College Park, 112 pp.
41. Zomlefer, M. R., Ho, R., Levine, W. S., and Zajac, F. E. (1975): A study of the coordination of cat hindlimb muscles during a maximal vertical jump. *Proceedings of the 1975 IEEE Conference on Decision and Control,* pp. 402–407. IEEE Control Systems Soc., Houston.
42. Zomlefer, M. R., Zajac, F. E., and Levine, W. S. (1977): Kinematics and muscular activity of cats during maximum height jumps. *Brain Res.*, 126:563–566.

Posture and Movement, edited by
R. E. Talbott and D. R. Humphrey.
Raven Press, New York © 1979.

Features of Segmental Motor Control Revealed in Single-Unit Recordings During Natural Movements

Douglas G. Stuart, Marc D. Binder, and Barry R. Botterman

Department of Physiology, College of Medicine, Arizona Health Sciences Center, University of Arizona, Tucson, Arizona 85724

This report is a selective review of the expanding body of literature on the firing patterns of single motor and sensory peripheral nerve fibers and of single motor units during natural and "near-natural" movements of humans and experimental animals. In particular, we have asked ourselves the extent to which experiments of this general kind have illuminated features of segmental motor control not discernible from more traditional experimental approaches, and the extent to which these new data have provided important confirmations or negations of hypotheses derived from more traditional approaches. The report will begin with a brief catalogue of the data base followed by four sections, each discussing a general issue to which the various *natural movement* studies have addressed themselves.

THE DATA BASE

Human Studies

The human studies referred to in this report include those which have related unitary EMG and neuronal signals to various kinematic or kinetic features of shortening, isometric, and lengthening contractions undertaken by fully conscious subjects. The recent technical progress in EMG recording and signal extraction has been reviewed elsewhere (19). Other progress includes measurement of single motor unit contractile properties (65,66), including their fatigability (79), and single-fiber microneurography which has produced recordings from single muscle afferents in muscle-nerve fascicles, as well as from single skin afferents and postganglionic sympathetic efferent fibers in different fascicles of the same peripheral nerves (reviewed in refs. 5,36,82).

The major advantage of these various procedures and approaches is that they can be undertaken in fully conscious and cooperative subjects during performance of finely graded and controlled tasks. Their major limitation, particularly for neuronal signals, is that movement must be relatively stereotyped and of quite limited displacement and force to ensure stable recording conditions. Even in the case of single-unit EMG recording, considerable difficulties are encountered once the level of muscle contraction exceeds about 20% of the maximum voluntary contraction (64; see however ref. 20).

Animal Studies

The animal literature covered in this report began to appear largely in the mid-1960s with studies of neural-unit activity associated with: respiratory movements during the quiet breathing of anesthetized cats (reviewed in refs. 22,71, 72,73); hind-limb muscles of unanesthetized decerebrate cats during controlled treadmill locomotion (reviewed in refs. 32,75,84); and, jaw-closing muscles of lightly anesthetized cats during reflexly induced swallowing movements (81). More recent studies have expanded upon such observations to include analysis of unitary activity associated with jaw-closing movements during natural chewing, swallowing, and licking activity in conscious cats (17) and monkeys (16, 25,56) and hind-limb movements of conscious cats during treadmill and overground locomotion, as well as during other forms of voluntary and/or automatically-controlled movements (50,51,68,69).

The main advantage of the respiratory system for the analysis of segmental motor control is that a near-natural movement can be studied in the anesthetized state. However, one must be cautious in extrapolating from such stereotyped conditions to those in which conscious voluntary movements are superimposed. For example, differences in the extent of α-γ coactivation have been observed by the same group for the jaw-closing muscles of cats when slightly anesthetized (81) and when fully conscious (17). An advantage of studies on jaw-closing muscles is that unitary recordings of spindle afferent discharge can be obtained from cell bodies located in the brainstem. Spindle activity can then be followed during natural movements of greater force and amplitude than is possible in other motor systems. The main limitation of the jaw-closing studies appears to be problems associated with relating unitary neural signals to complex triplanar chewing, swallowing, and licking movements (24,55), but this problem is not easily disposable in studies attempting similar correlations in any form of natural movement.

Apart from their major contribution to the understanding of the neural control of locomotion, studies on controlled and "fictive" locomotion of high decerebrate cats were particularly influential a decade ago with respect to several basic features of segmental motor control (reviewed in refs. 32,52,75,84). While this preparation remains appropriate for certain types of analysis (e.g., ref. 87), there has been concern in recent years regarding the extent to which restraints imposed upon the natural movements of the lower spinal cord and hips can influence the patterning of unitary afferent and efferent discharge (26,52,69). This limitation and, no doubt, the desire to explore motor control in the fully conscious and unrestrained state, has led to the development of alternative preparations by Prochazka and Loeb and their colleagues (50,68), which permit stable, chronic recording from primary afferent neurons of the freely moving cat coupled with kinematic analysis of the movements. Prochazka's technique provides recording periods of 1 to 4 hr largely from group I axons among the dorsal roots. Loeb's technique involves implanting electrodes within the dorsal root ganglion (DRG) to record from the cell bodies of both group I and II afferents for periods as long as 1 to 4 days. With both techniques, the recording conditions are far more stable than those achieved in human microneurography, such that movements of quite wide range, rate, and force can be analyzed. A limitation emphasized by Loeb *(personal communication)* is that the discharge of particularly sensitive afferents such as the primary ending of the muscle spindle can be correlated only with a somewhat indirect measurement of overall muscle length. This problem is also present in traditional acute experiments unless, of course, the receptor is actually removed from the parent muscle.

MOTOR UNIT RECRUITMENT

It is now well accepted that the graded development of muscle force involves usually, but not exclusively, an orderly recruitment of motor units whose functional thresholds are more or less in the order type S, FR, F (int), and FF, respectively (for details of this nomenclature system, see ref. 10). This phenomenon and its exceptions were first revealed in traditional animal experiments. However, data derived from experiments on conscious humans executing voluntary contractions have provided valuable corroborative evidence for orderly recruitment (for review see refs. 9,21).

Particularly following Henneman's elaboration of the "size principle" (38, 39,41,42,76), results from experiments performed on conscious humans became pivotal in the debate as to the "flexibility" of recruitment order in the course of voluntary contractions (e.g., refs. 33,40,48). These data combined with further traditional animal experiments have, in general, supported orderly recruitment, at least for slowly developing contractions up to approximately 40% of the maximum voluntary contraction (9,21). However, there remain several important questions regarding the extent to which this "strategy" is employed by the nervous system in response to its diverse motor control "problems."

At this stage, it is not entirely clear whether the normal recruitment pattern is retained during a rapid "ballistic" contraction, although the evidence for retention (20,21) appears more conclusive than that against it (e.g., refs. 35,37). A second issue concerns the extent to which functional thresholds of motor units may be altered under various circumstances. It has now been demonstrated that selective activation of motoneuronal inputs from the red nucleus (11),

Deiter's nucleus (13), and the sural nerve (46) in experimental animals and low-threshold cutaneous stimulation of the human hand during voluntary contractions (78) can produce reversals in normal recruitment order. In addition, there are reports of reversals in orderly recruitment in conscious humans by a presumed blockade or alteration of proprioceptive afferent activity (23,33,34). However, the connection between these results and recruitment-order alterations in the course of "normal" natural movements is not yet obvious. In any case, it would appear that further *natural movement* experiments on humans are required to provide the critical data.

In light of the demonstrations of recruitment reversals during activation of select inputs to a given motoneuronal pool mentioned above, any model proposing a mechanism(s) underlying the phenomenon of orderly recruitment must account for both intrinsic motoneuronal properties (e.g., cell size, membrane resistance, absolute threshold voltage, or their combination) and the organization of synaptic input to a given pool (i.e., extrinsic properties). Although a complete model is as yet unavailable, its basic framework has been formalized by Burke and co-workers (see in particular refs. 9,12). The development of a more complete model should take into account the potential significance of a recent finding from experiments on conscious humans. Stein and colleagues (62,63) have shown that normal recruitment patterns are retained in patients with partial depletion of a motoneuronal pool (following amyotrophic lateral sclerosis) or with partial nerve degeneration (following pressure or entrapment neuropathies) even though these conditions are associated with a substantial rearrangement of the muscular component of the surviving motor units. In contrast, in patients with a previous complete severance of the ulnar nerve followed by its self-reinnervation, the normal recruitment order is dramatically altered even though the twitch tensions of individual motor units are all within normal range. Although these data clearly support the investigators' suggestion that "a maintained physical continuity between nerve and muscle is a necessary condition for the preservation of the orderly recruitment of motor units during voluntary contraction," their bearing on *mechanisms* underlying orderly recruitment remain obscure. Nonetheless, as is pointed out by these investigators, any hypothesis based on their data is testable in both animal and human experimentation.

Orderly Recruitment for γ Motoneurons?

In introducing the "size principle," Henneman and his colleagues emphasized that the tendency of γ motoneurons to fire continuously in decerebrate preparations was apparently attributable to their small size (41). However, they did not suggest an orderly recruitment pattern within the γ component of the motoneuronal pool (39), even though, in a companion paper (86), a relatively broad spectrum of γ cell "sizes" (axonal conduction velocities) was reported for cat medial gastrocnemius (10 to 50 m/sec).

Spontaneous fusimotor activity has been observed in many other experimental preparations, but the issue of an orderly recruitment order for γ motoneurons has not been raised in the literature on traditional animal experiments. However, in a recent study on humans it was concluded by Burke et al. (7), that the threshold for activation of a spindle afferent in an isometric voluntary contraction appears to be determined by its fusimotor innervation, and that "fusimotor neurons probably have a recruitment order much as do the skeletal motor neurones." This is an intriguing possibility, but more direct evidence from analysis of fusimotor activity in traditional animal preparations is needed to test their hypothesis.

α-γ COACTIVATION DURING MOVEMENT

In the past 50 years, a wide variety of experiments performed on anesthetized or decerebrate preparations have characterized the response patterns of muscle spindles to: (a) stretch of the passive or active muscle; (b) α-γ coactivation under isometric conditions or during slow muscle shortening; (c) combinations of stretch-shorten epochs with α-γ coactivation; and (d) fusimotor activation of an otherwise passive muscle at a constant length (for review of these developments, see ref. 57). In toto, these data do not provide a singular profile of α-γ coactivation in segmental motor control but rather one which depends to a large extent on the role of a particular muscle in the generation of a specific movement. Moreover, a similar picture has emerged from recent studies of natural movements in animals. In most of these studies (17,25,51,68,69; see, however, ref. 56) there is evidence of considerable independence in the control of α and γ motor pathways. Nonetheless, generalizations that correlate muscle function to the extent of α-γ coactivation are indeed possible. For example, Loeb and Duysens (51) have noted that in the course of some stereotyped movements, muscle spindles in the associated "prime-mover" muscles display consistent patterns of afferent activity. The secondary endings of spindles (in the presence of some degree of fusimotor input) usually generate spike trains which are correlated with absolute muscle length. The primary endings are more strongly biased by α-γ coactivation and generally display activity during active muscle contractions. Moreover, in nonstereotyped movements which require more conscious attention (e.g., licking, postural adjustments, evasive maneuvers), there are generally higher levels of fusimotor activity in both active and passive muscles. However, isolated fusimotor activity, independent of α-motoneuron discharge, has been observed only in select muscles, and during movements for which these muscles are not primarily responsible. This qualification has prompted Loeb and Duysens (51) to suggest that alternative motor control strategies may be utilized in these cases.

In contrast to the recent animal literature, the majority of studies on conscious humans performing voluntary contractions have emphasized a somewhat rigid CNS "strategy" of α-γ coactivation (e.g., refs. 5,8,82). There are, however, a number of exceptions (reviewed in ref. 67). Resolution of these disparate views requires experiments which feature a sequence of behavioral and kinesiological situations that will reveal the extent to which the human CNS can make optimal use of its separate skeletomotor and fusimotor pathways.

SENSORY ACTIVATION DURING MOVEMENT

Although they have received considerable experimental attention, the central connections of muscle, joint, and cutaneous receptor afferents are still poorly understood (for review, see refs. 31,57,59). Of necessity, the experimental approaches have involved either a composite selective activation of afferents from the same group (e.g., refs. 18,47,54) or analysis of the motoneuronal connections of a single afferent axon (49,60,77,83). These approaches are fundamental to the study of segmental motor control, but one should not infer from their results that a selective activation of a single afferent species occurs necessarily during natural movements. The studies reviewed in this report have not addressed this issue directly, but there is now sufficient data on the firing patterns of different types of afferent fiber during the various active and passive phases of different movements to make inferences on the extent to which the different species of afferent fiber are either co- or selectively activated during the different phases of a movement.

Some generalizations which emerge from all the natural movement literature cited in this report include:

1. Muscle spindle (Ia and II) and tendon organ (Ib) afferents demonstrate discharge during both weak and strong muscle-contractions. In contrast, stretch of the passive muscle results in a relatively selective activation of spindle Ia and group II afferents until substantial rates and extents of stretch are attained. Under these latter conditions limited Ib firing is usually also observed.
2. Thus far, the natural movement experiments have not revealed a selective activation of Ia over spindle group II afferents for any phase of a movement. Sample sizes for spindle group II afferents are still quite small, however, and there are many naturally occurring types of fine muscle perturbation (both active and passive) that have yet to be studied.
3. Cutaneous and joint receptor afferents are as consistently activated as muscle afferents in the movements studied to date.

While these generalizations suggest that sensory coactivation is a feature of natural movements, the relative balance of input among the different afferent types is obviously changing as a movement evolves. This is particularly evident in locomotion where the distinct phases of the hind-limb step cycle (i.e., F, E^1, E^2, E^3) impose dramatically different tasks on the participant muscles (see, for example, Fig. 14 in ref. 80).

There have been few attempts in the *natural movement* literature to model the collective input of even a single afferent species (e.g., ref. 70) during even a single phase of a movement. This approach has proven particularly useful for assessing our current understanding of reflex and supraspinal contributions to locomotion (see, for example, Fig. 21 in ref. 84). It must be cautioned, however, that the presence of afferent input from a select afferent species (even Ia input) does not necessarily result in an obligatory segmental reflex output (25). This throws the burden of segmental reflex control back within the CNS itself, as has been emphasized so effectively by Lundberg and his colleagues (53).

NATURE OF SENSORY INFORMATION DURING MOVEMENT

The transducing properties of muscle, joint, and cutaneous receptors have been traditionally studied in immobile preparations where "controlled input" permits quantitative estimates of stimulus-response relationships. This approach has proven particularly fruitful in the study of muscle receptors, where reasonably consistent descriptions of receptor behavior are available from experiments performed on isolated receptors and receptors within an *in vivo* muscle that is either: (a) completely passive; (b) passive but for the electrical activation of one or more units; or, (c) active as in a decerebrate cat undergoing tonic stretch and crossed-extension reflexes. In the case of joint afferents a less consistent pattern has emerged, largely owing to variations in the different preparations (e.g., intracapsular tension, see refs. 28,29). Similarly, the analysis of cutaneous receptor responses in immobile preparations might have little bearing on the response of these receptors to the same stimuli during natural movements.

An inherent strength of the *natural movement* experiments is the provision for correlating mechanical events within muscle groups to the firing patterns of individual spindle, tendon organ, joint, and cutaneous afferents. The extent to which these data have contributed to our current understanding of the transducing properties of these receptors is discussed below.

Muscle Spindles

Excluding well-established differences between the primary and secondary afferents (reviewed in refs. 27,57), the key properties of muscle spindles that have emerged from studies on immobile preparations include: (a) an exquisite sensitivity to passive length (1 to 200 μ, see ref. 57) and active force (1,85) perturbations provided that there is an appropriate "control" stress on the receptor; (b) a relatively stronger response to small perturbations (58) but nonetheless a substantial response to physiological-range muscle-length perturbations (26); and (c) a firing rate that is determined at any moment by the interaction of muscle length and fusimotor stimuli (57).

The *natural movement* data support these generalizations and, in most instances, extend upon them by specifying the kinesiological conditions under which the receptor is activated. For example, the fact that even in a passive muscle, individual spindle discharge may not provide an accurate representation of whole muscle length, is not paradoxical provided that initial muscle length and passive tension are taken into account (51).

A feature of spindle responsiveness emerging from *natural movement* studies, is that in slow movements (involving lengthening, isometric, or shortening contractions of the parent muscle) against heavy loads, spindle discharge frequency increases with contraction strength (6). This finding might have been predicted from the immobile-preparation literature, but its clear demonstration was provided by human microneurographic studies. What have not been established in either immobile preparations or *natural movement* studies are the kinesiological conditions under which the spindle's small perturbation sensitivity becomes dominated by more global stresses (e.g., rate and extent of overall muscle lengthening or shortening, number of active motor units, extent of fusimotor drive).

Tendon Organs

Experiments on traditional immobile preparations (reviewed in ref. 44) have clearly demonstrated the very low threshold of Golgi tendon organs to "in-series" muscle force. These receptors exhibit vigorous discharge in active muscle contractions (when select muscle fibers exert forces directly on the receptor capsule) and, in general, relatively variable and feeble discharge during stretch of the passive muscle (when only a small component of the passive tension is actually coupled to the receptor). The *natural movement* literature tends to support these findings (see, e.g, ref. 82), although there is as yet no information available on the effects of single motor unit contractions on the firing patterns of these receptors during the course of natural movements. Such motor unit–muscle receptor analysis has proven particularly valuable in the immobilized-preparation work. For example, such studies have not only revealed the very low threshold of tendon organs to "in-series" force (2,45), but further, led to the view that the behavior of individual endings reflects the activity of a restricted sample of motor units in the muscle (70). Under current investigation and discussion is the relationship between these findings and the extent to which the discharge of a single tendon organ in an active muscle is a continuous or discontinuous function of parent muscle force (44,82). Further work along these lines is obviously necessary, in both immobile preparations and *natural movement* studies.

Joint Receptors

Our current understanding of the transducing properties of joint afferents is based almost exclusively on traditional immobile preparations (4,14,15,28, 29,30). Joint afferents appear to function as capsular stretch receptors, responding to any combination of mechanical events that exert forces on the parent capsule. When, for example, the knee joint of an anesthetized cat or monkey is rotated in a completely passive limb, most afferents from the medial and posterior articular nerves appear to respond only at the extremes of angular displacement (4,30). Alternatively, activation of muscles that insert on a joint capsule can evoke joint afferent discharge even in the absence of any angular displacement (28, 29,30,61). Presumably between these two extreme stimulus conditions, a variety of mechanical events produced by both muscle action and joint displacement occur in the course of most movements such that joint afferents are capable of signaling intermediate as well as extreme joint positions. Specific delineation of such mechanical combinations has not as yet been attempted in immobile preparations, and it would appear that *natural movement* experiments will prove more appropriate to this end (50). Once the technical difficulties are surmounted, human experimentation on finger-joint afferents would appear of particular value in that the mechanical combinations could be controlled very carefully by conscious action.

Loeb and his colleagues (50) have recently reported on the firing patterns of knee-joint afferents in the conscious, unrestrained cat during treadmill locomotion. Their results display remarkable variability in the behavior of these receptors, some being responsive to small fluctuations in ankle angle, with others displaying almost no activity at all during walking. Such results suggest that the functional significance of joint afferent discharge might be elucidated only by *natural movement* experiments, in that the number and complexity of mechanical variables are simply too great to be controlled and analyzed in traditional immobile preparations.

Cutaneous Afferents

Several recent reviews have outlined considerable progress on the analysis of cutaneous mechanoreceptors (e.g., refs. 3,43). It now appears that practically all "naturally occurring" tactile stimuli to the primate and cat forepaw excite a large and diverse population of mechano-sensitive receptor types. These findings have now been extended to the human hand by use of single-fiber microneurography with an emphasis on the psychophysics of tactile sensation (reviewed in refs. 82,88)

Cutaneous afferent input has been shown to exert relatively powerful segmental reflex effects on motoneurons (11,57,78), even though classic experiments revealed that the abolition of this input has little serious effect on movements such as stepping and climbing (74). Resolution of these conflicting views requires further information on the nature of cutaneous afferent input during natural movements and the subtleties of movement deficits produced in the absence of this input.

This issue has understandably not surfaced in the traditional approach to the transducing properties of cutaneous afferents, even though it has long been intuitively obvious that the unusually low threshold of many cutaneous mechanoreceptors must lead to their activation in movements, even for those innervating skin encountering no external perturbation during the movement (74). Again, new *natural movement* data are beginning to emerge on this issue (50).

SUMMARY

A great deal is yet unknown about segmental motor control and "how the CNS uses the continuous flow of information that it receives from the muscle afferents" (57). Results from *natural movement* experiments have now become particularly influential in both the confirmation of existing hypotheses and the elaboration of new ones. In some instances, such as motor-unit recruitment order, results of natural movement experiments are critical in establishing the boundary conditions of this principle and determining the functional significance of its exceptions. In other cases, it is only through the analysis of a natural movement that we might gain insight into CNS strategies such as those involved in α-γ coactivation, and it is clear that we have only begun to design experiments that challenge human and animal subjects on such strategies. Finally, it seems clear that future developments in our understanding of segmental motor control will involve a continuous interplay between the traditional and *natural movement* approaches.

ACKNOWLEDGMENTS

This work was supported in part by U.S.P.H.S. grants NS 07888 (to D.G.S.), NS 05871 (to B.R.B.), H 607249 (to Department of Physiology) and RR 05675 (to College of Medicine). M.D.B. was supported by a research fellowship of the Muscular Dystrophy Association of America. His present address is the Department of Physiology and Biophysics, University of Washington School of Medicine, Seattle, Washington 98195.

REFERENCES

1. Binder, M. D., Kroin, J. S., Moore, G. P., Stauffer, E. K., and Stuart, D. G. (1976): Correlation analysis of muscle spindle responses to single motor unit contractions. *J. Physiol. (Lond.),* 257:325–336.
2. Binder, M. D., Kroin, J. S., Moore, G. P., and Stuart, D. G. (1977): The response of Golgi tendon organs to single motor unit contractions. *J. Physiol. (Lond.),* 271:337–349.
3. Bovie, J. J. G., and Perl, E. R. (1975): Neural substrates of somatic sensation. In: *Neurophysiology, Vol. 3,* edited by C. C. Hunt, pp. 303–411. Butterworths, London.
4. Burgess, P. R., and Clark, F. J. (1969): Characteristics of knee joint receptors in the cat. *J. Physiol. (Lond.),* 203:315–335.
5. Burke, D. (1978): The fusimotor innervation of muscle spindle endings in man. In: *Trends in Neurosciences,* 1:89–92.
6. Burke, D., Hagbarth, K-E., and Löfstedt, L. (1978): Muscle spindle responses in man to changes in load during accurate position maintenance. *J. Physiol. (Lond.),* 276:159–164.
7. Burke, D., Hagbarth, K-E., and Skuse, N. F. (1978): Recruitment order of human spindle endings in isometric voluntary contractions. *J. Physiol. (Lond.),* 285:101–112.
8. Burke, D., Hagbarth, K-E., and Skuse, N. F. (1978): Voluntary activation of spindle endings in human muscles temporarily paralysed by nerve pressure. *J. Physiol. (Lond.),* 287:329–336.

9. Burke, R. E. (1978): Motor unit recruitment: What are the critical factors? In: *Progress in Clincal Neurophysiology, Vol. 9: Recruitment Patterns of Motor Units and the Gradation of Muscle Force,* edited by J. Desmedt. Karger, Basel *(in press).*
10. Burke, R. E., and Edgerton, V. R. (1975): Motor unit properties and selective involvement in movement. In: *Exercise and Sports Science Reviews, Vol. III,* edited by J. H. Wilmore and J. F. Keogh, pp. 33–81. Academic Press, New York.
11. Burke, R. E., Jankowska, E., and Bruggencate, G. ten. (1970): A comparison of peripheral and rubrospinal synaptic input to slow and fast twitch motor units of triceps surae. *J. Physiol. (Lond.),* 207:709–732.
12. Burke, R. E., and Rudomin, P. (1977): Spinal neurons and synapses. In: *Handbook of Physiology, Sect. 1, Vol. 1, The Nervous System: The Cellular Biology of Neurons,* edited by E. R. Kandel, pp. 877–944. American Physiological Society, Washington, D.C.
13. Clamann, H. P., and Kukulka, C. G. (1977): Reversals of recruitment order in medial gastrocnemius produced by stimulation of Deiter's nucleus. *Neurosci. Abstr.,* 3:269.
14. Clark, F. J. (1975): Information signalled by sensory fibers in medial articular nerve. *J. Neurophysiol.,* 38:1464–1472.
15. Clark, F. J., and Burgess, P. R. (1975): Slowly adapting receptors in cat knee joint: can they signal joint angle? *J. Neurophysiol.,* 38:1448–1463.
16. Clark, R. W., Luschei, E. S., and Hoffman, D. S. (1978): Recruitment order, contractile characteristics, and firing patterns of motor units in the temporalis muscle of monkeys. *Exp. Neurol.,* 61:31–52.
17. Cody, F. W. J., Harrison, L. M., and Taylor, A. (1975): Analysis of activity of muscle spindles of the jaw-closing muscles during normal movements in the cat. *J. Physiol. (Lond.),* 253:565–582.
18. Coppin, C. M. L., Jack, J. J. B., and McLennan, R. C. (1970): A method for the selective activation of tendon organ afferents from the cat soleus. *J. Physiol. (Lond.),* 210:18–20P.
19. De Luca, C. J. (1979): Physiological and mathematical basis of myoelectric signals. In: *Muscles Alive: Their Function Revealed by Electromyography,* 5th ed., edited by J. V. Basmajian. Williams & Wilkins, Baltimore *(in press).*
20. Desmedt, J. E., and Godaux, E. (1977): Ballistic contractions in man: Characteristic recruitment pattern of single motor units of the tibialis anterior muscle. *J. Physiol. (Lond.),* 264:673–693.
21. Desmedt, J. E., and Godaux, E. (1978): Ballistic skilled movements Load compensation and patterning of the motor commands. In: *Progress in Clinical Neurophysiol., Vol. 4: Cerebral Motor Control in Man: Long loop mechanisms,* edited by J. E. Desmedt, pp. 21–55. Karger, Basel.
22. Euler, C. von (1973): The role of proprioceptive afferents in the control of respiratory muscles. *Acta Neurobiol. Exp.,* 33:329–341.
23. Fukishima, K., Taniguchi, I., Kamishina, Y., and Kato, M. (1976): Peripheral factors contributing to the volitional control of firing rates of human motor units. *Neurosci. Lett.,* 3:33–36.
24. Goodwin, G. M., and Luschei, E. S. (1974): Effects of destroying spindle afferents from jaw muscles on mastication in the monkey. *J. Neurophysiol.,* 37:967–981.
25. Goodwin, G. M., and Luschei, E. S. (1975): Discharge of spindle afferents from jaw-closing muscles during chewing in alert monkeys. *J. Neurophysiol.,* 38:560–571.
26. Goslow, G. E., Stauffer, E. K., Nemeth, W., and Stuart, D. G. (1973): The cat step cycle II: Responses of muscle spindles and tendon organs to passive stretch within the locomotor range. *Brain Res.,* 60:35–54.
27. Granit, R. (1970): *The Basis of Motor Control.* Academic Press, New York.
28. Grigg, P. (1973): Mechanical factors influencing responses of joint afferent neurons from cat knee. *J. Neurophysiol.,* 38:1473–1483.
29. Grigg, P. (1976): Response of joint afferent neurons in cat medial articular nerve to active and passive movements of the knee. *Brain Res.,* 118:482–485.
30. Grigg, P., and Greenspan, B. J. (1977): The response of primate joint afferent neurons to mechanical stimulation of the knee joint. *J. Neurophysiol.,* 40:1–8.
31. Grigg, P., Harrigan, E. P., and Fogarty, K. E. (1978): Segmental reflexes mediated by joint afferent neurons in cat knee. *J. Neurophysiol.,* 41:9–14.
32. Grillner, S. (1975): Locomotion in vertebrates—Central mechanisms and reflex interaction. *Physiol. Rev.,* 55:247–304.
33. Grimby, L., and Hannerz, J. (1968): Recruitment order of motor units on voluntary contraction:

Changes induced by proprioceptive afferent activity. *J. Neurol. Neurosurg. Psychiatry,* 31:565–573.
34. Grimby, L., and Hannerz, J. (1976): Disturbances in voluntary recruitment order of low and high frequency motor units on blockades of proprioceptive afferent activity. *Acta Physiol. Scand.,* 96:207–216.
35. Grimby, L., and Hannerz, J. (1977): Firing rate and recruitment order of toe extensor motor units in different modes of voluntary contraction. *J. Physiol. (Lond.),* 264:865–879.
36. Hagbarth, K-E. (1976): Microneurography in man. In: *Sensory Functions of the Skin,* edited by Y. Zotterman, pp. 129–136. Pergamon Press, New York.
37. Hannerz, J. (1974): Discharge properties of motor units in relation to recruitment order in voluntary contraction. *Acta Physiol. Scand.,* 91:374–384.
38. Henneman, E. (1957): Relation between size of neurons and their susceptibility to discharge. *Science,* 126:1345–1346.
39. Henneman, E., and Olson, C. B. (1965): Relations between structure and function in the design of skeletal muscles. *J. Neurophysiol.,* 28:581–598.
40. Henneman, E., Shahani, B. T., and Young, R. R. (1976): Voluntary control of human motor units. In: *The Motor System: Neurophysiology and Muscle Mechanisms,* edited by M. Shahani, pp. 73–78. Elsevier, Amsterdam.
41. Henneman, E., Somjen, G., and Carpenter, D. O. (1965): Functional significance of cell size in spinal motoneurons. *J. Neurophysiol.,* 28:560–580.
42. Henneman, E., Somjen, G., and Carpenter, D. (1965): Excitability and inhibitability of motoneurons of different sizes. *J. Neurophysiol.,* 28:599–620.
43. Horch, K. W., Tuckett, R. P., and Burgess, P. R. (1977): A key to the classification of cutaneous mechanoreceptors. *J. Invest. Dermatol.,* 69:75–82.
44. Houk, J. C., Crago, P. E., and Rymer, W. Z. (1978): Functional properties of Golgi tendon organs. In: *Progress in Clinical Neurophysiology, Vol. 8: Motor Control in Man: Suprasegmental and Segmental Mechanisms,* edited by J. E. Desmedt. Karger, Basel *(in press).*
45. Houk, J. C., and Henneman, E. (1967): Responses of Golgi tendon organs to active contraction of the soleus muscle of the cat. *J. Neurophysiol.,* 30:466–481.
46. Kanda, K., Burke, R. E., and Walmsley, V. (1977): Differential control of fast and slow twitch motor units in the decerebrate cat. *Exp. Brain Res.,* 29:57–74.
47. Kato, M., and Fukushima, K. (1976): Selective activation of group II muscle afferents and its effects on cat spinal neurons. In: *Understanding the Stretch Reflex,* edited by S. Homma, pp. 185–196. Elsevier, Amsterdam.
48. Kato, M., and Tanji, J. (1972): Volitionally controlled single motor units in human finger muscles. *Brain Res.,* 40:345–357.
49. Kirkwood, P. A., and Sears, T. A. (1975): Monosynaptic excitation of motoneurones from muscle spindle secondary endings of intercostal and triceps surae muscles in the cat. *J. Physiol. (Lond.),* 240:64–66P.
50. Loeb, G. E., Bak, M. J., and Duysens, J. (1977): Long-term unit recording from somatosensory neurons in the spinal ganglia of the freely walking cat. *Science,* 197:1192–1194.
51. Loeb, G. E., and Duysens, J. (1978): Activity patterns in individual primary and secondary muscle spindle afferents during normal movements in unrestrained cats. *J. Neurophysiol.,* 42:420–440.
52. Lundberg, A. (1969): *Reflex Control of Stepping.* Nansen Memorial Lecture to Norwegian Academy of Sciences and Letters. Universitetsforlaget, Oslo.
53. Lundberg, A. (1969): Convergence of excitatory and inhibitory action on interneurones in the spinal cord. In: *The Interneurone,* edited by M. A. B. Brazier, pp. 231–256. University of California Press, Los Angeles.
54. Lundberg, A., and Winsbury, G. (1960): Selective activation of large afferents from muscle spindles and Golgi tendon organs. *Acta Physiol. Scand.,* 49:155–164.
55. Luschei, E. S., and Goodwin, G. M. (1974): Patterns of mandibular movement and jaw muscle activity during mastication in the monkey. *J. Neurophysiol.,* 37:954–966.
56. Matsunami, K., and Kubota, K. (1972): Muscle afferents of trigeminal mesencephalic tract nucleus and mastication in chronic monkeys. *Jap. J. Physiol.,* 22:545–555.
57. Matthews, P. B. C. (1972): *Mammalian muscle receptors and their central actions.* Arnold, London.
58. Matthews, P. B. C., and Stein, R. B. (1969): The sensitivity of muscle spindle afferents to small changes in length. *J. Physiol. (Lond.),* 200:723–743.

59. McIntyre, A. K. (1974): Central actions of impulses in muscle afferent fibers. In: *Handbook of Sensory Physiology, III/2,* edited by C. C. Hunt, pp. 235–288. Springer-Verlag, New York.
60. Mendell, L. M., and Henneman, E. (1971): Terminals of single Ia fibers: Location, density and distribution within a pool of 300 homonymous motoneurons. *J. Neurophysiol.,* 34:171–187.
61. Millar, J. (1973): Joint afferent fibres responding to muscle stretch, vibration and contraction. *Brain Res.,* 63:380–383.
62. Milner-Brown, H. S., Stein, R. B., and Lee, R. S. (1974): Pattern of recruiting human motor units in neuropathies and motor neurone disease. *J. Neurol. Neurosurg. Psychiatry,* 37:665–669.
63. Milner-Brown, H. S., Stein, R. B., and Lee, R. S. (1974): Contractile and electrical properties of human motor units in neuropathies and motor neurone disease. *J. Neurol. Neurosurg. Psychiatry,* 37:670–676.
64. Milner-Brown, H. S., Stein, R. B., and Lee, R. G. (1975): Synchronization of human motor units: Possible roles of exercise and supraspinal reflexes. *Electroencephalogr. Clin. Neurophysiol.,* 38:245–254.
65. Milner-Brown, H. S., Stein, R. B., and Yemm, R. (1973): The contractile properties of human motor units during voluntary isometric contractions. *J. Physiol. (Lond.),* 228:285–306.
66. Milner-Brown, H. S., Stein, R. B., and Yemm, R. (1973): The orderly recruitment of human motor units during voluntary isometric contractions. *J. Physiol. (Lond.)* 230:359–370.
67. Murthy, K. S. K. (1978): Vertebrate fusimotor neurones and their influences on motor behavior. *Prog. Neurobiol.,* 11:249–307.
68. Prochazka, A., Westerman, R. A., and Ziccone, S. P. (1976): Discharge of single hind limb afferents in the freely moving cat. *J. Neurophysiol.,* 39:1090–1104.
69. Prochazka, A., Westerman, R. A., and Ziccone, S. P. (1977): Ia afferent activity during a variety of voluntary movements in the cat. *J. Physiol. (Lond.),* 268:423–448.
70. Reinking, R. M., Stephens, J. A., and Stuart, D. G. (1975): The tendon organs of cat medial gastrocnemius: Significance of motor unit size and type for the activation of Ib afferents. *J. Physiol. (Lond.),* 250:491–512.
71. Sears, T. A. (1966): The respiratory motoneurone: Integration at spinal segmental level. In: *Breathlessness,* edited by J. B. L. Howell and E. J. M. Campbell, pp. 33–47. Blackwell, Oxford.
72. Sears, T. A. (1972): Breathing: A sensori-motor act. *Sci. Basis Med. Annu. Rev.,* pp. 129–147.
73. Sears, T. A. (1974): The afferent regulation of learnt movements. *Brain Res.,* 71:465–473.
74. Sherrington, C. S. (1910): Flexion-reflex of the limb, crossed extension-reflex, and reflex stepping and standing. *J. Physiol. (Lond.),* 40:28–121.
75. Shik, M. L., and Orlovsky, G. N. (1976): Neurophysiology of locomotor automatism. *Physiol. Rev.,* 56:465–501.
76. Somjen, G., Carpenter, D. O., and Henneman, E. (1965): Responses of motoneurons of different sizes to graded stimulation of supraspinal structures of the brain. *J. Neurophysiol.,* 28:958–965.
77. Stauffer, E. K., Watt, D. G. D., Taylor, A., Reinking, R. M., and Stuart, D. G. (1976): Analysis of muscle receptor connections by spike-triggered averaging: 2. Spindle secondary afferents. *J. Neurophysiol.,* 39:1393–1402.
78. Stephens, J. A., Garnett, R., and Buller, N. P. (1978): Reversal of recruitment order of single motor units produced by cutaneous stimulation during voluntary muscle contraction. *Nature,* 272:362–364.
79. Stephens, J. A., and Usherwood, T. P. (1977): The mechanical properties of human motor units with special reference to their fatigability and recruitment threshold. *Brain Res.,* 125:91–97.
80. Stuart, D. G., Mosher, C. G., and Gerlach, R. L. (1972): Properties and central connections of Golgi tendon organs with special reference to locomotion. In: *Research in Muscle Development and the Muscle Spindle,* edited by B. Q. Banker, R. J. Pryzbylsky, J. P. Van Der Meulen, and M. Victor, pp. 437–464. Exerpta Medica, Amsterdam.
81. Taylor, A., and Davey, M. R. (1968): Behavior of the jaw muscle stretch receptors during active and passive movements in cats. *Nature,* 220:301–302.
82. Vallbo, A. B., Hagbarth, K-E., Wallin, B. G., and Torebjork, H. E. (1979): Proprioceptive, somatosensory and sympathetic activity in human nerves. *Physiol. Rev. (in press).*
83. Watt, D. G. D., Stauffer, E. K., Taylor, A., Reinking, R. M., and Stuart, D. G. (1976): Analysis

of muscle receptor connections by spike-triggered averaging. 1. Spindle primary and tendon organ afferents. *J. Neurophysiol.,* 39:1375–1392.
84. Wetzel, M. C., and Stuart, D. G. (1976): Ensemble characteristics of cat locomotion and its neural control. *Prog. Neurobiol.,* 7:1–98.
85. Windhorst, U., and Meyer-Lohmann, J. (1977): The influence of extrafusal muscle activity on discharge patterns of primary muscle spindle endings. *Pfluegers Arch.,* 372:131–138.
86. Wuerker, R. B., McPhedran, A. M., and Henneman, E. (1965): Properties of motor units in a heterogeneous pale muscle (m. gastrocnemius) of the cat. *J. Neurophysiol.,* 28:85–99.
87. Zajac, F. E., and Young, J. L. (1975): Discharge patterns of motor units during cat locomotion and their relation to muscle performance. In: *Neural Control of Locomotion,* edited by R. M. Herman, S. Grillner, P. G. G. Stein, and D. G. Stuart, pp. 789–793. Plenum Press, New York.
88. Zotterman, Y., editor (1976): *Sensory Functions of the Skin.* Pergamon Press, New York.

Posture and Movement, edited by
R. E. Talbott and D. R. Humphrey.
Raven Press, New York © 1979.

Convergence on an Understanding of Motor Control

John M. Brookhart

Department of Physiology, School of Medicine, University of Oregon Health Sciences Center, Portland, Oregon 97201

The wide variety of experimental endeavors described by the authors of the preceding chapters in this volume might be confusing to the selective reader. Nevertheless, they all represent efforts to achieve a commonly shared goal. That goal is to understand how animals are able to use only a small fraction of their body weight, called a brain, to enable them to control their musculoskeletal systems and thus generate positions and changes of position in space—positions and changes of position that are inherently beautiful for their grace, accuracy, and effectiveness, which do not lose that beauty when directed toward mundane and common objectives, nor acquire additional beauty even though they may be life-preserving.

We are not new in the world by virtue of this common interest in a single goal. William Harvey's notebooks contain some of his musings on this topic (9). Only a few years later, Borelli (2) worked with great diligence and physical insights, and with woefully inadequate tools, to try to achieve some understanding of animal movement. In our present time-frame, it is hard to appreciate the importance of his inference that muscles only pull, and that they do this with equal forces on their origins and insertions. Lest we tend to derogate the importance of this insight, we should remember that it was in this same mid-17th-century period that Boyle and Toricelli were developing the concept of pressure—the *concept* of pressure differences and applying it to fluid mechanics.

If we leap to the last half of the 19th century, we find this remarkably clever, ingenious man named Etienne-Jules Marey (Fig. 1) devising ways to study men, horses, birds, fish, all in motion, with his multiple recording tambours and repeating photographic rifles. And then there was the intellectually gigantic small man named Sherrington, who taught us how to break large problems down into smaller, apparently simpler components long before Henry Kissinger[1] came on the scene. His student Magnus had the courage coupled with ingenuity to address larger components of the same whole (10). Still later there came

[1] Henry Kissinger was the U.S. Secretary of State (1969–1976) who applied the reductionist philosophy to diplomatic problems.

FIG. 1. Memorial statue of E. J. Marey in a small square near the Musée Marey in Beaune. The moving horse, in relief, and the bird in flight are the artist's reconstruction from Marey's records of motion studies still preserved in the museum. (Photo by author.)

James Gray, with his remarkable catholicity of interest in animal motion, no matter where in the phylogenetic scale (7). These are just some of the men whose shoulders support us, whose interests we perpetuate, and to whom we owe homage.

Today, we are in a state of capability of which these men could not have dreamed. We are in a technological revolution that can only dimly be appreciated by those not working in the 1930s. With the recognition of such an enrichment of tools, we now come to a question. Why, with the technological and technical capabilities available to us, have we not progressed further? Why are we still so ignorant? Is it because we have, individually and randomly, directed our attention to the same goal in a discoordinate and disorderly fashion?

I am persuaded that the answer does not lie in the fact that we are converging from different origins and directions. Our convergent state is rather an expression of the complexity of the task we have undertaken which, in turn, is a result of the complexity of the brain, its processes, and our own ignorance. This combination of complexity with ignorance is so overwhelming that some level of reductionism appears to be absolutely essential to even begin to make inroads on the puzzle.

On the other hand, I propose that at the base of our problem lies the fact that we do not have a specification of the end-product of the behaviors we seek to understand. We have not yet seen clearly how to relate the behavior of the components to the behavior of the whole. Over the four decades from 1920 to 1960, Nicholas Bernstein devoted his exceptional intellectual powers to the consideration of global problems of movement control (1). A Japanese otologist, Fukuda, took a different view of the problem and made, for his time and technical capability, a serious attempt to address it by using the simple process of observing with educated eyes (6). He had been persuaded to study Magnus, and, in 1943, published the first of several accounts of the ways he saw living, moving people using the reflexes described by Magnus. In his picture of a baseball player (Fig. 2) attempting to stop a high line-drive over shortstop, he saw the consequences of operation of the tonic neck reflex initiated by head rotation. He saw it again in the soccer player (Fig. 3), exhibiting extension on the chin side and flexion on the occiput side just after "heading" the ball. He saw it still again in Fred Astaire and a dancing partner (Fig. 4). Admittedly, this was a simple way to approach a complex problem, but it represented a real effort to see parts functioning in a holistic context. Workers in Maryland

FIG. 2. Photograph used by Fukuda to illustrate how head and neck reflexes described by Magnus are used in reinforcement of goal-directed volitional movements. In each instance, head and neck rotation contribute to the extension of the limbs on one side and their contralateral flexion. (From ref. 6, with permission.)

FIG. 3. See legend, Fig. 2.

FIG. 4. See legend, Fig. 2.

(see F. E. Zajac and W. S. Levine, *this volume*) have, in a sense, turned the problem upside down and started out, in a manner resembling some of Bernstein's early studies, with a specific movement pattern that can be characterized in quantitative terms. Within this known, predetermined framework, I expect they will be seeking evidences of component behavior. In this task of relating parts to the whole, our friends in the sensory physiology game have the jump on us, because they can compare the input that creates a sensation with measurements of the output as sensory experience quantitated through psychophysical methods.

There is another barrier, more dimly seen and of unknown magnitude, that lies in our converging paths. This barrier constitutes the probability that in dealing with the maintenance or change of body position, we are dealing with emergent properties of the total system that are not described by the sum of the properties of its parts. Bunge has laid this out very clearly for us in recent issues of *Neuroscience* (3) and the *American Journal of Physiology* (4). We must devise ways to determine whether or not parts of the system function with the whole in the same way they do in isolation. Several of the stories we have heard today represent attempts to get at the problem by examining subsystems operating in intact animals. However, learning how a system subcomponent behaves in a carefully defined and constrained set of circumstances gives us no guarantee that our subcomponent will behave according to the same rules if the circumstances were changed.

I would like to give an example of the manner in which the conditions under which measurements are made can alter the behavior of a part of a system. Talbott (14) described this phenomenon and Mirka (11a) has since enlarged upon it. The polar plot in Fig. 5 describes the gain and phase of a dog's body movements while she is making postural adjustments to linear optokinetic stimuli at a series of frequencies. The smaller inner loop is derived from the dog's pelvic movements during a purely optokinetic stimulation when her visual field alone was moved sinusoidally at increasing frequencies going clockwise. The larger outer loop was also generated by pelvic movements induced by that same visual field motion. However, the problem was made more complex by also moving the platform at a frequency 2.88 times that of visual field motion. It is clear that the visual input is somehow made more effective, by a factor sometimes approaching 10 when the two forms of perturbation are combined. I think it is extremely important that we keep possibilities of this kind in our minds when looking at the behavior of parts of systems and wondering how they might be used by the whole animal. Our method of display and the nature of an equation that describes such curves lead us to follow Merton (12) and others by calling this increase a gain change. If this is accepted, it must also be accepted that we don't know what this means in neurophysiological terms, nor do we know how the animal uses these kinds of information in her everyday life.

Let me turn now towards (Borelli's) attempts to understand limb motion

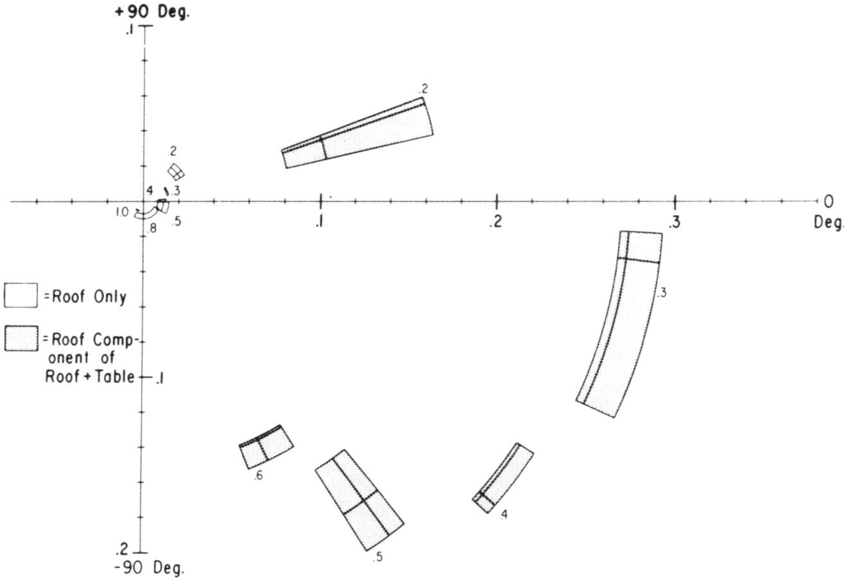

FIG. 5. Variable influence of visual input on postural adjustments executed by a trained dog. This polar plot describes the relative magnitude of sinusoidal pelvic movement by the length of a radius extending from the origin to the data point. The angle of this radial line expresses the phase angle between the sinusoidally driven visual stimulus and the resulting movement of the animal. The median value of 5 trials lies at the crossing points of the enclosed lines; the enclosure expresses the variability over the 5 trials. The *unshaded boxes* describe the characteristics of the postural compensation during the illusion of movement produced by moving a pattern in the animal's visual field at the frequencies (indicated in Hz). When the supporting table was also moved at a frequency 2.8 × higher than that used for visual stimulation, the magnitude and phase of the visually induced response were radically altered as indicated by the shaded boxes. Clearly, the remarkable changes in relative magnitude of pelvic movements induced by the visual stimulation emphasize the fact that visual influences on postural adjustments are sensitive to the environmental context in which the observations are made. (From ref. 14.)

through the application of mechanical principles. John Manter's studies also centered on this aspect of motion (11). Rack (13) and Grillner (8) in our time have reinforced the notion that we should keep the physical properties of moving muscles continually in our minds. And finally for some time now, and again today (see Partridge, *this volume*), we have been cautioned to think carefully about the way in which physical[2] properties of the muscle must put constraints on possible control mechanisms.

In 1973, Carlton Cross, then an engineering student at Oregon State, did his thesis research in our laboratory (5). He learned that some important portions

[2] I use the word *physical* in preference to the word *passive* that is often used in this context. The word *physical* can be used in reference to muscle which is either physiologically passive or active.

of the shear forces generated at the feet during a step function of platform displacement are determined by the physical constants of the moved limbs and do not depend upon neurally controlled adaptations. Since that time, Richard Talbott and Daniel Schuster have been bothered by their ignorance of the degree to which the behavior of our animals is conditioned by simple mass, elastic, and viscous properties. They have established the fact that our dogs can be likened to an inverted pendulum held upright by viscous and elastic constraints and supported on a movable base (13a). The behavior of such a system can be described on a two-dimensional linear gain-phase plot. If you should look down vertically on Fig. 6, that is what you would see. To the gain-phase plot, we have added a third dimension in the form of a dependent variable. This is the error angle. It measures the inclination of the dog's leg with respect to the moving platform surface. This third dimension creates for us a curved surface. The portion of this surface bounded by the curved left margin and the curved dotted line starting at 90° contains all the transfer functions of systems described only by the physical characteristics of mass, viscosity, and elasticity, whether

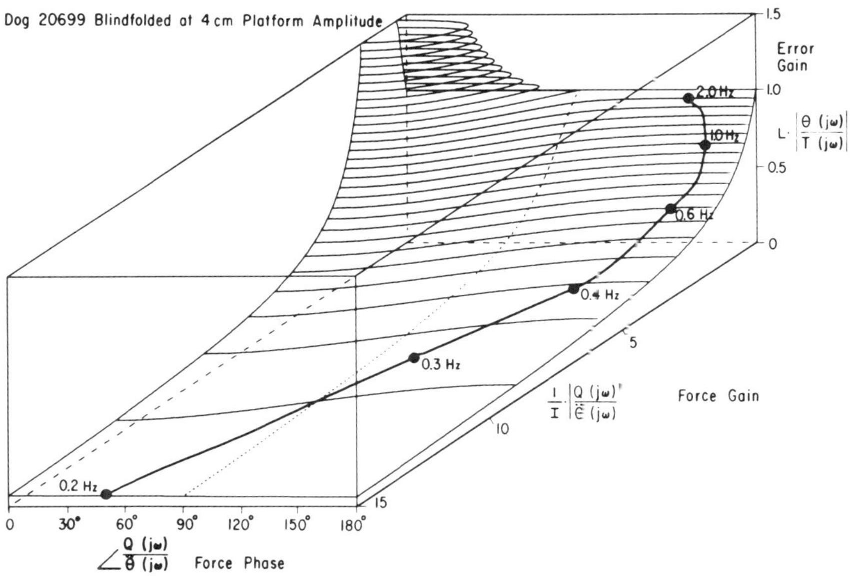

FIG. 6. A three-dimensional display of the relations between relative forces at the animal's feet (force gain), the phase shift relative to the moving surface, and the relative tilt of the supporting limbs (error gain) of a dog during sinusoidal movement at the indicated frequencies (in Hz). If the system behavior is determined only by mass, viscosity, and elasticity, its describing function would fall in the area between the curved margin to the left and the curved dotted line beginning at 90°. The *solid line* connects data points derived from a dog during table oscillations at 6 frequencies. The position of this line to the right of the dotted line for most of its extent demonstrates that the dog's behavior is governed by more than inherent physical traits—that neurally controlled muscle forces participate in the generation of this behavior. (From ref. 13a.)

these be constant or varying. The fact that the animal's behavior, as depicted by the heavy solid line, falls outside these limits shows that the active forces generated by muscles under control of the central nervous system (CNS) participate in determining the behavior. This form of analysis makes it clear that in the intermediate range of frequencies, the CNS is heavily involved in controlling the muscle forces required to make the system behave this way. At both upper and lower ends of the tested frequency range, *evidence* for CNS-guided muscle forces diminishes with a resulting increase in the probability that the system behavior may be dominated by physical characteristics. I am not proposing that the CNS and muscles quit working at the high and low frequencies. Indeed, EMG discharges occur cyclically in some muscles at low forcing frequencies, and at a high level and continuously at high forcing frequencies. By presenting this example, I am simply emphasizing the probability that no single set of mechanical properties will be adequate to cover all varieties and phases of movements.

Let me turn away from the systems type of studies toward another kind of endeavor. Over the past 10 to 15 years, out enthusiasm and hopes have been aroused by the promise that we can gain important information about neural mechanisms of control by looking at the behaviors of neurons, singly or in groups. This kind of information is essential to our understanding of the complete story. Several chapters in this volume deal with activities in this area. The promise is still there, but the road will not be smooth and easy. Speculations must be constrained by uncertainties, only one of which I will discuss. When we look at patterns of neural behavior, we must wonder where those impulses are going and what they do when they get there. There is no guarantee that impulses originating in one place are going to be delivered to a single recipient. Neither is there any guarantee that the recipient neurons are critically involved in the control processes of interest to us. How are we going to find out whether the action generated is a necessary control signal, whether it is involved in the kind of function called "corollary discharge," or whether it is simply contributing to sensory experience? These are serious questions that have to be answered sometime in the future.

But enough of specific uncertainties! They come in many disguises—mechanical, technical, interpretive, conceptual, and even philosophical. We all know we face them and must respond to them. The important thing is to keep going from every direction that offers promise. As our models are further refined, if we share our problems with others of different orientations, if we maintain our convergent efforts, I am confident that we, or our successors, will meet at our central goal.

ACKNOWLEDGMENT

The author gratefully acknowledges support by a grant from the National Institutes of Health (NS 04744).

REFERENCES

1. Bernstein, N. (1967): *The Coordination and Regulation of Movements.* Pergamon, London.
2. Borelli, G. A. (1680): *De Motu Animalium.* A. Bernato, Rome. (See also: *Founders of Experimental Physiology* (1971), edited by W. Blasius, J. Boylan, and K. Kramer. J. Lehmanns, Munich.)
3. Bunge, M. (1977): Emergence and the mind. *Neurosci.,* 2:501–509.
4. Bunge, M. (1977): Levels and reduction. *Am. J. Physiol.,* 233:R75–R82.
5. Cross, C. E. (1973): *Analysis of Postural Dynamics in the Dog.* Doctoral Dissertation, Oregon State University.
6. Fukuda, T. (1961): Studies on human dynamic postures from the viewpoint of postural reflexes. *Acta Otolaryngol. [Suppl.] (Stockh.),* 161:1–52.
7. Gray, James (1968): *Animal Locomotion.* Weidenfeld, London.
8. Grillner, S. (1975): The role for muscle stiffness in meeting the changing postural and locomotor requirements for force development by the ankle extensors. *Acta Physiol. Scand.,* 86:92–108.
9. Harvey, Wm. (1627): *De Motu Locali Animalium,* edited and translated by Gweneth Whitteridge. Cambridge University Press, London, 1959.
10. Magnus, R. (1924): *Körperstellung.* J. Springer, Berlin.
11. Manter, J. T. (1938): The dynamics of quadripedal walking. *J. Exp. Biol.,* 15:522–540.

11a. Mirka, A., Talbott, Richard E., Brookhart, John M. (1978): Role of the visual cortex in visual feedback for postural control. *Soc. Neurosci. (Abstr.),* 4:638.

12. Marsden, C. D., Merton, P. A., and Morton, H. B. (1976): Servo action in the human thumb. *J. Physiol. (Lond.),* 257:1–44.
13. Rack, P. M. H. (1970): The significance of mechanical properties of muscle in the reflex control of posture. In: *Excitatory Synaptic Mechanisms,* edited by P. Andersen and J. K. S. Jansen, pp. 317–321. Oslo Universitet Svorlaget.

13a. Schuster, Daniel, and Talbott, Richard E. (1979): Optimal and adaptive control in canine postural regulation *(submitted).*

14. Talbott, R. E., and Brookhart, J. M. (1976): Variable postural response of the dog to visual stimuli depending upon the environmental context. *Neurosci. Abstr.* 2:534.

Subject Index